Separation in Point-Free Topology

Jorge Picado • Aleš Pultr

Separation in Point-Free Topology

Birkhäuser

Jorge Picado
CMUC, Department of Mathematics
University of Coimbra
Coimbra, Portugal

Aleš Pultr
Department of Applied Mathematics
Charles University
Prague, Czech Republic

ISBN 978-3-030-53478-3 ISBN 978-3-030-53479-0 (eBook)
https://doi.org/10.1007/978-3-030-53479-0

Mathematics Subject Classification: 06D22, 18F70, 54D10

This book is published under the imprint Birkhäuser, www.birkhauser-science.com, by the registered company Springer Nature Switzerland AG.
The registered company address is: Gewerbestrasse 11, 6330 Cham, Switzerland

To our wives Jitka and Isabel and to our teacher and friend Bernhard Banaschewski

Preface

Point-free topology is a general discipline of geometry exploiting the algebraic properties of natural pieces of spaces. It originated in individual results in the late 1930s and early 1940s, developed to the current concept system in the 1960s and 1970s, and flourishes ever since. The reader can learn the basics (and certainly more than just that) in several monographs and chapters in handbooks and similar publications. There is, however, no systematic comprehensive presentation of special properties needed when treating concrete special problems. The aim of this book is to fill in this gap.

Similarly like in classical topology, when dealing with special problems the scope of the quite general concepts has to be restricted by specific conditions. In the classical theory, the most significant role in this respect is played by the so-called separation axioms of various strength. In the point-free context, the need to delimit and specify the objects suitable for particular purposes is perhaps even stronger, and it is natural to borrow and imitate the classical requirements (already in the early times of the theory, one discovered the benefits of the condition of regularity corresponding precisely to the homonymous axiom of classical topology; also, we should not forget to note the use of the so-called disjunctivity replacing the T_1-axiom in the point-free prehistory). Although the points seem to be crucial in the classical formulations of separation axioms and similar conditions, it has turned out that one can either produce exact counterparts or at least mimic them to advantage. And sometimes one discovers very useful conditions of essentially point-free nature.

This book bids a systematic study of this area. We present as much as possible of the broad range of separation type conditions, from very weak ones (in among of which the reader might be surprised by the benefits of the so far underestimated subfitness), over the pleiad of the Hausdorff type conditions, to very strong ones (the particularly strong scatteredness may be surprising to be found in this company, but in the point-free context it belongs, and plays a very interesting role). We discuss their interrelations and point out consequences.

We would like to acknowledge the support, excellent conditions and the active and friendly atmosphere in the institutions we are happy to work in—the Department of Mathematics and the Centre for Mathematics of the University of Coimbra and the Department of Applied Mathematics and IUUK of the Charles University in Prague. Further, our thanks go to our colleagues and collaborators and numerous members of the international community of categorical topology, point-free topology and related topics for their kind interest and encouragement. And we wish to express most warmly our thanks to our families for invaluable support, understanding and patience.

Coimbra, Portugal
Prague, Czech Republic
April 2020

Jorge Picado
Aleš Pultr

Introduction

1

In the nineteenth century, mathematical analysis flourished. The correct formulation of newly discovered facts needed precise thinking about phenomena like convergence, continuity (and different types of continuity) or approximation. A space was not any more just the Euclidean space or a part of it. More or less explicitly, people started to think in terms of fairly general metric spaces. This was a great progress, and this is how we treat with advantage a lot of questions of analysis (and not only of analysis) since.

But it did not take long to observe that for some basic notions, notably in particular for the most fundamental concept of continuity, the concrete metric structure was unimportant: although the definitions were naturally expressed in terms of distance (metric), replacing a concrete metric by others gave the same results. What, then, is the structure that makes a set to a space? The attempts to solve this problem culminated in Hausdorff's ingenious idea published in 1914 (*Grundzüge der Mengenlehre*, Leipzig) that laid the foundation for (set-theoretic) topology, the generalized geometry as we know it today. The idea is very transparent and intuitive: for an element (point) of a set, one has to make the difference between being surrounded by the set (like a boat in the middle of a lake, surrounded by water from all sides) or just being in the set, possible on the border (like a boat landed, still in water, but already bordering the shore). It turned out, perhaps surprisingly, that for understanding continuity and related phenomena, it suffices to understand a space as a set X with the additional information whether a subset $U \subseteq X$ containing an element x surrounds it (one speaks of being a *neighbourhood* of x) or not, with a few very simple and very intuitive axioms.

2

Instead of taking the idea of neighbourhood for the primitive concept, one can, equivalently, start with the concept of an *open set*. In a space defined as above, an open set is a set that is a neighbourhood of each of its points; conversely, if we start with open sets, we can define a neighbourhood V of x as a set such that there is an open U between x and V, that is, such that $x \in U \subseteq V$. This variant of the definition of a (topological) space was used very shortly after introducing the neighbourhood idea. The system of axioms needed here is even simpler than specifying the properties of neighbourhoods: it is only a simple property of a sublattice of the lattice of all subsets of the set of points. While it may be slightly less intuitive, it is technically much more expedient, and it is the definition mostly used.

In fact, it is less intuitive only from the point of view of a space as a suitably structured set of points. We can think of a space from another angle, though, and then the concept will become very natural. Points are constructs, not very realistic entities without extent. Forget about them and think of a space as about a system of places, spots of a typically nontrivial extent. They join to make bigger spots, and they "hold together" if they meet. This is all the geometry we need to start with, and this is the pivotal idea of point-free topology. The reader certainly knows that what we are speaking about is the so-called *frame* (or *locale*)—see [161, 220] or the Appendix below—the generalized space some aspects of which are the main topic of this book. Good and typical models of such generalized spaces are the lattices $\Omega(X)$ of open sets of topological spaces.[1] There are other frames, and the theory is a considerable extension of the classical one (we will speak of the point-based topological spaces as of the *classical ones*), but we will relate the explanatory examples in this Introduction mostly to the $\Omega(X)$. Just keep in mind that in the spaces one usually works with, the open sets are models of such realistic nontrivially extended easily understandable places, and that in the general theory such places have their own existence and do not have to consist of points.

3

Now, let us return to the classical topological spaces. The general concept is indeed very general, for many purposes too general (in one respect, however, the Hausdorff's axiomatics was not general enough; we will come to it shortly). Therefore, one often restricts the scope of spaces in question by various extra conditions, among which a particular role is played by the so-called *separation*

[1]The spaces X of a very broad class are determined and well represented by the lattice $\Omega(X)$. Therefore, one often speaks of the lattices as of the spaces.

axioms. Already the original Hausdorff's system of axioms contained such a separation condition, requiring that

(H) *any two distinct points have disjoint neighbourhoods*

(in other words, any two distinct points are *separated* by neighbourhoods). This is what we call today the *Hausdorff axiom*.

In the first years of topology, the interest was predominantly focused to spaces that were more and more geometric (or, rather, more and more like spaces with the structure determined by a metric). Thus, one considered the *regularity* requiring that

(reg) *any point x and a closed set $A \not\ni x$ have disjoint neighbourhoods*

(a point not contained in a closed set and the closed set can be *separated* by neighbourhoods), *complete regularity* where one separates such a pair of a point and a closed set by a continuous real function f (in the sense that $f(x) = 1$ and $f[A] = \{0\}$), or the *normality* requiring that

(norm) *any two disjoint closed subsets have disjoint neighbourhoods.*

Further, there are stronger types of normality like the *complete normality* where one assumes that every subspace of the given one is normal (for the plain normality this is not generally the case, unlike for the Hausdorff property or regularity) or *perfect normality* where one assumes that any two disjoint closed subsets A, B can be separated by a continuous real function f such that $f^{-1}[\{0\}] = A$ and $f^{-1}[\{1\}] = B$ (spaces satisfying this condition are already very similar to metrizable ones) or *full normality* where every open cover has a star refinement (in normality this holds for *finite* open covers).

Conversely, it turned out that there are applications of topology where the Hausdorff axiom is too strong. This led to the weaker assumption that

(T_1) *any point can be separated from any other by a neighbourhood*

or the even weaker (T_0) where for any two distinct points, at least one can be separated from the other.[2]

This is of course not an exhaustive list of separation axioms. Some more will be presented in Chap. I, for instance, the very important T_D, or the symmetry, and also the particularly important sobriety (which is in fact an axiom of a different nature, sometimes included into the separation area for quite good reasons), but the reader has certainly already an idea what we have in mind when speaking of conditions of this type.

4

In point-free topology, the need to specify the objects suitable for particular purposes is perhaps even stronger than in the realm of classical spaces. First, of

[2]For the T_k notation, see Chap. I (Sect. 4.4); T_1 is sometimes referred to as the *Fréchet axiom*.

course, there are the questions whether that or other classical result has a point-free counterpart. The spaces in such classical facts are mostly subjected to special conditions that have to be somehow satisfied also in the extended situation. But there are also (and often) phenomena that have no classical background or that are more interesting in the point-free context. Such facts very rarely hold in absolute generality. Hence, what are the (hopefully natural) conditions under which they do hold? All this calls for studying the point-free separation.

This need was obvious from the very beginning of the theory. In particular, already in the founding years of point-free topology, the role of regularity and complete regularity was recognized,[3] and very useful results on (completely) regular frames were obtained (to name just one, the choice-free variant of the Stone–Čech compactification presented in [43, 44]). Soon there also appeared an article [81] showing that one can formulate separation axioms or suitable replacements in the point-free setting. In [152], Isbell introduced and discussed important separation conditions specific for the point-free spaces.

In the recent decades, separation was studied with a growing intensity. It turned out, a.o., that some conditions (and some of them neglected for years, e.g., subfitness) play a much more important role than had been expected. Therefore, we think that the subject deserves a comprehensive treatment. It will be the main topic of this book.

Thus, what topics we wish to discuss? First of all, we will have to analyse the conditions that extend or mimic the standard ones from classical theory. Their adaptability to the point-free situation varies. Some of them can be directly or almost directly translated (such as normality and also regularity). Then there are translations based on a necessary and sufficient condition, not obvious, but better suited for the point-free context. And there are also such that are translated only seemingly: they can lead to very good analogues or sometimes to something quite different. We will also meet cases where the translation is precise but not very useful, because at a closer scrutiny they reduce to speaking on classical spaces only.

Secondly, but about this we will speak later, there are separation conditions specific for the point-free context, not making much sense for classical spaces but very useful in the general situation.

We will be interested, even in the case of spaces, only in the facts that can have a point-free interpretation. Therefore, ***we will mostly ignore the axiom*** $\boldsymbol{T_0}$. Two points that cannot be distinguished (that is, separated) by an open set have to be treated, in our perspective, as one. Therefore, for trivial reasons, there will be made no attempt to find an analogue of this axiom for frames, and, however, when speaking of classical spaces, the axiom T_0 will be automatically assumed (perhaps now and then recalled, only not to be forgotten).

Before discussing concrete examples: let us agree to call an extension of a concept *conservative* if, when applied to the frame $\Omega(X)$, it coincides with the

[3] We speak of the 1960s and later; a very important separation condition was discovered and used in 1938 (by Wallman [282]) and then neglected for decades, see paragraphs 7 and 8 below.

homonymous property of X. It can sometimes happen that a definition is not conservative, but so useful that it deserves keeping the name; however, there are also cases of conservative extensions that are practically useless.

5

From the conditions above, only one, namely normality, has an absolutely straightforward translation into the language of open sets. Replacing the closed sets A, B by their open complements C, D, we can reformulate the axiom (norm) for a space X by requiring that

for any two open sets C, D such that $C \cup D = X$, there are disjoint opens U, V such that $C \cup U = X = D \cup V$.

But also the regularity condition is very simple. For open sets, define the *rather below* relation

$$V \prec U \quad \equiv_{\mathrm{df}} \quad \overline{V} \subseteq U$$

(in the language of the lattice of open sets, $\overline{V} \subseteq U$ can be expressed, using the pseudocomplement $V^* = \operatorname{int}(X \setminus V)$ of V, as $V^* \cup U = X$). Then it is easy to see that a space is regular if and only if every open U is equal to the union $\bigcup\{V \mid V \prec U\}$.

This formula is an example of an *almost direct* translation mentioned above. Note that it comes immediately from a (very straightforward) characterization of regularity standardly used in classical topology, namely that regular spaces are those in which every neighbourhood of a point contains a closed one. (Complete regularity can be expressed similarly, with a variant of the relation $\prec$.[4] It needs however some more reasoning (see Chap. VI, Sects. 1 and 2) which brings a useful insight into the classical spatial condition as well.)

To give an example of a not quite straightforward characterization yielding a conservative extension, here is the Johnstone–Paseka–Šmarda–Sun Shu-Hao formula for the Hausdorff property (why so many authors: the formula merged from very differently motivated approaches by Johnstone and Sun Shu-Hao [171] and by Paseka and Šmarda [216]):

for any two open sets A, B such that $X \neq A \nsubseteq B$, there are disjoint opens U, V such that $U \nsubseteq A$, $V \nsubseteq B$.

[4]Namely, the *completely below* relation $\prec\!\prec$ introduced by Banaschewski in his 1953 doctoral dissertation [19], replacing a notion Alexandroff had introduced using real-valued continuous functions [2].

(It is very easy to prove that this is equivalent to the separation of points as in (H) above, one has only to reason just a bit. Note that incomparability of A and B would not do the job.)

6

The Hausdorff axiom, a point-free formulation of which we have presented in the previous paragraph, is fairly instructive. There are several approaches, two of them merging in the conservative formula we have recalled above.

Another one (Isbell [152]) was based on the following elegant (and obvious) classical characterization: a (T_0) topological space is Hausdorff if and only if the diagonal $\Delta = \{(x, x) \mid x \in X\}$ is closed in the product $X \times X$. Now this is nice: we know what the products $L \oplus L$ in the category of locales are, a diagonal subobject is well defined and well understood, and also closedness of subobjects is a straightforward concept (precisely corresponding to classical closedness of subspaces). Thus, we can say that a locale L is *Hausdorff* if the diagonal in the square $L \oplus L$ is a closed sublocale (generalized subspace). But beware: we have forgotten that the product in the category of locales does not quite correspond to the classical one: the category is bigger and hence the universality of the product is checked by many more objects than in the category of topological spaces, resulting in the fact that the product of locales $\Omega(X) \oplus \Omega(X)$ does not necessarily correspond to the product of spaces $X \times X$. This is an example of a definition that is translated only seemingly. And indeed thus defined Hausdorff property (call it (sH)) is not conservative, that is, a space can be classically Hausdorff while $\Omega(X)$ is not (sH). But it would be wrong to dismiss it. The situation is strange: locales satisfying this non-conservative definition behave like Hausdorff objects should. In many respects, they behave better than those with the conservative property (just to give an example, (sH) combined with compactness yields regularity, the conservative one does not). This shows that sometimes it can be advisable to mimic a classical property by something that is just analogous and not necessarily a precise extension.

There are several interesting facets of the Hausdorff property in the point-free context. Another one is an example of a contingency that was not mentioned yet. When extending a concept we may have a choice between two necessary and sufficient conditions that are (of course) equivalent for spaces but may diverge in the broader situation. Both the extended notions are then conservative, but different (see Chaps. III, 3.4 and 4.1).

7

We have seen that there are sometimes reasons for preferring a non-conservative analogue for a definition of an extended concept, even if we have a perfectly correct conservative one. The axiom T_1 will be an even better example. Consider the following necessary and sufficient condition:

> *A (sober[5]) space X is T_1 if and only if all the prime elements in $\Omega(X)$ are maximal.*

Here we have a nice statement in the language of open sets. Primeness is a lattice concept, not to speak of maximality. Yet, taking it for a definition of a frame separation axiom ("a frame is said to be T_1 if each of its prime elements is maximal"), which is sometimes done, is not particularly helpful. Why? It is because the primes represent the points of the spectrum (see A.3.4) so that one actually speaks only about the spectrum ΣL, the spatial part of L. Defining a property $\widetilde{P}$ for a frame L by requiring P for ΣL is somewhat cheap, but first of all it is practically useless: it cannot say anything new in the extended point-free area.[6]

This leads to an example of a separation condition that is inherently point-free (although it does make perfect sense for spaces as well). Consider the following requirement on a (complete) lattice L:

$$\forall a, b \in L \ \ (\text{if } a \nleq b \text{ then there is a } c \in L \text{ such that } a \vee c = 1 \neq b \vee c).$$

This property is called *subfitness*. It appeared, as a suitable substitute for T_1 already in 1938, in one of the first papers using point-free reasoning (Wallman [282]), under another name, long before point-free topology started to be cultivated. (It obviously holds in T_1-spaces: If $A \nsubseteq B$ takes an $x \in A \smallsetminus B$; then, we have the open $C = X \smallsetminus \{x\}$ with $A \cup C = X \neq B \cup C$. Moreover, under a very weak condition T_D, it is equivalent with T_1.)

Subfitness is strictly weaker than T_1 in classical spaces (see II.2.2). For frames it is extremely useful. It is not only a very good replacement of T_1; we will meet it again and again as a sufficient (often, necessary and sufficient) condition for important point-free facts.

8

Although subfitness makes a good sense in classical topological spaces (see Sect. 2.1 in Chap. II), we cannot view it as a translation of a classical property. It appeared in the aforementioned Wallman's article [282] as an inherent lattice theoretic concept

---

[5]For sobriety see I.7. We think now of the typical primes $X \smallsetminus \overline{\{x\}}$; if $\overline{\{x\}} \neq \{x\}$, we have two such open sets, both prime, comparable.

[6]Admittedly, this is a rather harsh rebuff. Thus defined T_1 has its role in analysing analogues C of stronger classical properties. The question whether such C implies T_1 is certainly legitimate (see, e.g., IV.1). But we are speaking on the property in se.

(called *disjunctivity*[7]) that could serve as a helpful *replacement* of T_1. When it reappeared after more than three decades, in a different but equivalent form, it was as a specific property of point-free spaces, already studied as objects in their own right.

Indeed, in 1972, Isbell [152] defined *subfitness* as the property that

(sfit) *every open sublocale (generalized subspace) is a join of closed ones,*

together with the *fitness* where he required that

(fit) *every closed sublocale (generalized subspace) is a meet of open ones.*

These two properties look, deceptively, as being somehow dual to each other. They are not. Actually, the latter can be characterized as hereditary subfitness (subfitness itself is not inherited by subobjects) and it is much stronger. While in spaces subfitness is weaker than T_1, fitness is almost as strong as regularity. The discrepancy may be surprising, but keep in mind that in the point-free context a space can have a lot of new sublocales, and requiring a given property for all of them is a rather strong claim.

The open-closed formulas for subfitness and fitness are good examples of conditions that look like translations from the classical theory but are indeed not so. In fact, requiring in spaces that every open subset is a union of closed ones makes a good, but different, sense (namely, it is equivalent with the so-called *symmetry*, which the subfitness is not), and by De Morgan formula the same is obtained requiring that every closed subset is an intersection of open ones. (Although open resp. closed sublocales are in perfect correspondence with open and closed subsets, the lattice structure of the system of sublocales is more complex; also, one has just a one-sided De Morgan formula.)

The formula for fitness is, perhaps surprisingly, equivalent with the formally stronger

(fit') *every sublocale whatsoever is a meet of open ones.*

We have already stated that fitness and subfitness are, despite the appearance, far from being dual to each other. Hence, one can hardly expect that the equivalence of fitness with (fit') should have a full analogue for subfitness. And indeed the condition that

every sublocale whatsoever is a join of closed ones

is much stronger than subfitness (actually, even much stronger than fitness). It turns out that it is equivalent with the well-known *scatteredness*, a property that is perhaps more important in point-free topology than in the classical one (in particular, *scattered spaces* are precisely those for which subspaces and sublocales coincide).

---

[7] Wallman had in mind the lattice of *closed* sets. Much later, when the opposite perspective of *open* sets was already firmly established, the authors of, e.g., [192, 254, 255] started to speak of *conjunctivity*.

Note that the analogous claims about spaces and subspaces as above are again the same, every subset is a union of closed ones if and only if every open subset is an intersection of open ones, and it is nothing else but T_1. Here we can see the pitfalls of extending notions by analogy: we have two necessary and sufficient conditions for T_1 and if we extend them with general subobjects instead of subspaces, we eventually obtain two very strong (distinct) conditions.

We have got acquainted with the setting. Now we can briefly outline the contents.

In Chap. I, we summarize the well-known separation axioms in classical topological spaces. We add some more: symmetry, the very important T_D, and sobriety. We start to think in terms of the lattice of open sets.

Chapter II is devoted to the basics of subfitness and fitness. It starts with T_1 and the point-free T_U (which differs from the spatial case where it coincides with symmetry). Some consequences of subfitness are presented: in particular, a somewhat surprising useful formula for the Heyting operation (and pseudocomplement) and a spatialization theorem. Isbell's formulas for subfitness and fitness are introduced, and fitness is shown to be the hereditarily modified subfitness. Specific properties of congruences under these conditions are presented.

In Chap. III, we turn to axioms of Hausdorff type. We introduce five different approaches one encounters in literature and discuss the resulting four concepts (two of the approaches merge) and show how they relate. Then, after introducing the necessary techniques, we prove some particularly nice properties of the "strong Hausdorff axiom" (the closed diagonal one) in which the others fail or most probably fail: the density theorem, the facts that under this condition compact sublocales are closed, and that strongly Hausdorff compact frames have stronger separation properties.

Chapter IV summarizes the "low separation properties". First, relations between T_1, T_U and subfitness are briefly discussed. Then, two more characterizations of the strong Hausdorff property are introduced, and the merits of the individual variants of point-free Hausdorff axioms are pointed out. After presenting some implications and (sometimes surprising) non-implications, we outline the tangle of relations in the low separation world.

In Chap. V, we discuss regularity, the first separation axiom that had been widely technically used in point-free theory and also the previously defined fitness from the perspective of its closedness to regularity. We show that regularity implies the strong Hausdorff property and all the lower separation properties. The historical role is emphasized and proofs originally used when regularity substituted weaker assumptions recalled. Some properties of fitness presenting this property as a relaxed regularity are proved, together with discussing other related assumptions.

In [220], we had an incorrect proof of reflexivity of fitness; we pay the debt now by rendering a correct one.

Chapter VI is concerned with complete regularity. First, we briefly discuss the relation "completely below" (to the analysis of which we return at the end of the chapter). Complete regularity is compared with regularity; we recall examples proving it is strictly stronger. The role of complete regularity in uniformization is recalled. The structure of cozero elements is discussed, and a Lindelöf reflection (a specifically point-free phenomenon) is presented. Finally, a choice-free variant of complete regularity is introduced (an application of which is, e.g., the choice-free Stone–Čech compactification).

In Chaps. VII and VIII, we treat the question of normality and various stronger properties (omitting paracompactness, also known as full normality, which had its own chapter in [220]). We start with plain normality and its basic properties. Next, we discuss the behaviour of finite covers and present the Wallman compactification of a subfit frame, showing that, under normality, it coincides with the Stone–Čech compactification. We conclude Chap. VII with a discussion of complete normality, the hereditary version of normality.

In Chap. VIII, we have two more variants of normality. There is the perfect normality, which turns out to be a conjunction of the classical perfectness (that is slightly different in the point-free context due to the different behaviour of sublocales and subspaces) and normality. Next, we deal with the technically important collectionwise normality, weaker than paracompactness. Then we present point-free real-valued functions and prove, in the penultimate section, the Katětov–Tong insertion theorem, using to advantage the techniques of the point-free real line. We finish with a unified view of several weaker variants of normality and a glimpse of the parallel between normality and extremal disconnectedness.

When comparing fitness and subfitness in 8 above, we mentioned that unlike fitness, where we can think of *all* sublocales as intersections of open ones, arbitrary sublocales being joins of closed ones is a property much stronger than subfitness. This and the theory of an ensuing envelope is the topic of Chap. IX. First, we recall the concept of scatteredness, both in the point-free and in the classical contexts, and show that for frames it coincides with the stronger formula from 8. For spaces, we present the Simmons (and Niefield–Rosenthal) theorems on sublocales in scattered spaces and frames. The frame of joins of closed sublocales $\mathsf{S}_{\mathfrak{c}}(L)$ is introduced. It is shown that for subfit frames it is a Boolean algebra, and more (for instance, that it is the maximal essential extension).

The last chapter contains some facts that did not exactly fit into the individual chapters.

For the convenience of the reader, we add a concise appendix containing some definitions and facts they may prefer to have at hand rather than having to look for them elsewhere. Besides repeating facts from our previous monograph [220], it also presents facts that are not quite so easy to look for in the literature (like, e.g., the frame coproduct $\oplus$ as a tensor product) and simpler proofs of known facts.

Contents

Chapter I
Separation in Spaces

We will start by recalling some standard separation axioms in topological spaces X and discussing how to cope with them without points. In the point-free setting we think of classical spaces in terms of the complete lattices $\Omega(X)$ of their open sets, and of general spaces as of complete lattices of similar nature. Hence, the fact that the classical separation is expressed by statements in which individual points (and non-open subsets) play a prominent role seems to be a principal obstacle for extending them to the more general context. But the situation is much more favourable than what one may expect. First of all, some of the separation conditions can be replaced by obviously equivalent statements using the calculus of the lattice only: such is an absolutely straightforward reformulation of normality, very easy (and classically transparent) reformulation of regularity, and a translation of complete regularity which needs some more explanation (but this explanation concerns the role of real functions which calls for explanation in the classical situation as well). This will be presented already in this chapter. Then there are reformulations and replacements that are more involved (in particular the Hausdorff axiom that will come in variants with nontrivial relations). And then there are specific point-free separation conditions that are of particular interest: some of them akin to classical ones but not quite equivalent, some of them naturally arising from lattice theoretic requirements. They will be discussed and analysed in the individual chapters below (to be more exact: regularity, complete regularity and normality will also have their individual chapters; what we have in mind is that their reformulations can be presented right away while the others will be postponed).

Besides the most standard classical separation axioms (those of the T_k-sequence, $k = 0, 1, 2, 3, 3\frac{1}{2}, 4$), we will introduce in this chapter also the symmetry and two more, T_D and sobriety. The last two named are extremely important for linking classical and point-free theories. Sobriety (which is, properly spoken, not really a separation axiom, although sometimes, for good reasons, included in the class) takes care for agreement of classical continuity and its algebraic expression in an appropriate type of lattice homomorphisms; T_D has several aspects, the

J. Picado, A. Pultr, *Separation in Point-Free Topology*,
https://doi.org/10.1007/978-3-030-53479-0_1

most important is in representing classical subspaces in among the generalized subobjects.

1 T_0: Not So Much for Point-Free Purposes, But with at Least One Important Aspect

1.1 The axiom T_0 This is the following very weak separation assumption.

T_0: *if $x, y \in X$, $x \neq y$, then there is an open U such that either $y \notin U \ni x$ or $x \notin U \ni y$.*

Equivalently, T_0 amounts to the very expedient implication

$$\overline{\{x\}} = \overline{\{y\}} \quad \Rightarrow \quad x = y. \tag{T_0}$$

1.2 Note and convention See what happens if the assumption from T_0 is not satisfied for two distinct x, y in X. Then we have that

$$\text{for every open } U \subseteq X, \ \ x \in U \text{ iff } y \in U.$$

Thus, using the language of the lattice of open sets (about which more will be said in 5), such two points are undistinguishable. As we have already agreed in the Introduction,

- we will not try to find a point-free replacement of the T_0-axiom,
- and working with classical spaces we will assume it (sometimes tacitly, sometimes explicitly, just not to forget).

Our convention of automatically assuming T_0 when working with classical spaces is of utmost importance, and the reader should keep it in mind. In particular it follows that the representation of continuous maps by their preimage functions will be in this book always automatically one-to-one.

1.3 Observation *Let Y be a T_0-space and let $f, g\colon X \to Y$ be continuous maps. If $f^{-1}[U] = g^{-1}[U]$ for every open U in Y, then $f = g$.*

(Indeed, let $f(x) \neq g(x)$ for some x. Choose an open U such that, say, $f(x) \notin U \ni g(x)$. Then $x \in g^{-1}[U] \smallsetminus f^{-1}[U]$.)

2 T_1 and Symmetry

2.1 T_1 is the following assumption.

T_1: *if $x, y \in X$, $x \neq y$, then there is an open U such that $y \notin U \ni x$.*

This is obviously equivalent to

$$\forall x \in X, \quad \overline{\{x\}} = \{x\}, \tag{T_1}$$

that is, to the fact that one-point sets are closed, and consequently

all finite sets are closed.

2.1.1 Note The closedness of one-point sets is a fairly "geometric assumption" and, particularly in the first formulation above, it may seem to be the first possible strengthening of T_0 (we just remove the choice of the order of the points x, y). It is not so: we will see shortly very natural and useful conditions that are weaker than T_1 (T_D still in this chapter, subfitness in the next).

2.2 Observation Since $X \smallsetminus \{x\}$, *if it is open, that is, if* $\overline{\{x\}}$ *is closed*, is a maximal element in the lattice $\Omega(X)$, and since *always* $X \smallsetminus \overline{\{x\}}$ is prime, one may think of the "point-free" representation of T_1 as requiring that

each prime element is maximal.

This does not seem to mention points, but, although it indeed is what is substituted for T_1, it is not really very satisfactory. See II.1.1 below.

2.3 Symmetry The following useful assumption played an important role in the enrichment of topology generalizing uniform structures [144, 146]. A space is *symmetric*[1] if we have

(sym): $\forall x, y \in X, \quad x \in \overline{\{y\}}$ *iff* $y \in \overline{\{x\}}$.

In other words,

$$x \in \overline{\{y\}} \quad \Rightarrow \quad \overline{\{x\}} = \overline{\{y\}}. \tag{sym}$$

We will see, more than once, that it plays a role of its own.

2.3.1 Proposition *Under T_0, symmetry is equivalent to T_1.*

Proof Suppose (sym) holds. Let $x \in \overline{\{y\}}$. Then by (sym), $\overline{\{x\}} = \overline{\{y\}}$ and, by T_0, $x = y$. On the other hand, if we have T_1 and $x \in \overline{\{y\}}$, then $x = y$ and hence trivially $y \in \overline{\{x\}}$. □

2.3.2 Proposition *A T_0-space Y is symmetric iff for any X and continuous $f, g\colon X \to Y$ we have $f^{-1}[U] \subseteq g^{-1}[U]$ for all open $U \subseteq Y$ only if $f = g$.*

Proof Let Y be symmetric and let $f^{-1}[U] \subseteq g^{-1}[U]$ for all open $U \subseteq Y$. Let for some U, $f^{-1}[U] \subsetneq g^{-1}[U]$. Choose an $x \in g^{-1}[U] \smallsetminus f^{-1}[U]$. Then $f(x) \notin U \ni g(x)$, hence $g(x) \notin \overline{\{f(x)\}}$ and by symmetry $f(x) \notin \overline{\{g(x)\}}$. Hence there

[1] Symmetric spaces were introduced by Shanin in [250] and rediscovered by Morita in [203] (as *weakly regular spaces*) and by Davis in [77] (as *R_0-spaces*).

is a V such that $g(x) \notin V \ni f(x)$, and we see that $f^{-1}[V] \nsubseteq g^{-1}[V]$. Thus, $f^{-1}[U] = g^{-1}[U]$, and, by T_0, $f = g$.

On the other hand, let the implication hold and let $x \in \overline{\{y\}}$. Define X as the subspace $\{x, y\}$ of Y, $g \colon X \to Y$ as the embedding, and $f \colon X \to Y$ as the constant $x, y \mapsto x$. Then, obviously, $f^{-1}[U] \subseteq g^{-1}[U]$ for all open U and hence $f = g$, and consequently $x = y$. □

Just as there are many examples of T_0-spaces which are not T_1, there are also many examples of symmetric spaces that are not T_0. Consider, for instance, the plane endowed with the topology given by the pseudometric $\delta((x, y), (x', y')) = |x - x'|$.

3 Hausdorff Axiom and Regularity

3.1 A space is said to be *Hausdorff*, or T_2, if

T_2: *for any two $x, y \in X$, $x \neq y$, there are disjoint open U, V such that $x \in U$ and $y \in V$.*

This condition was formulated (in the language of neighbourhoods) already in the original pioneering Hausdorff article of [136]. Hausdorff spaces are already quite "geometric", behave nicely in view of compactness (compact subspaces are closed, compact Hausdorff spaces have automatically much stronger properties, etc.) and admit a nice theory of convergence. The point-free treatment is not quite straightforward and we will see in Chap. III that it can be approached in several distinct ways.

3.2 A space is said to be *regular* if

(reg): *for any $x \in X$ and any closed $A \subseteq X$ such that $x \notin A$, there are disjoint open U, V such that $x \in U$ and $A \subseteq V$.*

Regularity looks at the first sight as a property stronger than T_2 but there is the hitch that a point does not have to be closed (and hence we cannot apply (reg) for $A = \{y\}$). Hence it is usually considered combined with T_1. But one can assume less.

3.2.1 Proposition *If a regular space is T_0 (and hence under our convention of 1.2 always), it is T_1, and hence Hausdorff.*

Proof Let $x \neq y$. If there is an open W with $y \notin W \ni x$ we have $x \notin \overline{\{y\}}$ and we can apply (reg) for x and $\overline{\{y\}}$. Similarly with $x \notin W \ni y$ we can apply (reg) for y and $\overline{\{x\}}$. □

3.3 Unlike the Hausdorff property, regularity can be very easily formulated in terms of the lattice $\Omega(X)$. For $U, V \in \Omega(X)$ say that V is *rather below* U and write

$$V \prec U$$

if $\overline{V} \subseteq U$. This is very intuitive:

This is still not quite in the language of $\Omega(X)$ but can be easily mended: $\Omega(X)$ has pseudocomplements V^*, that is, largest U such that $U \cap V = \emptyset$, and one sees easily that

$$V^* = X \setminus \overline{V}.$$

Thus we have

$$V \prec U \quad \text{iff} \quad \overline{V} \subseteq U \quad \text{iff} \quad V^* \cup U = X.$$

3.3.1 Proposition *A space X is regular if and only if*

$$\forall U \in \Omega(X), \quad U = \bigcup\{V \mid V \prec U\}. \tag{$*$}$$

Proof

I. Let X be regular and let $U \in \Omega(X)$. For $x \in U$ take $A = X \setminus U$ and, using (reg), open $V_x \ni x$ and $W \supseteq A$ such that $V_x \cap W = \emptyset$. Then $W \cup U = X$, that is, $x \in V_x \prec U$ and

$$U = \bigcup\{V_x \mid x \in X\} \subseteq \bigcup\{V \mid V \prec U\} \subseteq U.$$

II. Let $(*)$ hold and let $x \notin A$ for a closed A. Then $x \in U = X \setminus A$ and hence $x \in V$ for some $V \prec U$ and we have $A \subseteq W = X \setminus \overline{V}$ and $V \cap W = \emptyset$. □

4 Complete Regularity, Normality and the T_k Sequence

4.1 Usually, a space is said to be *completely regular* if

(creg): *for any $x \in X$ and any closed $A \subseteq X$ such that $x \notin A$, there is a continuous mapping $f\colon X \to \mathbb{I}$ ($\mathbb{I}$ is the closed interval $[0, 1] \subseteq \mathbb{R}$) such that $f(x) = 0$ and $f[A] = \{1\}$.*

Obviously, complete regularity implies regularity: consider the disjoint open $[0, \frac{1}{2})$ and $(\frac{1}{2}, 1]$ in $[0, 1]$ and their preimages under f.

4.2 The formulation in (creg) is expedient but the use of the real line already prefabricated beforehand makes it suspect. When one learns, say, of the Tychonoff's embedding of a completely regular (T_0) space X into a product $\mathbb{R}^M$ (and hence, of a representation of X as a subspace of an "Euclidean space of (possibly) infinite dimension") one cannot escape the feeling that here is a property sort of ready-made for such representation rather than a property of topological spaces only slightly more general than the regular ones.

In fact, however, the use of continuous real-valued functions is only a technical simplification. Let us explain what is really happening.

4.2.1 The interpolative modification of $\prec$: completely below One may expect that the "rather below" relation, $\prec$, is *interpolative*, that is, if $U \prec V$, then there is an open W such that $U \prec W \prec V$. But it is not.[2]

Complete regularity can be obtained by mending this flaw.

It is easy to see that every transitive relation R contains a largest interpolative one, $\widetilde{R}$ (the union of all interpolative $S \subseteq R$). Let us, then, concentrate on the relation *completely below* defined as

$$\prec\!\!\prec = \widetilde{\prec}.$$

Here is the standardly used construction.

Denote by D the set of dyadic rationals in the closed unit interval $\mathbb{I} \subseteq \mathbb{R}$,

$$D = \left\{ \frac{k}{2^n} \mid n \in \mathbb{N},\ k = 0, 1, \dots, 2^n \right\}.$$

We say that U is *completely below* V in $\Omega(X)$ and write

$$U \prec\!\!\prec V$$

if there are open U_d with $d \in D$ such that

$$U_0 = U,\ \ U_1 = V, \quad \text{and} \quad U_d \prec U_e \ (\text{that is, } \overline{U}_d \subseteq U_e) \ \text{ for } d < e.$$

Note that

- if $U' \subseteq U \prec\!\!\prec V \subseteq V'$, then $U' \prec\!\!\prec V'$,
- $\prec\!\!\prec$ is interpolative, and
- it contains each interpolative S contained in $\prec$.

---

[2] Actually, the intuition is not quite so deceptive. There is a famous (and standard) theorem stating that every regular space with a countably generated topology is normal, which makes the relation $\prec$ interpolative (see VII.1.2). Hence, a construction of a counterexample has to be essentially uncountable. Finding an *easy* example of a regular but not completely regular space was a hard task for the topology of the thirties and took a considerable time to solve (by Mysior, in 1981 [207]). Before Mysior's example, such a construction seemed quite complicated (the first such example, the *Tychonoff corkscrew*, can be found in Tychonoff's 1930 paper [274]).

Thus, $<<$ is the largest interpolative subrelation of $<$. Note that to prove this one needs a very weak choice principle. Later (in VI.8) we will show that the idea of the largest interpolative relation can be used for a fully choice-free treatment of complete regularity.

4.2.2 Lemma *Let there be open U_d with $d \in D$ in a topological space X such that*

$$U_0 = U, \quad U_1 = V, \quad \text{and} \quad U_d < U_e \ \text{for}\ d < e.$$

Define $\phi\colon X \to \mathbb{I}$ by setting

$$\phi(x) = \inf\{d \mid x \in U_d\}.$$

Then ϕ is continuous.

Proof

I. Let $\phi(x_0) = 1$. Choose $\varepsilon > 0$ and $d, e \in D$ with $1 - \varepsilon < d < e < 1$. Then $x_0 \notin U_e$ and hence $x_0 \in X \smallsetminus \overline{U_d}$. If $x \in V = X \smallsetminus \overline{U_d}$ is arbitrary, then $x \notin U_d$ and $\phi(x) \geqslant d > 1 - \varepsilon$. Hence we have a neighbourhood V of x_0 such that $\phi[V] \subseteq (1 - \varepsilon, 1]$. (Note that this reasoning is valid also for $x_0 \notin U_1$.)
II. If $\phi(x_0) = 0$ choose $d \in D$, $0 < d < \varepsilon$ such that $x_0 \in U_d$. Then for any $x \in U_d$, $\phi(x) \leqslant d < \varepsilon$.
III. If $\phi(x_0) = a$, $0 < a < 1$, choose $d_1, d_2 \in D$ such that $a - \varepsilon < d_1 < a < d_2 < a + \varepsilon$. Then $x_0 \in V = (X \smallsetminus \overline{U_{d_1}}) \cap U_{d_2}$ and $\phi[V] \subseteq (a - \varepsilon, a + \varepsilon)$ (for $X \smallsetminus \overline{U_{d_1}}$ use the same reasoning as in I). □

Note This is the well-known construction from Urysohn's Lemma, see, e.g., [179], standardly used for separating closed sets in a normal space by real-valued functions.

4.2.3 Theorem *A topological space X is completely regular iff for every $U \in \Omega(X)$,*

$$U = \bigcup\{V \mid V << U\}. \qquad \text{(creg)}$$

Proof

I. Let X be completely regular. Let U be open and $x \in U$. Then $x \notin X \smallsetminus U$ and hence there is a continuous $f\colon X \to \mathbb{I}$ such that $f(x) = 0$ and $f(y) = 1$ for $y \notin U$. For $d \in D$ set

$$U_d = f^{-1}\Big([0, \frac{1+d}{2})\Big) \quad \text{and} \quad B_d = f^{-1}\Big([0, \frac{1+d}{2}]\Big).$$

Then U_d is open, B_d is closed and, for $d < e$, $B_d \subseteq U_e$, and hence $U_d < U_e$. Thus the system witnesses for $U(x) = U_0 << U_1$ and since $x \in U(x)$ and

$U_1 \subseteq U$ we have $x \in U(x) << U$ and

$$U = \bigcup\{U(x) \mid x \in X\} \subseteq \bigcup\{V \mid V << U\} \subseteq U.$$

II. Let (creg) hold, let A be closed and let $x \notin A$. If we set $U = X \setminus A$ we have $x \in U$ and, by (creg), a $V << U$ such that $x \in V$. If we now take a system $(U_d)_{d \in D}$ witnessing for $V = U_0 << U_1 = U$ and the continuous map ϕ from 4.2.2 we see that $\phi(x) = 0$ (x is in all the U_d) and for $y \in A$, $\phi(y) = 1$ because here $\{d \mid y \in U_d\} = \emptyset$. □

4.3 Normality This is the following assumption.

(norm): *If $A, B \subseteq X$ are disjoint closed subsets, then there are disjoint open U, V such that $A \subseteq U$ and $B \subseteq V$.*

It is a very useful one. It should be noted, however, that in constructions, normality does not behave very "normal", unlike the other mentioned axioms.

4.3.1 Normality is already formulated without points. Of course it can be modified to the language of $\Omega(X)$, requiring that

(norm): *if $S, T \in \Omega(X)$ are such that $S \cup T = X$, then there are disjoint open U, V such that $S \cup U = X$ and $T \cup V = X$.*

4.4 The sequence T_k The conditions above are in a way more and more exacting, but they do not really increase in strength. One-point sets are not necessarily closed, and hence regularity does not generally imply the Hausdorff property, and normality does not imply regularity. Therefore one usually considers the combined assumptions

$$T_3 = (\text{reg})\ \&\ T_1, \quad T_{3\frac{1}{2}} = (\text{creg})\ \&\ T_1, \quad \text{and} \quad T_4 = (\text{norm})\ \&\ T_1.$$

This gives a sequence of axioms strictly increasing in strength:[3]

$$T_0 \Longleftarrow T_1 \Longleftarrow T_2 \Longleftarrow T_3 \Longleftarrow T_{3\frac{1}{2}} \Longleftarrow T_4.$$

We have seen in 3.2.1 that T_1 in fact follows from regularity and T_0 (and hence from complete regularity and T_0). Therefore, in our convention from 1.2, after all, regularity is basically the same as T_3 and complete regularity is basically the same as $T_{3\frac{1}{2}}$.

On the other hand, normality and T_0 does not imply T_1. Consider the trivial example of $X = \{0 < 1 < 2\}$ with up-sets for open sets: then two closed sets are disjoint only if at least one of them is void. But even here we have T_1 implied

[3]In the well-known terminology of Alexandroff and Hopf [4], Kolmogorov spaces $\supseteq$ Fréchet spaces $\supseteq$ Hausdorff spaces $\supseteq$ Vietoris spaces $\supseteq$ Tychonoff spaces $\supseteq$ Tietze spaces.

if we assume, together with normality, something weaker (namely subfitness). See VII.1.

4.5 The sequence R_k Recall the symmetric spaces from 2.3. In [77] Davis calls them R_0-*spaces* and speaks also of R_1-*spaces*. A space X is R_1 if

R_1: $\forall x, y \in X,\ \overline{\{x\}} \neq \overline{\{y\}} \ \Rightarrow\ \overline{\{x\}} \subseteq U$ *and* $\overline{\{y\}} \subseteq V$ *for some disjoint open* U, V.

Similarly as in 2.3.1 we have

4.5.1 Proposition *Under T_1, R_1 is equivalent to T_2.*

Proof Suppose R_1 holds. Let $x \neq y$. By T_1, $\overline{\{x\}} = \{x\} \neq \{y\} = \overline{\{y\}}$ and hence by R_1 there exist disjoint open U, V such that $x \in U$ and $Y \in V$. The converse is obvious. □

4.5.2 Now, denoting (reg) by R_2 and (creg) by R_3 we have

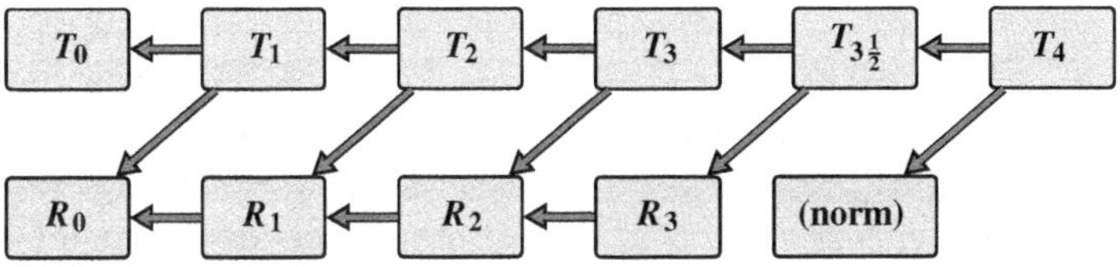

and the combined assumptions in 4.4 are augmented to

$$T_i = R_{i-1} \mathbin{\&} T_{i-1} = R_{i-1} \mathbin{\&} T_0 \quad (i = 1, 2, 3), \qquad T_{3\frac{1}{2}} = R_3 \mathbin{\&} T_3 = R_3 \mathbin{\&} T_0$$

and, not quite following the pattern,

$$T_4 = (\text{norm}) \mathbin{\&} T_{3\frac{1}{2}} = (\text{norm}) \mathbin{\&} T_1.$$

5 A Few Notes About the Lattices $\Omega(X)$ and Frames

5.1 The time has come to prepare a few concepts for further investigations and summarize some of the so far observed facts. First of all, we have that *the lattice $\Omega(X)$ of open sets of a space X is a complete distributive lattice with joins given as unions, and* ***finite*** *meets given as intersections. In fact, this lattice is* *more than just distributive. Namely, we have the* ***frame law***

$$(\bigcup_{i\in J} U_i) \cap V = \bigcup_{i\in J}(U_i \cap V) \tag{frm}$$

for any $\{U_i\}_{i\in J} \subseteq \Omega(X)$ *and any* $V \in \Omega(X)$.

(Note, however, that although infinite meets in $\Omega(X)$ do not coincide with intersections they are sometimes also of interest.)

5.2 Frame homomorphisms A mapping $h\colon \Omega(X) \to \Omega(Y)$ is a *homomorphism* (more precisely, a *frame homomorphism*, but there will be no other homomorphisms in this chapter) if it preserves all joins (unions), all binary meets (intersections), and the top.

A typical homomorphism is $\Omega(f)\colon \Omega(X) \to \Omega(Y)$ obtained from a continuous map $f\colon Y \to X$ by setting $\Omega(f)(U) = f^{-1}[U]$.

5.3 Frame congruences In accordance with 2.2 we will consider the *congruences* on the lattices $\Omega(X)$ as the equivalence relations respecting all joins and all finite meets (again, more precisely, one should speak of *frame congruences*).

Obviously, given a homomorphism $h\colon \Omega(X) \to \Omega(Y)$, we have a congruence

$$E_h = \{(U, V) \mid h(U) = h(V)\}.$$

In particular, an embedding $j\colon Y \subseteq X$ of a subspace yields the congruence

$$E_Y = \{(U, V) \mid U \cap Y = V \cap Y\}.$$

Such congruences can be used as representations of subspaces, and such a representation is correct under very weak conditions (see next section). On the other hand, the reader certainly knows (see also IX.3) that $\Omega(X)$ can have (and typically has) natural subobjects that are not induced by subspaces, and that this is actually a very useful fact.

5.4 Filters, in particular completely prime ones Similarly like in any distributive lattice, we consider filters in $\Omega(X)$. In this chapter we will be particularly interested in the *completely prime* ones, that is, the filters $\mathcal{F}$ such that

$$\text{if } \bigcup_{i\in J} U_i \in \mathcal{F} \text{ then there is a } k \in J \text{ such that } U_k \in \mathcal{F}.$$

For instance, for any $x \in X$ we have the completely prime filter of open neighbourhoods of x,

$$\mathcal{F}(x) = \{U \in \Omega(X) \mid x \in U\}.$$

5.5 Frames: some separation axioms in lattice terms In this book we will work with *frames* (recall A.3), complete lattices satisfying the distributivity law

$$(\bigvee_{i\in J} a_i) \wedge b = \bigvee_{i\in J} (a_i \wedge b),$$

of which the $\Omega(X)$ are a special case (recall 5.1). We have seen that some of the classical separation axioms can be easily translated to the lattice (frame, point-free) language. Thus in particular we can define $a \prec b$ in a frame L by setting

$$a \prec b \quad \equiv_{\mathrm{df}} \quad \exists c,\ a \wedge c = 0 \text{ and } c \vee b = 1$$

and say that L is *regular* if

(reg): *for every* $a \in L,\ a = \bigvee\{x \mid x \prec a\}$

(compare with Sect. 3) and following Sect. 4 define $a \prec\prec b$ in an obvious way and declare L for being *completely regular* if

(creg): *for every* $a \in L,\ a = \bigvee\{x \mid x \prec\prec a\}$.

Or, we can say that L is *normal* if

(norm): *for every* $a, b \in L$ *such that* $a \vee b = 1$,
$\exists u, v \in L$ *such that* $a \vee u = b \vee v = 1$ *and* $u \wedge v = 0$.

These concepts will be discussed in detail in Chaps. V, VI, VII and VIII.

We have seen a point-free reformulation of T_1 in 2.2. It was not very satisfactory but we will do better in Chap. II. We have not said anything so far about the Hausdorff axiom; it is an intriguing topic and will be discussed in Chaps. III and IV.

In the text to follow, there will be, of course, numerous other conditions discussed, some of them specific for the point-free context.

6 One More Axiom into the T_k Sequence: T_D

6.1 In [5], the authors discussed separation conditions stronger than T_0 but weaker than T_1. In particular they defined the following.

T_D: *For every* $x \in X$ *there is an open* U *such that* $x \in U$ *and* $U \setminus \{x\}$ *is open.*

Obviously, T_1 implies T_D and T_D implies T_0. A typical example of a T_D-space that is not T_1 is the *Alexandroff space* of a nontrivial poset: take a partially ordered set $(X, \leqslant)$ with its *Alexandroff topology*, that is, with the $\Omega(X, \leqslant)$ consisting precisely of the up-sets (i.e. the U such that $\uparrow U = U$). The resulting space is T_D: indeed, for every $x \in X$ we have the open $\uparrow\{x\} \ni x$ and $\uparrow\{x\} \setminus \{x\}$.

(Note that it was essential that $\leqslant$ was a partial order. If it were just a preorder, that is, if we could have $x \leqslant y$ and $y \leqslant x$ for some $x \neq y$, the space of up-sets would not be T_D—not even T_0, for that matter.)

T_D axiom turned out to be very useful. In this section we will prove two important consequences (in particular the 6.3 is crucial) but there will be others throughout the text.

6.2 Already in 1962, one of the authors of [5] proved that, under T_D, the spaces with isomorphic lattices of open sets are homeomorphic [271]. This was the first result of this kind, and although there came more expedient ones later (see 7.4 below), it is worth presenting here.

6.2.1 Lemma *A space X satisfies T_D iff for every $x \in X$, $(X \smallsetminus \overline{\{x\}}) \cup \{x\}$ is open.*

Proof Let X satisfy T_D. Choose an open U as in the definition. Each $V = (X \smallsetminus \overline{\{x\}}) \cup \{x\}$ is obviously a neighbourhood of each $y \in V$, $y \neq x$. But it is also a neighbourhood of x: we have $U \smallsetminus \overline{\{x\}} = U \smallsetminus \{x\}$ and hence $x \in U \subseteq V$.

On the other hand, if V is open, we have $x \in V$ and $V \smallsetminus \{x\} = X \smallsetminus \overline{\{x\}}$ is also open. □

6.2.2 Theorem *Let X and Y satisfy T_D and let the lattices $\Omega(X)$ and $\Omega(Y)$ be isomorphic. Then X and Y are homeomorphic.*

Proof Let $\varphi\colon \Omega(X) \to \Omega(Y)$ be a lattice isomorphism. For each $x \in X$ set

$$U(x) = X \smallsetminus \overline{\{x\}} \quad \text{and} \quad V(x) = U(x) \cup \{x\}.$$

They are distinct open sets (the latter by the Lemma) and hence we have nonempty difference $\varphi(V(x)) \smallsetminus \varphi(U(x))$.

Claim *The difference $D = \varphi(V(x)) \smallsetminus \varphi(U(x))$ consists of precisely one-point.*

Proof of Claim Suppose there are two points y_1, y_2 in D. Since Y is T_D it is, in particular, T_0 and therefore we have, say, $y_2 \notin \overline{\{y_1\}}$. Since y_1 is not in the open $\varphi(U(x))$ we have $\overline{\{y_1\}} \cap \varphi(U(x)) = \emptyset$. Hence

$$\varphi(U(x)) \subsetneq \varphi(V(x)) \smallsetminus \overline{\{y_1\}} \subsetneq \varphi(V(x))$$

and denoting $W = \varphi(V(x)) \smallsetminus \overline{\{y_1\}}$ and applying the inverse isomorphism φ^{-1} to the inequalities we obtain a contradiction

$$X \smallsetminus \overline{\{x\}} \subsetneq \varphi^{-1}(W) \subsetneq (X \smallsetminus \overline{\{x\}}) \cup \{x\}.$$

Hence, denoting by $f(x)$ the single element of D, we have

$$\varphi(V(x)) = \varphi(U(x)) \cup \{f(x)\}.$$

We have already observed above that $\overline{\{y_1\}} \cap \varphi(U(x)) = \emptyset$ and hence

$$\varphi(U(x)) \subseteq Y \smallsetminus \overline{\{f(x)\}}.$$

We cannot have $\varphi(U(x)) \subsetneq Y \smallsetminus \overline{\{f(x)\}}$: indeed, in such a case we would have

$$U(x) \subsetneq \varphi^{-1}(Y \smallsetminus \overline{\{f(x)\}})$$

but then $V(x) \subseteq \varphi^{-1}(Y \smallsetminus \overline{\{f(x)\}})$ and $\varphi(V(x)) \subseteq Y \smallsetminus \overline{\{f(x)\}}$, contradicting $f(x) \in \varphi(V(x))$. Thus we see that

$$\varphi(X \smallsetminus \overline{\{x\}}) = Y \smallsetminus \overline{\{f(x)\}}.$$

Similarly for φ^{-1}, we have a mapping $g: Y \to X$ such that

$$\varphi^{-1}(Y \smallsetminus \overline{\{y\}}) = X \smallsetminus \overline{\{g(y)\}}.$$

The mappings $f: X \to Y$ and $g: Y \to X$ are inverse to each other (indeed,

$$X \smallsetminus \overline{\{x\}} = \varphi^{-1}(\varphi(X \smallsetminus \overline{\{x\}})) = \varphi^{-1}(Y \smallsetminus \overline{\{f(x)\}}) = X \smallsetminus \overline{\{g(f(x))\}}$$

and hence $x = gf(x)$, and similarly $y = fg(y)$). We will prove that they are homeomorphisms, for which it suffices to show that for any $U \in \Omega(X)$, $f[U] = \varphi(U)$. Now if $x \notin U$, $U \subseteq X \smallsetminus \overline{\{x\}}$ and hence

$$\varphi(U) \subseteq \varphi(X \smallsetminus \overline{\{x\}}) = Y \smallsetminus \overline{\{f(x)\}}$$

so that $f(x) \notin \varphi(U)$. Thus

$$x \notin U \;\Rightarrow\; f(x) \notin \varphi(U)$$

and similarly

$$y \notin \varphi(U) \;\Rightarrow\; g(y) \notin \varphi^{-1}(\varphi(U)) = U$$

so that, considering $y = f(x)$, $x \notin U$ iff $f(x) \notin \varphi(U)$, and finally $f[U] = \varphi(U)$. □

6.3 In the sequel we will encounter T_D in various roles (cf. [57]). Here is one of its features that is particularly important for point-free topology, namely that this is precisely the condition under which subspaces are correctly represented by frame congruences (and hence by sublocales, see A.5.5). This we will present right away.

Recall the congruences

$$E_Y = \{(U, V) \mid U \cap Y = V \cap Y\}$$

from 5.3. We have

Theorem *$E_Y \neq E_Z$ for any two distinct $Y, Z \subseteq X$ iff X is a T_D-space.*

Proof

I. Let X be T_D and suppose, say, there is an $a \in Y \smallsetminus Z$. Take an open $U \ni a$ such that $V = U \smallsetminus \{a\}$ is also open. Then $U \cap Y \neq V \cap Y$ while $U \cap Z = V \cap Z$. Thus, $E_Y \neq E_Z$.

II. If distinct Y, Z yield distinct E_Y, E_Z we have, in particular, $E_{X \smallsetminus \{x\}} \neq E_X = \Delta_{\Omega(X)}$. Hence there have to be open $U \nsubseteq V$ such that

$$U \cap (X \smallsetminus \{x\}) = U \smallsetminus \{x\} = V \cap (X \smallsetminus \{x\}) = V \smallsetminus \{x\}$$

which is possible precisely if $x \in U$ and $V = U \smallsetminus \{x\}$. □

7 Sobriety: Not Quite Separation, But Important

7.1 Sobriety, the condition we will discuss in this section, is not really a separation axiom. Rather, it is a sort of completeness. In 7.4 the reader will see the analogy with the completeness in uniform (or just metric) spaces: it is a claim that a filter of open sets that behaves like a system of neighbourhoods of a point really is a system of neighbourhoods of a point.

Nevertheless, it is closely related to genuine separation axioms (for instance, it is sort of complementary to T_D, and it is naturally implied by T_2). But, first of all, it is a fundamental link between the classical and point-free topology. Namely, it is precisely the condition under which continuous maps are correctly represented by frame homomorphisms (consequently, we also have that for a sober space X the lattice $\Omega(X)$ determines the space X, similarly like in 6.2.2, but this time we have much more).

7.2 Recall that an element $p \neq 1$ in a distributive lattice is said to be *prime* if

$$p = a \wedge b \;\Rightarrow\; p = a \text{ or } p = b$$

or, equivalently,

$$a \wedge b \leqslant p \;\Rightarrow\; a \leqslant p \text{ or } b \leqslant p.$$

Let X be a space and let $x \in X$. Then

$$X \smallsetminus \overline{\{x\}} \tag{$*$}$$

is prime in $\Omega(X)$.

(Indeed, if $X \smallsetminus \overline{\{x\}} = U \cap V$, then x cannot be in both U and V, hence, say, $x \notin U$, $x \in X \smallsetminus U$, hence $\overline{\{x\}} \subseteq X \smallsetminus U$, and $U \subseteq X \smallsetminus \overline{\{x\}}$.)

A space is said to be *sober* (in the formulation of Grothendieck and Dieudonné [118]) if

(sob): *it is T_0 and all the primes in $\Omega(X)$ are the sets of the form* $(*)$.

(The role of T_0 is that one wishes for the x in $X \smallsetminus \overline{\{x\}}$ to be uniquely determined. Recall that we had already agreed that the spaces in this book will be T_0; thus, $x \mapsto X \smallsetminus \overline{\{x\}}$ is always a one-to-one correspondence.)

7.3 Proposition *Every Hausdorff space is sober, and* (*sob*) *is incomparable with* T_1.

Proof

I. Let X be Hausdorff and let $P \in \Omega(X)$ be prime. Suppose there are $x, y \notin P$, $x \neq y$. Choose disjoint U, V such that $x \in U$ and $y \in V$. Then $P = (P \cup U) \cap (P \cup V)$ and neither $P \cup U$ nor $P \cup V$ is P.
II. The *cofinite topology* on an infinite X (where $U \subseteq X$ is open if either it is empty or $X \smallsetminus U$ is finite) is T_1 but not sober, because $\emptyset$ is prime.
III. Consider the Sierpiński space $X = (\{0, 1\}, \{\emptyset, \{1\}, \{0, 1\}\})$ (or, for that matter, any finite chain endowed with the Alexandroff topology). Then the primes are $\emptyset = X \smallsetminus \overline{\{1\}}$ and $\{1\} = X \smallsetminus \overline{\{0\}}$, hence X is sober (but not T_1). □

7.4 Theorem *A T_0-space X is sober iff the completely prime filters in $\Omega(X)$ are precisely the neighbourhood filters*

$$\mathcal{F}(x) = \{U \in \Omega(X) \mid x \in U\}.$$

Proof

I. Let X be sober and let $\mathcal{F}$ be a completely prime filter. Consider $U_0 = \bigcup\{U \mid U \notin \mathcal{F}\}$. Then by the complete primeness, $U_0 \notin \mathcal{F}$, hence it is the largest element of $\Omega(X)$ that is not in $\mathcal{F}$. Thus, since filters are up-sets we see that

$$U \in \mathcal{F} \quad \text{iff} \quad U \nsubseteq U_0.$$

U_0 is prime: it is obviously not X (else $\mathcal{F}$ would be trivial), and if $U \cap V \subseteq U_0$ then $U \cap V \notin \mathcal{F}$ and we cannot have both U and V in the filter $\mathcal{F}$.

Thus $U_0 = X \smallsetminus \overline{\{x\}}$ for some x and we conclude that

$$U \in \mathcal{F} \quad \text{iff} \quad U \nsubseteq U_0 \quad \text{iff} \quad U \nsubseteq X \smallsetminus \overline{\{x\}} \quad \text{iff} \quad x \in U.$$

II. Let the statement on complete prime filters hold and let P be prime in $\Omega(X)$. Set

$$\mathcal{F} = \{U \in \Omega(X) \mid U \nsubseteq P\}.$$

Then $\mathcal{F}$ is a filter: obviously it is an up-set, and if $U, V \nsubseteq P$, then $U \cap V \nsubseteq P$ by primeness. It is trivially completely prime: if $U_i \subseteq P$ for all $i \in J$, then also $\bigcup_{i \in J} U_i \subseteq P$. Thus $\mathcal{F} = \mathcal{F}(x)$ for some x and we have

$$U \nsubseteq P \text{ iff } x \in U, \quad \text{that is,} \quad U \subseteq P \text{ iff } \{x\} \cap U = \emptyset \text{ iff } \overline{\{x\}} \cap U = \emptyset$$

and we conclude that $P = X \smallsetminus \overline{\{x\}}$. □

7.4.1 Corollary *A sober space X can be reconstructed from the lattice $\Omega(X)$.*

Proof In the lattice $L = \Omega(X)$ consider the set

$$\widetilde{X} = \{\mathcal{F} \mid \mathcal{F} \text{ is a completely prime filter on } L\}.$$

By 7.4 this set is in a one-to-one correspondence $x \mapsto \mathcal{F}(x)$ with X. Then, if we define for a $U \in \Omega(X)$ a subset $\widetilde{U} = \{\mathcal{F} \mid U \in \mathcal{F}\} \subseteq \widetilde{X}$ we have the $\widetilde{U}$ determined in terms of the lattice L, without reference to the original points of X, and we have obtained a topological space

$$\left(\widetilde{X}, \{\widetilde{U} \mid U \in \Omega(X)\}\right)$$

homeomorphic with the original X, since $\mathcal{F}(x) \in \widetilde{U}$ iff $x \in U$. □

7.5 Theorem *Let X be sober. Then every homomorphism*

$$h \colon \Omega(X) \to \Omega(Y)$$

is $h = \Omega(f)$ (recall 5.2) for a continuous map $f \colon Y \to X$.

On the other hand, if every homomorphism $h \colon \Omega(X) \to \Omega(Y)$ is $h = \Omega(f)$ for a continuous map $f \colon Y \to X$, then X is sober.

Proof

I. Let X be sober and let $h \colon \Omega(X) \to \Omega(Y)$ be a homomorphism. For each $y \in Y$ set

$$\mathcal{F}_y = \{U \in \Omega(X) \mid y \in h(U)\}.$$

Then $\mathcal{F}_y$ is a completely prime filter: obviously it is an up-set, if $U, V \in \mathcal{F}_y$, then $y \in h(U) \cap h(V) = h(U \cap V)$, hence $U \cap V \in \mathcal{F}_y$, and if $\bigcup_{i \in J} U_i \in \mathcal{F}_y$ we have $\bigcup_{i \in J} h(U_i) = h(\bigcup_{i \in J} U_i) \ni y$ and hence $y \in h(U_j)$ for some j, so that $\mathcal{F}_y$ is completely prime. By 7.4, $\mathcal{F}_y = \mathcal{F}(x)$ for some (in fact, uniquely determined) x. If we set $x = f(y)$ we see that

$$y \in f^{-1}[U] \quad \text{iff} \quad x = f(y) \in U \quad \text{iff} \quad U \in \mathcal{F}(x) = \mathcal{F}_y \quad \text{iff} \quad y \in h(U),$$

so that f is continuous and $h = \Omega(f)$.

II. In particular, every homomorphism $h\colon \Omega(X) \to \Omega(P)$, where $P = \{p\}$ is a one-point space, is $\Omega(f)$. If $\mathcal{F}$ is a completely prime filter in $\Omega(X)$ define $h\colon \Omega(X) \to \Omega(P)$ by setting

$$h(U) = \begin{cases} P \text{ if } U \in \mathcal{F}, \\ \emptyset \text{ if } U \notin \mathcal{F}. \end{cases}$$

We easily check we have a homomorphism, hence $h = \Omega(f)$ for an $f\colon P \to X$. There is, however, only one such f, sending p to some $x \in X$. Now we have

$$U \in \mathcal{F} \quad \text{iff} \quad h(U) = f^{-1}[U] = P \quad \text{iff} \quad x = f(p) \in U.$$

□

7.6 Sober modification For a general T_0-space X construct a space $\widetilde{X}$ as in 7.4.1, that is,

$$\widetilde{X} = \Big(\{\mathcal{F} \mid \mathcal{F} \text{ is a completely prime filter on } \Omega(X)\},\ \{\widetilde{U} \mid U \in \Omega(X)\}\Big),$$

where $\widetilde{U} = \{\mathcal{F} \mid U \in \mathcal{F}\}$.

Since our filters are nontrivial, we have $\widetilde{\emptyset} = \emptyset$ and $\widetilde{X}$ is the whole of the space (the notation $\widetilde{X}$ is used in two senses, but the result coincides). Further we have

$$\widetilde{U} \cap \widetilde{V} = \{\mathcal{F} \mid \mathcal{F} \ni U \ \&\ \mathcal{F} \ni V\} = \{\mathcal{F} \mid \mathcal{F} \ni U \cap V\} = \widetilde{U \cap V},$$

and

$$\widetilde{\bigcup_{i \in J} U_i} = \{\mathcal{F} \mid \bigcup_{i \in J} U_i \in \mathcal{J}\} = \{\mathcal{F} \mid \exists i,\ U_i \in \mathcal{F}\} = \bigcup_{i \in J} \widetilde{U_i}.$$

Finally, denoting again $\mathcal{F}(x) = \{U \mid x \in U\}$, we see that if $U \nsubseteq V$ there is an $x \in U \smallsetminus V$ and hence $\mathcal{F}(x) \in \widetilde{U} \smallsetminus \widetilde{V}$. Thus,

$$U \mapsto \widetilde{U} \textit{ constitutes an isomorphism between } \Omega(X) \textit{ and } \Omega(\widetilde{X}). \qquad (*)$$

7.6.1 Proposition *The space $\widetilde{X}$ is sober.*

Proof Let $\mathfrak{F}$ be a completely prime filter in $\Omega(\widetilde{X})$. Define

$$\mathcal{F} = \{U \in \Omega(X) \mid \widetilde{U} \in \mathfrak{F}\}.$$

Then,

I. $\mathcal{F}$ is a completely prime filter: if $U, V \in \mathcal{F}$, then $\widetilde{U \cap V} = \widetilde{U} \cap \widetilde{V} \in \mathfrak{F}$, similarly if $U \in \mathcal{F}$ and $V \supseteq U$, then $V \in \mathcal{F}$, and finally if $\bigcup_{i \in J} U_i \in \mathcal{F}$, then

$$\widetilde{\bigcup_{i \in J} U_i} = \bigcup_{i \in J} \widetilde{U_i} \in \mathfrak{F}$$

and there is an i with $\widetilde{U_i} \in \mathfrak{F}$ and $U_i \in \mathcal{F}$.

Second,

II. we have $\widetilde{U} \in \mathfrak{F}$ iff $U \in \mathcal{F}$ iff $\mathcal{F} \in \widetilde{U}$. Thus,

$$\mathfrak{F} = \{\widetilde{U} \mid \widetilde{U} \in \mathcal{F}\}$$

and the statement follows. □

7.6.2 Proposition *A T_0-space X has the property that $\Omega(X) \cong \Omega(Y)$ only for Y homeomorphic with X iff it is sober.*

Proof If X is sober recall 7.4.1. If it is not, consider $\widetilde{X}$ which is sober and hence not homeomorphic to X; we have, nevertheless, $\Omega(X) \cong \Omega(\widetilde{X})$ by $(*)$. □

8 A Few Notes About Categories

This book is not a book about categories. But now and then it is in order to pay attention to the categorical aspects of some phenomena. The question whether this or another condition is, say, inherited by products or subobjects is certainly of interest. But it should not be overestimated: look at the following comparison of sobriety and the axiom T_D.

8.1 One easily sees that

- a space X is sober iff one cannot *add* a point without changing the lattice $\Omega(X)$, and
- a space X is T_D iff one cannot *subtract* a point without changing the lattice $\Omega(X)$.

Thus, sobriety and T_D are properties in a sense dual to each other. Now, however,

- the subcategory of sober spaces is very well behaved as a subcategory of the category **Top** of topological spaces and continuous mappings. It is reflective, with a very natural reflection (the *sobrification* of a space [220]), and hence it is closed under all limit constructions (in particular, under products and subspaces).
- On the other hand, the subcategory of T_D-spaces is not categorically well-behaved at all; to obtain a reflection one has to radically change the morphism

structure replacing the category **Top** by something we were originally not interested in (see [56]).

If we should decide by the categorical nicety what to study and what not we would have to dismiss one half of a pair of dual phenomena. And there are aspects of point-free topology in which T_D is the important element, and where sobriety does not help.

8.2 From classical topology we know examples of useful conditions that do not behave well under constructions, e.g. normality or paracompactness (the latter improves in the point-free context, but in spaces it cannot behave worse; it is another story).

We will discuss shortly the property of subfitness. It is not inherited by subobjects, and it is not preserved by products. In consequence, it was for decades taken for uninteresting. Only relatively recently it turned out it belonged to the most useful ones [147, 223, 224, 257].

8.3 Nevertheless, good categorical properties of some of the separation properties will be sometimes pointed out, in particular if they yield a reflective or coreflective subcategory. We have here a slightly confusing situation due to the algebraic and geometric aspects of our reasoning. The (more algebraic) category of frames **Frm** is dual to the (more geometric) category of locales **Loc** which is the natural covariant extension of the category of spaces (more precisely, of *sober* spaces see A.3.3 but this is not important at this moment).

Thus, the fact that a subcategory of **Loc** is reflective (often in parallel with the fact in spaces) appears as a construction of a *coreflection* in **Frm**. We sometimes say that the property in question is (co)reflective; the reader will understand what we mean.

Chapter II
Subfitness and Basics of Fitness

We can only agree with Peter Johnstone who wrote in [162] that

> *the first person (apart of Stone) to exploit the possibility of applying lattice theory to topology was Henry Wallman.*

In his article [282] published in 1938 (already briefly mentioned in the Introduction), Wallman presented a compactification technically based on lattice theoretic principles, and proved that to determine the homology type of a space X one needs only the lattice of closed sets. When doing that, he needed a lattice formula substituting a sufficiently weak topological separation. His ingenious idea of the "disjunctive property", namely the requirement that

> *if* $a \neq b$ *then there is a* c *such that precisely one of* $a \wedge c$ *and* $b \wedge c$ *is zero*

worked very well. Thus defined concept (now called, in the dual form, the *subfitness*) turned out to be one of the most important weak separation properties suitable for the point-free context.

In this chapter we will present a few of its properties, equivalents and consequences, some of them fairly surprising. More will come in the later chapters.

In one of its reappearances (in Isbell [152]), subfitness was introduced as the weaker one of a pair of related concepts. The stronger one, the fitness from the title, might have looked at the first sight as a technical amendment overcoming some categorical flaws of subfitness. As it has turned out, however, it is actually so much stronger that it can hardly be thought about as a "low separation axiom"; some of its properties will be, more properly, discussed in the chapter on regularity, and later. On the other hand, it is technically very closely connected with subfitness; this makes it appropriate, even necessary, to introduce the basics already here.

J. Picado, A. Pultr, *Separation in Point-Free Topology*,
https://doi.org/10.1007/978-3-030-53479-0_2

1 T_1, T_U and Subfitness

1.1 The axiom T_1 Let us recall the observation I.2.2. T_1 was there rewritten as the statement that every prime element in the lattice $\Omega(X)$ is maximal. Since this is a formula in the lattice language, it suggests itself as a point-free extension of this axiom. That is, we may say that

a frame L is T_1 if every prime element in L is maximal. (∗)

This convention is sometimes really adopted, but it is not quite as useful as it may sound. With each frame L one has naturally associated the "closest classical topological space", the spectrum ΣL (see A.3.4),[1] and (∗) just claims T_1 property for this classical space. Hence it can be said, in a way, that such an axiom is not really point-free at all. Nevertheless, it should not be quite dismissed. It can have some use when compared with other axioms, more point-free and more interesting (for instance, the question whether an axiom implies (∗) or not is sometimes of importance).

1.1.1 Note It may be of some interest to explain the (∗) as a classical T_1 in some detail. Recall the definition of a sublocale (A.5.2). The smallest sublocale, representing the void generalized subspace, is $\mathsf{O} = \{1\}$. Hence, thinking of points (or, one-point spaces) we think of the smallest nontrivial sublocales, that is of those consisting of precisely *two* elements. Now if $\{p, 1\}$ is a sublocale then for each b we have $b \to p$ equal to p or 1. In other words, if $a \wedge b \leqslant p$ and $b \nleqslant p$, then $a \leqslant p$, that is, p is a prime; thus, points are precisely the $\{p, 1\}$ with p prime. Now p is maximal iff $\{p, 1\}$ is a closed sublocale (A.6.1), precisely as the T_1 in classical spaces claims that each one-point space $\{x\}$ is closed.

1.2 The axiom T_U On the other hand we can try to extend the axiom of symmetry which, by I.2.3.1, is under T_0 (assumed anyway) equivalent with T_1. The statement in I.2.3.2 suggests the following formula (in which $h_1 \leqslant h_2$ means that $h_1(a) \leqslant h_2(a)$ for all $a \in L$).

T_U: *For any two frame homomorphisms $h_1, h_2 \colon L \to M$ such that $h_1 \leqslant h_2$ we have $h_1 = h_2$.*

This is indeed an interesting axiom[2] and will play an important role. But keep in mind that it is not conservative (and in the general context captures neither T_1 nor symmetry). Proposition I.2.3.2 heavily depends on the target object being a space; the frame M being general radically changes the situation. Yet, this T_U did appear in the literature as T_1 (e.g. in [85, 110]).

[1] There are two equivalent constructions presented there, both of them, basically extending the reconstruction of a space X from $\Omega(X)$. In this section we have in mind the variant from A.3.4.2.

[2] Isbell introduced it in [153] and refers to the locales satisfying it as *unordered locales*. The name T_U (abbreviation for *totally unordered*) is due to Johnstone [161, III.1.5].

1.3 "Conjunctivity" (subfitness) The disjunctivity in [282] was envisaged for working with the lattice of closed sets. For reasons explained in the Introduction, the point-free intuition naturally prefers the reasoning in terms of open sets. Hence it now appears in the dual form.

Thus in Simmons' papers [254, 255] in 1978 (and then in [192] and [163]) it is the *conjunctivity*:

(sfit): *if* $a \nleq b$, *then there is a* c *such that* $a \vee c = 1 \neq b \vee c$

and this is the condition we will adopt. Note that we have labelled it as (sfit), short for *subfit*, which is the term commonly used nowadays.

It should be noted that in 1978 the disjunctive property was probably already a folklore; the term "conjunctivity" seems to be an allusion to that, although not explicitly mentioned (cf. [257]).

The disjunctive property is not mentioned in the Isbell's pioneering article [152] of 1972 either. There, however, subfitness comes in a different (second order) form (see Sect. 4).

2 Some Basic Facts on Subfitness

2.1 Subfitness makes a good sense in classical spaces. It amounts to a condition only slightly weaker than T_1.

Theorem *A space is subfit (that is,* $\Omega(X)$ *satisfies (sfit)) if and only if for each* $x \in X$ *and each open* $U \ni x$ *there is a* $y \in \overline{\{x\}}$ *with* $\overline{\{y\}} \subseteq U$.

Proof Let X be subfit and let $U \ni x$ be open. Since $U \nsubseteq U \smallsetminus \overline{\{x\}}$ there is a W such that

$$U \cup W = X \quad \text{and} \quad (U \smallsetminus \overline{\{x\}}) \cup W \neq X. \tag{$*$}$$

Choose a $y \notin (U \smallsetminus \overline{\{x\}}) \cup W$. Then

$$y \notin W \quad \text{and} \quad y \notin U \smallsetminus \overline{\{x\}} \text{ so that } y \in \overline{\{x\}}. \tag{$**$}$$

Let $z \notin U$ for some $z \in \overline{\{y\}}$. Then, by ($*$), $z \in W$ and since W is open, $y \in W$ contradicting ($**$).

On the other hand let the condition hold and let $U \nsubseteq V$. Take an $x \in U \smallsetminus V$ and the y from the condition. Then for $W = X \smallsetminus \overline{\{y\}}$ we have $W \cup U = X$ and $y \notin W \cup V$ because $y \in V$ makes $x \in V$. □

2.2 Subfitness and T_D Subfitness is weaker than T_1 (a trivial example of a space that is subfit but not T_1 is

$$\omega + 1 = \{0, 1, 2, \dots\} \cup \{\omega\}$$

with $U \subseteq \omega+1$ open if it is void or contains ω). But it is not much weaker. We have

Proposition *T_D & (sfit) coincides with T_1.*

Proof Obviously T_1 implies T_D & (sfit). On the other hand let both T_D and (sfit) hold. We will show that each $\{x\}$ is closed. By T_D choose an open $U \ni x$ such that $U \smallsetminus \{x\}$ is open, and by (sfit) a W such that $W \cup U = X$ and $W \cup (U \smallsetminus \{x\}) \neq X$. Then

$$W \cup (U \smallsetminus \{x\}) = X \smallsetminus \{x\}$$

and $X \smallsetminus \{x\}$ is open. □

2.2.1 Since both T_D and (sfit) are strictly weaker than T_1 we immediately obtain

Corollary *T_D and (sfit) are incomparable.*

2.3 Note The example above also shows that subfitness (even under T_0) does not imply symmetry. On the other hand, symmetry (with T_0) is T_1 and hence it implies subfitness in spaces. But in the general context the situation radically changes. Although T_U is much stronger than symmetry, it does not imply subfitness: subfitness and T_U are in general incomparable (see IV.1).

But symmetry in spaces and subfitness in general frames do have an important feature in common. In their contexts they are precisely the conditions for admitting a nearness structure; see Sect. 7.

2.4 In I.4.4 we mentioned that normality plus T_0 does not imply regularity but that one can formally require less than T_1. Indeed we have

Proposition *A subfit normal space is regular.*

Proof Let X satisfy both (sfit) and (norm), and let

$$U \neq \bigcup\{V \mid V \prec U\}.$$

Then by (sfit) there is an open W such that $W \cup U = X$ and

$$W \cup \bigcup\{V \mid V \prec U\} \neq X.$$

By (norm) we have open U_1, U_2 such that $U \cup U_1 = W \cup U_2 = X$ and $U_1 \cap U_2 = \emptyset$. Hence $U_2 \prec U$ so that $U_2 \subseteq \bigcup\{V \mid V \prec U\}$ and since $W \cup U_2 = X$ we have a contradiction $W \cup \bigcup\{V \mid V \prec U\} = X$. □

Note This holds generally for point-free normality and regularity. See Chap. VII.

2.5 Isbell's spatialization theorem When introducing fitness and subfitness in [152], John Isbell was rather sceptical concerning the importance of the latter (probably because of its bad categorical behaviour). The one merit of this property he approbated, though, was the spatialization theorem in 2.5.4 below. It stands somewhat apart from the other facts presented in this chapter, but we think it is in order to place it here, as one of the first results proved about subfitness.

2.5.1 Recall that a frame L is *spatial* if it is isomorphic to an $\Omega(X)$ which holds precisely if

every element of L is a meet of primes.

A somewhat stronger property is the T_1-*spatiality* (being isomorphic to an $\Omega(X)$ with a T_1-space X) where we require that

every element of L is a meet of maximal ones.

2.5.2 A frame L is said to be *max-bounded* if for each $x \neq 1$ in L there is a maximal $p < 1$ such that $x \leqslant p$.

2.5.3 Proposition *A max-bounded frame is T_1-spatial iff it is subfit.*

Proof $\Leftarrow$: Let $a \nleqslant b$. Take a c such that $a \vee c = 1 \neq b \vee c$, and a maximal $p < 1$ such that $p \geqslant b \vee c$. Then $p \ngeqslant a$ (else $p \geqslant a \vee c = 1$) so that $a \nleqslant p \geqslant b$. Thus, maximal elements distinguish distinct elements.

$\Rightarrow$: Let $a \nleqslant b$. Choose a maximal p such that $a \nleqslant p \geqslant b$. Then $a \vee p > p$ and hence, by maximality, $a \vee p = 1$, and $b \vee p = p \neq 1$. □

2.5.4 Theorem *A compact subfit frame is T_1-spatial.*

Proof Let C be a chain in $L \smallsetminus \{1\}$. Then its join is not 1 because by compactness there would otherwise be an element $c = 1$ in C. Hence, by Zorn's lemma, each element in $L \smallsetminus \{1\}$ is majorized by an element in this set, that is, it is max-bounded, and by 2.5.3 it is T_1-spatial. □

2.5.5 Note As a consequence of 2.5.4 we have that

every finite subfit frame is a Boolean algebra.

It can be easily proved directly. In a finite L consider for each a the $b = \bigwedge\{x \mid x \vee a = 1\}$. By (finite) distributivity, $a \vee b = 1$. Now let L be subfit and $b \wedge a \neq 0$. Then there is a $c \neq 1$ such that $1 = c \vee (a \wedge b) = (c \vee a) \wedge (c \vee b)$. But then $c \vee a = 1$, hence $c \geqslant b$, and $c = c \vee b = 1$, a contradiction. Thus, b is the complement of a.

3 Two Somewhat Unexpected Formulas

3.1 In the proof in 2.5.5 above we used subfitness restricted to $b = 0$ only. This condition, namely

(wsfit): $\forall a \neq 0\, \exists c \neq 1\colon\ a \vee c = 1$

(see [151], but it appeared already in 1942, in [278], as property Π_0) is called *week subfitness* and is sometimes surprisingly useful.[3]

For a space X, the frame $\Omega(X)$ is weakly subfit iff each nonempty open set contains a nonempty closed set; in other words,

$$\forall U \in \Omega(X),\ U \neq \emptyset, \quad \exists x \in U \text{ such that } \overline{\{x\}} \subseteq U.$$

Weak subfitness is really strictly weaker than subfitness:

3.1.1 Example Let $X = \mathbb{N} \cup \{\omega_1, \omega_2\}$ endowed with the topology

$$\big\{A \mid A \subseteq \mathbb{N} \text{ with } \mathbb{N} \smallsetminus A \text{ finite}\big\} \cup \big\{A \cup \{\omega_1\} \mid A \subseteq \mathbb{N} \text{ with } \mathbb{N} \smallsetminus A \text{ finite}\big\} \cup$$
$$\cup \big\{A \cup \{\omega_1, \omega_2\} \mid A \subseteq \mathbb{N} \text{ with } \mathbb{N} \smallsetminus A \text{ finite}\big\} \cup \big\{\emptyset\big\}.$$

Then

$$\overline{\{n\}} = \{n\} \text{ for all } n \in \mathbb{N}, \quad \overline{\{\omega_1\}} = \{\omega_1, \omega_2\} \quad \text{and} \quad \overline{\{\omega_2\}} = \{\omega_2\}.$$

$\Omega(X)$ is clearly weakly subfit (since each nonempty open set contains a closed set $\{n\}$ for some $n \in \mathbb{N}$) but it is not subfit since it does not satisfy the characterizing condition of Theorem 2.1: for $U = \mathbb{N} \cup \{\omega_1\} \in \Omega(X)$ and $x = \omega_1 \in U$ we have $\overline{\{x\}} = \{\omega_1, \omega_2\}$ but neither $\overline{\{\omega_1\}} = \{\omega_1, \omega_2\}$ nor $\overline{\{\omega_2\}} = \{\omega_2\}$ are contained in U.

3.1.2 Proposition *L is subfit iff each closed sublocale of L is weakly subfit.*

Proof We use the fact that, obviously, in a closed sublocale $\mathfrak{c}(b) = {\uparrow}b$ (see A.6.1) the nontrivial joins coincide with those in L.

If $a \neq b$ in ${\uparrow}b$ we have $a \vee b > b = 0_{\uparrow b}$ and hence there is a $c \neq 1$, $c \geqslant b$, such that (here we use 2.5.1) $(a \vee b) \vee c = a \vee c = 1$, and $b \vee c = c \neq 1$. □

3.2 Theorem *In a weakly subfit frame we have for pseudocomplement the formula*

$$a^* = \bigwedge\{x \mid a \vee x = 1\}.$$

On the other hand, if this formula holds in a frame L, then L is weakly subfit.

Proof

I. Set $u = \bigwedge\{x \mid a \vee x = 1\}$. If $a \vee x = 1$, then $a^* = a^* \wedge (a \vee x) = a^* \wedge x$ so that $a^* \leqslant u$.

Now suppose that $a \wedge u \neq 0$. By (wsfit) there is an $x \neq 1$ such that $(a \wedge u) \vee x = (a \vee x) \wedge (u \vee x) = 1$. Hence $(a \vee x) = 1$ so that $x \geqslant u$, and $(u \vee x) = 1$ so that $x = x \vee u = 1$, a contradiction. Hence $u \wedge a = 0$ and $u \leqslant a^*$.

[3] For example, one can find some interesting consequences in [204] where it appeared under the name of *jointfit*.

II. If L is not weakly subfit there is an $a \neq 0$ such that $a \vee x = 1$ only if $x = 1$. Then $u = \bigwedge\{x \mid a \vee x = 1\} = 1$ and $a \wedge u = a \neq 0$. □

3.2.1 Remark Note that it is as if we computed the *supplement* (the smallest b such that $a \vee b = 1$, see A.2.1) which of course in a general frame (even a general subfit frame) need not exist. In this concrete situation, the resulting meet itself does not have to join with a to make the top 1: for that one would need the coframe distributivity $a \vee \bigwedge b_i = \bigwedge(a \vee b_i)$.

But we can conclude the following.

Theorem *Let L be a weakly subfit frame that is also a coframe. Then L is a Boolean algebra.*

In particular, a finite distributive lattice is Boolean iff it is weakly subfit.

3.2.2 Perhaps more important is the consequence of Theorem 3.2 for the co-linearity (recall A.2.3) of individual elements.

Proposition *In a subfit frame, an element is co-linear iff it is complemented.*

Note The assumption of subfitness is essential: in a finite frame all elements are co-linear, but not all of them are complemented.

3.3 Heyting operation as a system of pseudocomplements Recall the basics on sublocales (see A.5–6). Obviously the meets and the Heyting operation in S coincide with the meets and Heyting operation in L. In particular, pseudocomplements in $\mathfrak{c}(b) = {\uparrow}b$ are (for $a \geqslant b$)

$$a^{*b} = a \to b.$$

3.4 Theorem *In a subfit frame we have for the Heyting operation the formula*

$$a \to b - \bigwedge\{x \mid a \vee x - 1,\ x \geqslant b\}.$$

On the other hand, if this formula holds in a frame L, then L is subfit.

Proof

I. Let L be subfit. Then, by 3.1.2, ${\uparrow}b$ is weakly subfit. We have

$$a \to b = (a \to b) \wedge (b \to b) = (a \vee b) \to b$$

and since $a \vee b \geqslant b$ we have $(a \vee b) \to b = (a \vee b)^{*b}$ and we can use the formula from 3.2 (we require $x \geqslant b$ because the formula is computed in ${\uparrow}b$) to obtain

$$a \to b = (a \vee b)^{*b} = \bigwedge\{x \mid a \vee b \vee x = 1,\ x \geqslant b\} = \bigwedge\{x \mid a \vee x = 1,\ x \geqslant b\}.$$

II. Let L not be subfit. By 3.1.2 some of its closed sublocales $\uparrow b$ is not weakly subfit and hence by 3.2 there has to be an instance of an $a \geqslant b$ such that

$$a^{*b} \neq \bigwedge\{x \mid a \vee x = 1,\ x \in \uparrow b\},$$

that is, $a \to b \neq \bigwedge\{x \mid a \vee x = 1,\ x \geqslant b\}$. □

3.5 The formula from 3.4 is in a way peculiar. In particular, as we already observed, the special case from 3.2 is strongly reminiscent of the formula for supplement. Of course, the pseudocomplement a^* is not also a supplement (and hence a complement) in general weakly subfit frames: this would need the coframe distributivity. Thus, the consequence of the facts in this direction is simply the assertion in 3.2.1 that

a weakly subfit frame that is also a coframe is a Boolean algebra.

It is good to know and has useful consequences; but the reader may be slightly disappointed by only a very special fact resulting from such beautiful formulas.

There are, however, very important consequences also in the context of plain frames (with no extra distributivity assumptions). Here is a very important one.

3.5.1 Theorem *Let L be a subfit frame. Then every complete homomorphism $h\colon L \to M$ preserves the Heyting operation.*

Proof Set $H(u, v) = \{x \mid x \vee u = 1,\ x \geqslant v\}$. Then in any frame, and for any frame homomorphism h,

$$u \to v \leqslant \bigwedge H(u, v) \quad \text{and} \quad h[H(u, v)] \subseteq H(h(u), h(v)). \tag{$*$}$$

(Indeed, for any $x \in H(u, v)$, $u \to v = (x \vee u) \wedge (u \to v) \leqslant x \vee (u \wedge (u \to v)) = x \vee v = x$, and if $x \in H(u, v)$, then $h(x) \vee h(u) = h(x \vee u) = 1$ and $h(x) \geqslant h(v)$.)

Now, by 3.4, $a \to b = \bigwedge H(a, b)$. Thus, by $(*)$,

$$h(a \to b) = h(\bigwedge H(a, b)) = \bigwedge h[H(a, b)] \geqslant \bigwedge H(h(a), h(b)) \geqslant h(a) \to h(b),$$

and $h(a \to b) \leqslant h(a) \to h(b)$ because $h(a) \wedge h(a \to b) = h(a \wedge (a \to b)) \leqslant h(b)$ always. □

3.5.2 By the famous Joyal–Tierney Theorem [173], open continuous maps are represented in the point-free context by the frame homomorphisms that, moreover, preserve all meets and the Heyting operation (geometrically, it corresponds, as one would desire, to localic maps for which images of open sublocales are open). Thus, by 3.5.1,

for subfit frames L, the open frame homomorphisms $h\colon L \to M$ are precisely the complete lattice homomorphisms.

This has further impact on the study of the geometry of completeness, that is, in particular, what is the role of subfitness in comparison of open sublocales and the complete ones. It will be discussed in Chap. X.

4 Isbell's Approach: Fitness

We will use the technique of the coframe of sublocales, including a basic fact on nuclei. The reader wishing to refresh it may consult A.5.

4.1 Proposition *Let L be subfit. Then for every sublocale $S \neq L$ there is a nonempty closed sublocale $\mathfrak{c}(a)$ such that $\mathfrak{c}(a) \cap S = \mathsf{O}$.*

Proof Let $S \subseteq L$ be disjoint from no nonempty closed sublocale. Let ν_S be the nucleus associated with S (that is, $\nu_S(x) = \bigwedge\{s \in S \mid x \leqslant s\}$). Then

$$\nu_S(a) = 1 \;\Rightarrow\; a = 1$$

(else we would have $\uparrow a \cap S = \mathsf{O}$ with $\uparrow a \neq \mathsf{O}$). Now let $x \in L$ be arbitrary and let $c \vee \nu_S(x) = 1$. Since $\nu_S(c \vee x) \geqslant c \vee \nu_S(x) = 1$ we have $c \vee x = 1$. Thus, by (sfit), $\nu_S(x) \leqslant x$, that is, $\nu_S(x) = x$ and since x is arbitrary, $S = L$. □

4.2 Theorem *The following statements about a frame L are equivalent.*

(1) *L is subfit.*
(2) *The only sublocale of L that is disjoint from no nonempty closed sublocale is L itself.*
(3) *Every open sublocale in L is a join of closed sublocales.*

Proof (1)⇒(2) is in 4.1.

(2)⇒(3): For an open sublocale $\mathfrak{o}(a)$ set $S = \bigvee\{\mathfrak{c}(b) \mid \mathfrak{c}(b) \subseteq \mathfrak{o}(a)\}$. Let $\mathfrak{c}(x)$ be a closed sublocale disjoint from $\mathfrak{c}(a) \vee S$. Then $\mathfrak{c}(x \vee a) = \mathfrak{c}(x) \cap \mathfrak{c}(a) = \mathsf{O}$ (and hence $\mathfrak{c}(x) \subseteq \mathfrak{o}(a)$ and $\mathfrak{c}(x) \subseteq S$), and $\mathfrak{c}(x) \cap S = \mathsf{O}$; hence we see that $\mathfrak{c}(x) = \mathsf{O}$ and, by (2), $\mathfrak{c}(a) \vee S = L$. Since $S \subseteq \mathfrak{o}(a)$ we conclude that $S = \mathfrak{o}(a)$.

(3)⇒(1): We have

$$\mathfrak{c}(a) \subseteq \mathfrak{o}(b) \quad \text{iff} \quad a \vee b = 1 \tag{$*$}$$

(if $\mathfrak{c}(a) \subseteq \mathfrak{o}(b)$, then $\mathfrak{c}(a \vee b) = \mathfrak{c}(a) \cap \mathfrak{c}(b) \subseteq \mathfrak{o}(b) \cap \mathfrak{c}(b) = \mathsf{O}$, and if $\mathfrak{c}(a \vee b) = \mathfrak{c}(a) \cap \mathfrak{c}(b) = \mathsf{O}$, then $\mathfrak{c}(a) \subseteq \mathfrak{o}(b)$).

Now, if $a \nleqslant b$, then $\mathfrak{o}(a) \nsubseteq \mathfrak{o}(b)$ and there is an x such that $\mathfrak{c}(x) \subseteq \mathfrak{o}(a)$ and $\mathfrak{c}(x) \nsubseteq \mathfrak{c}(b)$, that is, $x \vee a = 1$ and $x \vee b \neq 1$. □

4.3 The condition that

every open sublocale is a join of closed sublocales (op-as-joins) was the original definition of Isbell for subfitness in [152]. For reasons which will become apparent soon it came together with a formally similar *fitness*:

every closed sublocale is a meet of open sublocales. (cl-as-meets)

(The definitions in [152] were formulated in terms of open and closed *congruences*, but the present ones say the same—see A.6; mostly it is much easier to work with meets and joins of sublocales.)

The term "subfit" is now commonly used and it would be hard to change. We keep it, not very gladly (the concept turned out to be the basic one rather than "sub"-something).

4.4 We can think of 4.2 as of a translation of Isbell's subfitness (op-as-joins) definition to the first order formula (sfit) of 1.3. Now we will similarly translate the (cl-as-meets) definition of fitness.

4.4.1 Theorem *Every closed sublocale in L is an intersection of open ones iff we have*

(fit): $\forall a, b \in L, \quad a \nleqslant b \;\Rightarrow\; \exists c,\ a \vee c = 1 \text{ and } c \to b \neq b.$

Proof We will use the formula for $\mathfrak{c}(a) \subseteq \mathfrak{o}(b)$ from (∗) in the proof of 4.2. The equality

$$\mathfrak{c}(a) = \bigcap\{\mathfrak{o}(x) \mid a \vee x = 1\}$$

is obviously equivalent with the inclusion

$$\mathfrak{c}(a) \supseteq \bigcap\{\mathfrak{o}(x) \mid a \vee x = 1\}$$

which, explicitly, is the same as saying (recall the formula for an open sublocale) that if $a \vee x$ and $x \to b = b$ then $a \leqslant b$, that is, when viewing this implication negatively, that if $a \nleqslant b$, then there is an x such that $a \vee x = 1$ and $x \to b \neq b$. □

Observation 4.4.2 (Fitness implies subfitness) (This immediately follows from the first order formulas. Take the c from (fit). The $c \vee b = 1$ would yield $b = 1 \to b = (c \vee b) \to b = (c \to b) \wedge (b \to b) = (c \to b)$.)

4.5 The statements (op-as-joins) and (cl-as-meets) look, deceptively, dual to each other. Actually, however, fitness is much stronger, so much stronger that we cannot think of it as a low separation axiom (see Chap. IV). Also, (cl-as-meets) is equivalent with a formally much stronger statement about general sublocales, namely that

every sublocale whatsoever is an intersection of open ones

(see 6.6 below) for which there is no parallel concerning subfitness (see Chap. IX). At this moment we will just show what happens in a simple example.

4.5.1 An easy example: the cofinal topology We have an infinite set X and the topology τ consisting of $\emptyset$ and all the $X \smallsetminus A$ with $A \subseteq X$ finite. The Heyting operation is as follows.

$$U \to \emptyset = \begin{cases} X & \text{if } U = \emptyset \\ \emptyset & \text{if } U \neq \emptyset \text{ (because then } V \cap U = \emptyset \text{ only if } V = \emptyset), \end{cases}$$

$$\emptyset \to U = X \text{ for any } U,$$

$$(X \smallsetminus B) \to (X \smallsetminus C) = X \smallsetminus (C \smallsetminus B),$$

the last because

$$(X \smallsetminus A) \cap (X \smallsetminus B) = X \smallsetminus (A \cup B) \subseteq X \smallsetminus C \text{ iff } C \subseteq A \cup B \text{ iff } C \smallsetminus B \subseteq A.$$

Now $\tau = \Omega(X, \tau)$ is obviously subfit since (X, τ) is T_1. But it is not fit. We have $X \smallsetminus \{x\} \nsubseteq \emptyset$; if $(X \smallsetminus \{x\}) \cup U = X$, then U is nonempty, in which case $U \to \emptyset = \emptyset$.

Note The fact that in this space every subset is an intersection of sufficiently many open sets does not contradict 4.4.1. The intersections $\bigcap \mathfrak{o}(U_i)$ of open sublocales do not need to correspond to the subspaces $\bigwedge U_i$ (although each of the $\mathfrak{o}(U_i)$ correctly represents the individual U_i).

4.6 What happens in spaces The conditions (op-as-joins) and (cl-as-meets) applied for open and closed subspaces of spaces are different than subfitness and fitness.

4.6.1 Proposition *The following statements about a space X are equivalent.*

(1) *Every open $U \subseteq X$ is a union of closed subsets.*
(2) *Every closed $A \subseteq X$ is an intersection of open subsets.*
(3) *X is symmetric.*

Proof (1)⇔(2) by De Morgan formulas.

(1)⇒(3): If $x \notin \overline{\{y\}}$, then $x \in X \smallsetminus \overline{\{y\}}$ and, by (1), $\overline{\{x\}} \subseteq X \smallsetminus \overline{\{y\}}$ so that $\overline{\{x\}} \cap \overline{\{y\}} = \emptyset$ and $y \notin \overline{\{x\}}$.

(3)⇒(1): Use the formula (sym) from 2.3. If U is open, then $U = \bigcup\{\overline{\{x\}} \mid x \in U\}$. □

4.6.2 Proposition *The following statements about a space X are equivalent.*

(1) *Every $M \subseteq X$ is a union of closed subsets.*
(2) *Every $M \subseteq X$ is an intersection of open subsets.*
(3) *X is a T_1-space.*

Proof (1)⇔(2) by De Morgan formulas again.

(1)⇒(3): Every $\{x\}$ is a union of closed subsets, hence it has to be closed.

(3)⇒(1): $M = \bigcup\{\{x\} \mid x \in M\}$. □

5 Subfitness and Fitness in Sublocales

5.1 An important fact about computing in sublocales $S \subseteq L$ Recall from A.5.5.3 that

- *if $s \in S$ then for the associated nucleus ν_S and any $a \in L$, $a \to s = \nu_S(a) \to s$.*

5.2 Theorem *Each sublocale of a fit frame is fit.*

Proof If $a \nleq b$ in S, then $a \nleq b$ in L. Then there is a $c \in L$ such that $c \vee a = 1$ and $c \to b \neq b$. Set $c' = \nu_S(c)$. We have

$$c' \overset{S}{\vee} a \geqslant c' \vee a \geqslant c \vee a = 1,$$

and by 5.1, $c' \to b = c \to b \neq b$. □

5.3 Theorem *A frame L is fit iff each of its sublocales is subfit.*

Proof ⇒ follows from 5.2.

⇐: Suppose not. Then there are $a \nleq b$ such that for every u such that $a \vee u = 1$, $u \to b = b$. Set

$$S = \{x \mid a \vee u = 1 \ \Rightarrow u \to x = x\}.$$

This is a sublocale. Indeed:

1. Trivially $1 \in S$ and if $x_i \in S$ and $a \vee u = 1$, then $u \to x_i = x_i$ for all i and hence $u \to \bigwedge x_i = \bigwedge(u \to x_i) = \bigwedge x_i$ so that $\bigwedge x_i \in S$.
2. If $x \in S$ and y is arbitrary, and if $a \vee u = 1$, then

$$u \to (y \to x) = y \to (u \to x) = y \to x.$$

Then, by hypothesis, S is subfit. Note that b is in S, as well as a (if $a \vee u = 1$, then $a = (a \vee u) \to a = (a \to a) \wedge (u \to a) = u \to a$). Therefore, there is a $c \in S$ such that

$$a \overset{S}{\vee} c = 1 \neq b \overset{S}{\vee} c.$$

Now in general the join in S can be bigger than when taken in L. Here, however, by 5.1, if $a \vee u = 1$, then

$$u \to (a \vee c) = \bigwedge\{z \in L \mid u \vee z = 1,\ x \geqslant a \vee c\} = a \vee c$$

(since $u \vee a \vee z = 1$) and hence $a \vee c \in S$ and then it is equal to $a \overset{S}{\vee} c$. Thus $a \vee c = 1$ and hence $1 = c \to c = c$, a contradiction. □

5.4 Corollary *Subfitness is not a hereditary property.*

(Take a subfit frame that is not fit like the one of 4.5.1. In this particular example, the space X—that is, the subfit frame $\Omega(X)$—has plenty of non-subfit sublocales; using the formulas for $U \to V$, it is easy to check that all the 3-chains

$$\{\emptyset, X \smallsetminus \{x\}, X\}$$

are sublocales.)

5.4.1 Note This is one of the flaws that led to the disregard of subfitness after the definition in [152], but also later. The other is that subfitness is not preserved by (co)products. On the other hand, fitness makes for a reflective subcategory of the category of locales, and hence can be thought of as a satisfactory categorical rounding up the subfitness. Only, it turned out that it is actually too strong, while subfitness works very well as a "low separation axiom".

5.5 However, subfitness is inherited in some important cases.

5.5.1 Theorem *Let S be a complemented sublocale of a subfit frame L. Then S is subfit.*

Proof We will use A.6.5 and A.2.

Let $\mathfrak{o}_S(a)$ be open in S. Then $\mathfrak{o}_S(a) = \mathfrak{o}(a) \cap S$ with $\mathfrak{o}(a)$ open in L. We have $\mathfrak{o}(a) = \bigvee_{i \in J} \mathfrak{c}(b_i)$ for some closed sublocales $\mathfrak{c}(b_i)$ in L. Since S is complemented and hence linear we see that

$$\mathfrak{o}_S(a) = \mathfrak{o}(a) \cap S = (\bigvee_{i \in J} \mathfrak{c}(b_i)) \cap S = \bigvee_{i \in J} (\mathfrak{c}(b_i) \cap S) = \bigvee_{i \in J} \mathfrak{c}_S(\nu_S(b_i)).$$ □

5.5.2 Note We have used the second order formula. It might be of interest to see how this fact follows for the important cases of open and closed sublocales using the first order reasoning. Here it is.

Closed sublocales Let $\mathfrak{c}(u) = \uparrow u$ be a closed sublocale of L and let $a \nleq b$ in $\uparrow u$. Then $a \nleq b$ in L and there is a $c \in L$ such that $c \vee a = 1 \neq c \vee b$. We will use, again, the fact that nontrivial joins in $\uparrow u$ coincide with those in L. Set $c' = c \vee u$. Then trivially $c' \vee a = 1$ and $c' \vee b = c \vee u \vee b = c \vee b \neq 1$ since $b \geqslant u$.

Open sublocales Let $a \nleq b$ in $\mathfrak{o}(u)$, hence $a = u \to a$ and $b = u \to b$, and $a \nleq u \to b$. Then $a \wedge u \nleq b$ and hence there is a c such that $c \vee (a \wedge u) = 1$ and $c \vee b \neq 1$. Thus we have

$$c \vee a = c \vee u = 1 \neq c \vee b. \qquad (*)$$

Now consider the associated nucleus ν_S for $S = \mathfrak{o}(u)$ (recall from A.6.1.2 that $\nu_S(x) = u \to x$) and take $c' = \nu_S(c) = u \to c$. Then, first,

$$a \overset{\mathfrak{o}(u)}{\vee} c' \geqslant a \vee c = 1.$$

Secondly,

$$b \overset{\mathfrak{o}(u)}{\vee} c' = (u \to b) \overset{\mathfrak{o}(u)}{\vee} (u \to c) = u \to (b \vee c).$$

Now $u \to (b \vee c) \neq 1$ because $u \to (b \vee c) = 1$ would imply $u \leqslant b \vee c$ and consequently $b \vee c \geqslant u \vee c = 1$, a contradiction. □

We can make the last proof shorter by proving instead that each frame $\downarrow u$ is subfit (and using the fact from A.6.1.2 that the frames $\mathfrak{o}(u)$ and $\downarrow u$ are isomorphic). Let $a \nleqslant b$ in $\downarrow u$. Then $a \nleqslant b$ in L and thus there exists $c \in L$ such that $c \vee a = 1 \neq c \vee b$. Let $c' = c \wedge u \in \downarrow u$. Clearly, $c' \vee a = u$ (that is, $c' \vee a = 1$ in $\downarrow u$). Moreover, $c' \vee b \neq u$ (otherwise, $u = (c \vee b) \wedge u$, that is, $u \leqslant c \vee b$, would imply $1 = a \vee c = u \vee c \leqslant c \vee b \nleqslant 1$).

5.6 Weak subfitness in sublocales It follows immediately from 3.1.2 and 5.3 that

Corollary *A frame L is fit iff each of its sublocales is weakly subfit.*

5.6.1 Proposition *A frame L is weakly subfit iff each of its open sublocales is weakly subfit.*

Proof Assume L is weakly subfit and let $b \in \mathfrak{o}(a)$ such that $b \neq 0_{\mathfrak{o}(a)} = a^*$. Then $b \wedge a \neq 0$ and by weak subfitness there is a $c \neq 1$ such that $(b \wedge a) \vee c = 1$. Then $1 = b \vee c \leqslant b \vee (a \to c)$ and $a \to c \neq 1$ (otherwise, $a \to c = 1$ would imply $a \leqslant c$, that is, $1 = a \vee c = c$, a contradiction). Hence $c' = a \to c \in \mathfrak{o}(a)$ and the join of b and c' in $\mathfrak{o}(a)$ is given by $a \to (b \vee c') = a \to 1 = 1 \neq c'$.

Summing up, we have:

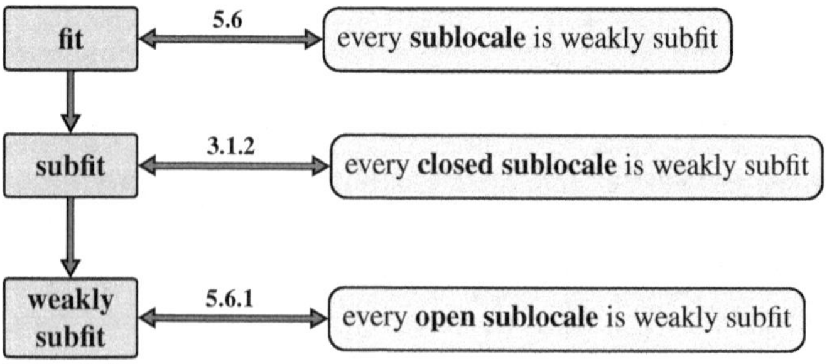

5.6.2 Corollary *Weak subfitness is not a hereditary property.*

6 Subfitness, Fitness and Congruences

To avoid misunderstanding: since congruences are in natural one-to-one correspondence with sublocales, a certain similarity of the titles of this and the previous section may suggest that the topics might be similar. They are not. Section 5 was concerned with the question how the properties are inherited. Here we will be concerned with concrete properties of congruences as algebraic phenomena.

6.1 A crucial role will be played by the sets

$$\downarrow(S \smallsetminus \{1\})$$

obtained from sublocales S. For the nucleus ν_S and the congruence E_S associated with S we obtain

6.1.1 Observation $\downarrow(S \smallsetminus \{1\}) = \{x \in L \mid \nu_S(x) \neq 1\} = L \smallsetminus E_S 1$.

6.2 Theorem *A frame L is subfit iff each congruence E on L such that $E1 = \{1\}$ is trivial.*

Proof Recall the equivalence (1)≡(2) from 4.2: we have a nonempty $\mathfrak{c}(a) = \uparrow a$ disjoint from a sublocale S iff there is an $a \in L$, $a \neq 1$ such that $a \notin \downarrow S$. Now apply this to the sublocale associated with E. □

6.3 In the following we will need some Heyting computing. Recall from A.4.3 that $x \leqslant a \to x$, that $x \to y = 1$ iff $x \leqslant y$, and in particular the formula

$$x = (x \vee c) \wedge (c \to x).$$

6.4 Proposition *A frame L is fit if and only if for any two sublocales $S, T \subseteq L$,*

$$\downarrow(S \smallsetminus \{1\}) = \downarrow(T \smallsetminus \{1\}) \;\Rightarrow\; S = T.$$

Proof ⇒: Let $\downarrow (S \smallsetminus \{1\}) = (T \smallsetminus 1)$ and let $b \in T$. Set $a = \nu_S(b)$. If $a \vee c = 1$ we have $\nu_S(b \vee c) \geqslant a \vee c = 1$ so that

$$b \vee c \notin \downarrow (S \smallsetminus \{1\}) \quad \text{and hence} \quad b \vee c \notin \downarrow (T \smallsetminus \{1\}).$$

By the formula from 6.3, $(b \vee c) \wedge (c \to b) \leqslant b$ and hence $b \vee c \leqslant (c \to b) \to b$ and since $b \vee c \notin \downarrow(T \smallsetminus \{1\})$ and $(c \to b) \to b \in T$, we see that $(c \to b) \to b = 1$ so that $c \to b \leqslant b$ and finally $c \to b = b$. Thus we have deduced that

$$a \vee c = 1 \quad \Rightarrow \quad c \to b = b$$

and since L is fit we conclude that $a = \nu_S(b) \leqslant b$, that is, $b \in S$.

$\Leftarrow$: Take a sublocale $S \subseteq L$ and consider the intersection

$$T = \bigcap \{\mathfrak{o}(x) \mid S \subseteq \mathfrak{o}(x)\}.$$

If $s \in S$ and $\nu_S(x) = 1$, then $x \to s = \nu_S(x) \to s = 1 \to s = s$, that is, $s \in \mathfrak{o}(x)$. Thus, if $\nu_S(x) = 1$, then $S \subseteq \mathfrak{o}(x)$ and we see that

$$S \subseteq T = \bigcap \{\mathfrak{o}(x) \mid \nu_S(x) = 1\}.$$

Hence for every $a \in T$, $x \to a = a$ whenever $\nu_S(x) = 1$, and if $a \neq 1$, since $a \to a = 1 \neq a$, $\nu_S(a)$ cannot be 1, so that $a \in (S \smallsetminus \{1\})$. Hence

$$T \smallsetminus \{1\} \subseteq \downarrow(S \smallsetminus \{1\}) \quad \text{and consequently} \quad \downarrow(T \smallsetminus \{1\}) \subseteq \downarrow(S \smallsetminus \{1\})$$

and since the other inclusion is trivial we conclude that $S = T$. In particular this holds for every closed S, and L is fit. □

Now we obtain a counterpart of 6.2, a very useful fact, and another comparison of fitness with subfitness.

6.5 Theorem *A frame L is fit if and only if for any two congruences E, F on L holds the implication*

$$E1 = F1 \quad \Rightarrow \quad E = F.$$

Proof It suffices to interpret the equality in view of the Observation 6.1.1. □

6.6 Note that in proving the implication "$\Leftarrow$" in 6.4 we have actually showed that $S = \bigcap \{\mathfrak{o}(x) \mid S \subseteq \mathfrak{o}(x)\}$ for *any sublocale* $S \subseteq L$, not only for the closed ones. Thus we have the following somewhat surprising fact (whose proof was promised in 4.5 above).

Theorem *A frame is fit if and only if every sublocale $S \subseteq L$ whatsoever is a meet of open ones.*

On the other hand the assumption that

every sublocale is a join of closed ones

is much stronger than subfitness (in fact, much stronger than fitness). It will be discussed in Chap. IX.

7 (Generalized) Nearness

7.1 A *nearness*[4] in classical topology [144, 146] is a generalization of uniformity omitting the requirement of a star refinement. That is, it is a nonempty system $\mathcal{A}$ of covers of a set such that if A refines B and A is in $\mathcal{A}$, then B is in $\mathcal{A}$, and that any two $A, B \in \mathcal{A}$ have a common refinement in $\mathcal{A}$.

7.2 In the point-free context this topic is usually approached as follows. A subset C of a frame L is a *cover* if $\bigvee C = 1$. A cover A *refines* a cover B if for every $a \in A$ there is a $b \in B$ such that $a \leqslant b$ (this is expressed by writing $A \leqslant B$), and we have the obvious common refinement

$$A \wedge B = \{a \wedge b \mid a \in A,\ b \in B\}$$

of A and B. Thus a *nearness* on L is a nonempty system $\mathcal{A}$ of covers of L such that

(N1) if $A \in \mathcal{A}$ and $A \leqslant B$, then $B \in \mathcal{A}$, and
(N2) if $A, B \in \mathcal{A}$, then $A \wedge B \in \mathcal{A}$.

But this is not all. In the classical context the covers are simply systems of subsets, and a nearness produces a topology by the obvious procedure. In a frame the topology is already inherent, and we have to take care that our system does not clash with that. This is done by means of the *admissibility* defined as follows. For a cover A and an element b we set

$$Ab = \bigvee \{x \in A \mid x \wedge b \neq 0\}$$

and for a system $\mathcal{A}$ of covers we write $b \lhd_{\mathcal{A}} a$ if there is an $A \in \mathcal{A}$ such that $Ab \leqslant a$. A nearness $\mathcal{A}$ has to be *admissible* in the sense that

$$\forall a \in L,\ a = \bigvee \{b \in L \mid b \lhd_{\mathcal{A}} a\}.$$

(For spatial frames $\Omega(X)$ this is precisely what one needs to conform the already given and the newly induced topologies.)

For more about point-free nearnesses see, e.g., [29, 51, 111, 147].

7.3 The nearness as described in 7.2 in fact does not extend the notion of a classical general nearness. At closer scrutiny one sees that when applied to spaces it yields a special case, the *regular* nearness [146]. To obtain a counterpart to the general concept we have to extend the notion of admissibility.

[4] *T-uniformity* in [203].

For a cover $A \subseteq L$ and a sublocale S write

$$AS = \bigvee\{a \in A \mid \mathfrak{o}(a) \cap S \neq \mathsf{O}\} \tag{$*$}$$

and call a nearness *quasi-admissible* if

$$\forall a \in L,\ \mathfrak{o}(a) = \bigvee\{S \mid \exists A \in \mathcal{A},\ AS \leqslant a\}.$$

It turns out that the counterpart of the classical general nearness is the *quasi-admissible nearness* [147, 221].

7.4 Theorem *A frame quasi-admits a nearness iff it is subfit.*

Proof $\Rightarrow$: See ($*$). We obviously have $AS = A\overline{S}$ (where $\overline{S}$ is the closure of the sublocale S) and hence if $\mathfrak{o}(a) = \bigvee\{S \mid \exists A \in \mathcal{A},\ AS \leqslant a\}$ we also have $\mathfrak{o}(a) = \bigvee\{\overline{S} \mid \exists A \in \mathcal{A},\ AS \leqslant a\}$, a join of closed sublocales.

$\Leftarrow$: We will show that if L is subfit, then the system of all covers (which, of course, satisfies (N1) and (N2)) is quasi-admissible, that is, for all $a \in L$

$$\mathfrak{o}(a) = \bigvee\{\mathfrak{c}(b) \mid \exists \text{ cover } C \text{ such that } C\mathfrak{c}(b) \leqslant a\}.$$

We have $\mathfrak{o}(c) \cap \mathfrak{c}(b) \neq \mathsf{O}$ iff $\mathfrak{o}(c) \nsubseteq \mathfrak{o}(b)$ iff $c \nleqslant b$, and since the inclusion $\supseteq$ is trivial we have to prove that

$$\mathfrak{o}(a) \subseteq \bigvee\{\mathfrak{c}(b) \mid \exists \text{ cover } C,\ (c \in C,\ c \nleqslant b)\ \Rightarrow\ c \leqslant a\}.$$

Let $x \in \mathfrak{o}(a)$. Then $x = a \to x$ and thus, by 3.4,

$$x = a \to x = \bigwedge\{y \mid y \vee a = 1,\ y \geqslant x\}.$$

By the formula for joins in the coframe of sublocales (see A.5.4.1) it then suffices to show that

$$\{y \mid y \vee a = 1,\ y \geqslant x\} \subseteq \bigcup\{\uparrow b \mid \exists \text{ cover } C,\ (c \in C, c \nleqslant b) \Rightarrow c \leqslant a\}.$$

This is easy: for each such y, take the cover $C = \{y, a\}$ and notice that $b = y$ satisfies the condition

$$(c \in C,\ c \nleqslant b)\ \Rightarrow\ c \leqslant a.$$

□

Chapter III
Axioms of Hausdorff Type

The axiom T_1 has not a very satisfactory conservative counterpart, but has a very satisfactory replacement, the subfitness.

On the other hand, translating the higher separation axioms, starting with regularity, creates no problems (although it sometimes opens new vistas).

With the Hausdorff axiom the situation is more complicated. There are classical facts that we can adopt for the point-free purposes, but they lead to different concepts, none of them without merit. Let us state right away that some of them are conservative (although distinct), and two of them not (but, nevertheless, at least one of the non-conservative ones behaves in some respect better when compared with spatial facts).

For technical reasons we will not be able to follow the chronological order in which the ideas appeared when discussing them in detail. Therefore we will start with a short historical survey.

1 Five Approaches and Four Concepts

1.1 C. H. Dowker and D. Papert Strauss In 1972 appeared two articles containing (among many other results) two very different mindsets. In "Separation axioms for frames" [81], Dowker and Papert Strauss followed a *geometric intuition* mimicking with spots, elements of a frame, the setting of open sets separating points. The concept was not conservative, but it sufficed to reinforce it with subfitness (under a different name) and it became so.

1.2 J. Isbell In the same year, Isbell in his "Atomless parts of spaces" [152] presented a quite different definition. This time it was based on a *categorical fact*, cum grano salis a necessary and sufficient condition. Why *cum grano salis*: the necessary and sufficient condition concerns product of spaces, and the product in

J. Picado, A. Pultr, *Separation in Point-Free Topology*,
https://doi.org/10.1007/978-3-030-53479-0_3

the category of locales is not precisely the same as in spaces. Therefore, the concept is not fully conservative. For spaces, Isbell's Hausdorff property implies the classical one, but a space can be Hausdorff without satisfying Isbell's condition.

1.3 J. Paseka and B. Šmarda In their preprint of 1987, "T_2-frames and almost compact frames", published in 1992 [216], Paseka and Šmarda used a quite different approach. Similarly like the regular spaces that can be characterized by the formula

$$U = \bigcup\{V \mid V \prec U\}$$

(I.3.3 above, and Chap. V below), Hausdorff spaces can be characterized by a certain formula

$$U = \bigcup\{V \mid V \sqsubset U\}$$

with suitably defined relation $\sqsubset$. Similarly like in the case of regularity this results in a conservative definition of point-free Hausdorffness. Furthermore, this property is preserved in sublocales and products of locales.

1.4 P. Johnstone and S. Shu-Hao About the same time, Johnstone and Shu-Hao [171] presented quite differently motivated Hausdorff formulas. Their approach was in *mending the categorical approach* as in 1.2 by suitably modifying the product $L \oplus L$ (or, rather, the diagonal in that). The authors themselves state that they were surprised by a resulting first order formula about which they were able to prove it to be equivalent with the one from 1.3. One can only agree with their claim that two so distinct approaches leading ultimately to the same concept indicate that this concept must be a natural one.

1.5 J. Rosický and B. Šmarda And here is another necessary and sufficient condition for classical Hausdorffness. In "T_1-locales" [243] Rosický and Šmarda made a different *order-theoretic* observation. While the (sober) T_1-spaces X are characterized by the fact that all the *primes* in $\Omega(X)$ are maximal (which is not quite so interesting, because it is, after all an immediate statement about points; recall II.1.1), Hausdorff spaces are characterized by all the *semiprimes* being maximal. The merit of this approach is certainly in paralleling the step from T_1 to T_2. But there is more to it. The primes are precisely the spectral points so that stating that primes are maximal is really just requiring that the spectrum is a T_1-space. But semiprimes are not, without other assumptions, closely connected with points. Hence, unlike the maximality of primes used for T_1 this order-theoretic condition has no direct relation to the spectrum.

1.6 Summing up it might seem that the Hausdorff property from 1.3 should be adopted as the basic point-free Hausdorff one. Its central role is expressed in the terminology—one speaks of such frames as of Hausdorff ones. But in fact the concepts from 1.1 and 1.2 are very important. The property from 1.1, that we will refer to as the

weak Hausdorff property,

is really rather weak and in concrete cases it is good to know that one can do just with that (like in the spatiality of frames satisfying the Raney identity—[183, 220]—, see 4.3. below).

Even more important is the Isbell's axiom; we will speak about the

strong Hausdorff property.

Its behaviour is in some respects in better parallel with classical topology: compact strongly Hausdorff frames are regular—compact Hausdorff frames are not necessarily even subfit— compact sublocales of strongly Hausdorff frames are closed. Such facts are fairly involved and deserve a detailed treatment. We will devote to them a good part of this chapter.

2 Weakly Hausdorff Frames

2.1 In the pioneering paper [81], Dowker and Papert Strauss suggested point-free counterparts for the standard T_k-axioms of classical topology. In fact, for T_2 there are several variants out of which the probably most widely adopted has been the following

(wH): *if* $a \vee b = 1$ *and* $a, b \neq 1$, *then there are* u, v *such that* $u \nleq a$, $v \nleq b$ *and* $u \wedge v = 0$.

We suggest to call it, in the point-free context, the *weak Hausdorff axiom.*

2.2 The open-hereditary modification Somehow more practical is the slightly stronger (in fact it is nothing but (wH) assumed for every open sublocale)

(wH'): *If* $a \nleq b$ *and* $b \nleq a$, *then there are* u, v *such that* $u \nleq a$, $v \nleq b$ *and* $u \wedge v = 0$.

2.2.1 An even stronger (and geometrically fairly intuitive) is

(wH''): *If* $a \nleq b$ *and* $b \nleq a$, *then there are* u, v *such that* $u \nleq a$, $v \nleq b$, $u \leqslant b$, $v \leqslant a$ *and* $u \wedge v = 0$.

We will see that under (sfit) the three conditions coincide (and without (sfit) they are not "Hausdorff enough").

2.3 These three conditions are not conservative. This is obvious: in a chain the premise $a \nleq b$ and $b \nleq a$ is void and hence the conditions are trivially satisfied.

Therefore, Dowker and Papert Strauss suggested, as an expedient Hausdorff axiom, the combination

$$\text{(wH) \& (sfit).}$$

This is indeed a conservative property (no need to prove it now: we will get if for free in 3.2 and 3.3.2 below). Now we will, rather, prove the following

2.3.1 Proposition *For subfit frames the conditions* (wH)*,* (wH')*and* (wH")*are equivalent.*

Proof Let L be subfit and let it satisfy (wH). First we will show that it also satisfies (wH').

Let $a \nleqslant b$ and $b \nleqslant a$. Then there is a c such that $a \vee c = 1 \neq b \vee c$. Hence $a \nleqslant b \vee c$ (else $1 = a \vee c \leqslant b \vee c$) and $a \vee (b \vee c) = 1$ so that we have $u \nleqslant a$ and $v \nleqslant b \vee c$ (and hence $v \nleqslant b$) such that $u \wedge v = 0$.

Now, again, take $a \nleqslant b$ and $b \nleqslant a$. Then there are c_1, c_2 such that

$$a \vee c_1 = 1 \neq b \vee c_1 \quad \text{and} \quad a \vee c_2 \neq 1 = b \vee c_2.$$

We have

$$b \vee c_1 \nleqslant a \vee c_2 \quad \text{and} \quad a \vee c_2 \nleqslant b \vee c_1$$

(indeed, if $b \leqslant a \vee c_2$, then $1 = b \vee c_2 \leqslant a \vee c_2$; the other statement by symmetry). We already know that L satisfies (wH') and hence we have u', v' such that $u' \nleqslant a \vee c_2$, $v' \nleqslant b \vee c_1$ and $u' \wedge v' = 0$. Then $u = u' \wedge b \nleqslant a$ since otherwise $u' = u' \wedge (b \vee c_2) \leqslant a \vee (u' \wedge c_2) \leqslant a \vee c_2$; in a similar way, we also get $v = v' \wedge a \nleqslant b$. □

3 Hausdorff Frames

In this section we will discuss the formulas arising from Paseka and Šmarda's view of the Hausdorff property as a "weak regularity", and on the other hand from Johnstone and Shu-Hao's mending the categorical diagonality approach.

In the former case the authors replace the relation $\prec$ by a weaker one (with an exception on the top).

3.1 In a topological space X define

$$U \sqsubset V \quad \equiv_{\text{df}} \quad U \subseteq V \text{ and } \overline{U} \cup V \neq X.$$

We have

3.2 Proposition *A T_0-space X is Hausdorff iff*

$$\forall V \in \Omega(X),\ V \neq X, \qquad V = \bigcup\{U \mid U \sqsubset V\}.$$

Proof $\Rightarrow$: Let $x \in V$ and $y \notin V$. There are open $U \ni x$ and $W \ni y$ such that $U \cap W = \emptyset$ and hence $\overline{U} \cap W = \emptyset$; obviously we can assume $U \subseteq V$. Then $y \notin \overline{U} \cup V$.

$\Leftarrow$: Here we use the T_0-property. Thus, say, $x \notin \overline{\{y\}}$ and we have

$$x \in X \smallsetminus \overline{\{y\}} = \bigcup\{U \mid U \sqsubset X \smallsetminus \overline{\{y\}}\}$$

and hence there is a $U \ni x$ such that $U \subseteq X \smallsetminus \overline{\{y\}}$, and a $z \notin \overline{U} \cup (X \smallsetminus \overline{\{y\}})$. In particular $z \notin \overline{U}$ and there is a $V \ni z$ such that $V \cap U = \emptyset$. But also $z \notin X \smallsetminus \overline{\{y\}}$, that is, $z \in \overline{\{y\}}$ and hence $y \in V$. □

3.2.1 We have obviously $\overline{U} \cup V \neq X$ iff $X \smallsetminus \overline{U} \nsubseteq V$ so that the formula from 3.1 can be written as

$$U \sqsubset V \quad \equiv_{\mathrm{df}} \quad U \subseteq V \text{ and } U^* \nsubseteq V.$$

This can be extended to general frames introducing the condition (Paseka and Šmarda [216], 1992)

(T_2): *If $a \neq 1$, then $a = \bigvee\{u \mid u \sqsubset a\}$ where $u \sqsubset a \equiv_{\mathrm{df}} (u \leqslant a \;\&\; u^* \nleqslant a)$.*

Obviously $\bigvee\{u \mid u \sqsubset a\} \leqslant a$ and the condition is equivalent with stating that if $1 \neq a \nleqslant b$ then there is a $v \sqsubset a$ such that $v \nleqslant b$. Setting $u = v^*$ we obtain a more explicit

(T_2): *If $1 \neq a \nleqslant b$, then $\exists u, v$ such that $u \nleqslant a$, $v \nleqslant b$, $v \leqslant a$ and $u \wedge v = 0$.*

3.3 In 1987, Johnstone and Shu-Hao [171] modified the Isbell's Hausdorff type condition to obtain a conservative one by slightly altering the product and the resulting diagonal. They showed that the result could be, somewhat surprisingly, reformulated to

(S_2): *If $1 \neq a \nleqslant b$, then $\exists u, v$ such that $u \nleqslant a$, $v \nleqslant b$ and $u \wedge v = 0$.*

Then they observed that this is the same as the Paseka and Šmarda's (T_2).

3.3.1 Proposition *(S_2) and (T_2) are equivalent.*

Proof Trivially, (T_2) implies (S_2). Now let (S_2) hold. If $a = a \wedge a \nleqslant b$, then $a \nleqslant a \to b$ and there are u, v with $u \wedge v = 0$, $u \nleqslant a$ and $v \nleqslant a \to b$. Thus, $v' = v \wedge a \nleqslant b$ and we have $v' \leqslant a$. □

3.3.2 (S_2) is strictly stronger than the weak Hausdorff property (which holds, for instance, for linear frames). But we have

Proposition (S_2) *implies (wH') and (wH') & (sfit) implies* (S_2).

Proof The former is trivial: If $b \nleqslant a$, then $a \neq 1$.

Now let L be subfit with (wH') and let $1 \neq a \nleqslant b$. Then we have a c such that $a \vee c = 1 \neq b \vee c$. Then $b' = b \vee c \nleqslant a$ because if $b \vee c \leqslant a$ then $a = a \vee c = 1$, and $a \nleqslant b'$ because if $a \leqslant b \vee c$, then $1 = a \vee c \leqslant b \vee c$. Apply (wH') for a, b': if $v \nleqslant b \vee c$, then $v \nleqslant b$. □

3.4 Hausdorff frames Thus we have here two equivalent conditions arriving from quite different directions. This confirmed (as stated in [171] and we cannot but agree) that one has here a good motivation for installing this formula as a canonical axiom. Hence, let us agree to call a frame *Hausdorff* if

(H): *for any* $1 \neq a \nleqslant b$ *there are* u, v *such that* $u \nleqslant a$, $v \nleqslant b$ *and* $u \wedge v = 0$.

In other words, if

(H): *for any* $1 \neq a \nleqslant b$ *there is a* u *such that* $u \nleqslant a$ *and* $u^* \nleqslant b$.

We have already spoken of *weakly Hausdorff frames* (and (wH) is really strictly weaker than (H)); there will be *strongly Hausdorff* ones in Sect. 5 below.

And there will be still another concept of Hausdorff type (see Sect. 4), perhaps not quite so expedient, but interesting at least because its source differs from all the others.

3.4.1 Good behaviour of Hausdorff frames First, from the approach in 3.3 above we immediately see that the concept is conservative, that is, for a (T_0)-space,

X is Hausdorff iff $\Omega(X)$ *is Hausdorff.*

Secondly, this property is inherited by sublocales and products:

Proposition

(1) *A sublocale of a Hausdorff locale is Hausdorff.*
(2) *A product of Hausdorff locales is Hausdorff.*

Proof

(1) Let $a, b \in S$ with $1 \neq a \nleqslant b$. This also holds in L, so we may find $x, y \in L$ with $x \wedge y = 0$, $x \nleqslant a$, $y \nleqslant b$. Recall the nucleus ν_S from A.5.5. Let $u = \nu_S(x)$ and $v = \nu_S(y) \in S$. Clearly, $u \nleqslant a$, $v \nleqslant b$ and $u \wedge v = \nu_S(x \wedge y) = 0_S$.
(2) Recall the terminology and notation from A.8.6 (cp-ideals, $\bigoplus_\alpha a_\alpha$, etc.). Let $(L_\alpha \mid \alpha \in \Lambda)$ be a family of Hausdorff locales, and let I, J be cp-ideals in $\prod_\alpha L_\alpha$ such that $1 \neq I \nsubseteq J$. Let

$$\widetilde{a} = (a_\alpha)_\alpha \in I \smallsetminus J,$$

and let $\{\alpha_1, \alpha_2, \dots, \alpha_n\}$ be the finite set of all the indices such that $a_{\alpha_i} \neq 1$. For each $i = 1, 2, \dots, n$, let

$$\widetilde{a}^{(i)} \in \prod_{\alpha} L_\alpha$$

be the Λ-tuple obtained from $\widetilde{a}$ by replacing each of $a_{\alpha_1}, \dots, a_{\alpha_i}$ by 1. Then

$$\widetilde{a}^{(0)} = a \in I \text{ but } \widetilde{a}^{(n)} = (1)_\alpha \notin I,$$

so there must exist i with $\widetilde{a}^{(i-1)} \in I$ but $\widetilde{a}^{(i)} \notin I$. For any $x \in L_{\alpha_i}$, let $\widetilde{x} = (x_\alpha)_\alpha$ denote the Λ-tuple obtained by replacing the α_i entry of $\widetilde{a}^{(i)}$ by $x_{\alpha_i} = x$. Now introduce

$$b = \bigvee\{x \in L_{\alpha_i} \mid \widetilde{x} \in I\} \text{ and } c = \bigvee\{x \in L_{\alpha_i} \mid \widetilde{x} \in J\}.$$

Since I and J are cp-ideals, we have $\widetilde{b} \in I$ and $\widetilde{c} \in J$. Moreover, $a_{\alpha_i} \leqslant b \neq 1$ and $a_{\alpha_i} \nleqslant c$ so that $b \nleqslant c$. Applying (H) in L_{α_i}, we obtain elements x, y with $x \wedge y = 0$, $x \nleqslant b$ and $y \nleqslant c$. Then $\widetilde{x} \notin I$ and $\widetilde{y} \notin J$, and therefore

$$\oplus_\alpha x_\alpha \nsubseteq I,\ \oplus_\alpha y_\alpha \nsubseteq J \text{ and } (\oplus_\alpha x_\alpha) \cap (\oplus_\alpha y_\alpha) = \mathsf{O}. \qquad \square$$

In fact, one has that

the subcategory of Hausdorff locales is reflective in the category **Loc**.[1]

3.4.2 Recall the regularity from Chap. I, 5.5 (it will be treated in more detail in Chap. V, in this chapter we need the definition only). Since $a \prec b \neq 1$ implies $a \sqsubset b$ (if $a^* \vee b = 1$ then $a^* \leqslant b$ yields $b = 1$) we immediately see that

each regular frame is Hausdorff.

3.5 A slightly peculiar example On the other hand, here is an example from [216] showing that Hausdorff frames do not always behave like Hausdorff spaces.

Example For a frame L define

$$\widetilde{L} = \{(x, y) \mid x \leqslant y\} \subseteq L \times \mathfrak{B}(L)$$

(recall from A.6.7 the *Booleanization*

$$\mathfrak{B}(L) = \{x \in L \mid x = x^{**}\} = \{x^* \mid x \in L\}$$

of L).

[1] This was originally proved in [171]; it is also mentioned in [216]. In [171, p. 191] it is, moreover, shown that the conjunction (sfit) & (H) is not inherited by sublocales, with an example of a non-subfit locale which is embeddable in a Hausdorff spatial locale.

3.5.1 Lemma *If $x^* = 0$ for every maximal element $x \in L$ and if L is Hausdorff, then also $\widetilde{L}$ is Hausdorff.*

Proof Suppose $(1, 1) \neq (a_1, a_2) \nleqslant (b_1, b_2)$ in $\widetilde{L}$. Then we have either (α) $1 \neq a_1 \nleqslant b_1$, or (β) $1 \neq a_1 \leqslant b_1$ and $1 \neq a_2 \nleqslant b_2$, or, finally, (γ) $1 \neq a_1 \leqslant b_1$ and $1 = a_2 \nleqslant b_2$.

Case (α) We have an $x \in L$ such that $x \nleqslant a_1$ and $x^* \nleqslant b_1$. Then $(x, x^{**}) \nleqslant (a_1, a_2)$ and $(x^*, x^*) \nleqslant (b_1, b_2)$ (and, of course, $(x, x^{**}), (x^*, x^*) \in \widetilde{L}$ and $(x, x^{**}) \wedge (x^*, x^*) = (0, 0)$).

Case (β) Consider an $x \in L$ such that $x \nleqslant a_2$ and $x^* \nleqslant b_2$. Now we have $(0, x^{**}), (x^*, x^*) \in \widetilde{L}$, and $(0, x^{**}) \nleqslant (a_1, a_2)$ and $(x^*, x^*) \nleqslant (b_1, b_2)$.

Case (γ) Here we have $a_1 \leqslant b_1 \leqslant b_2 \neq 1$. If $b_2 \nleqslant a_1$, since $b_2 \neq 1$ we have $b_2^* \nleqslant b_2$ and hence $(0, b_2^*) \nleqslant (b_1, b_2)$ and $(b_2, b_2) \nleqslant (a_1, a_2)$. Finally if $a_1 = b_2$, that is, $a_1 = b_1 = b_2 \neq 1$ we have $b_2^* \neq 0$, and by the assumption b_2 is not maximal. Choose a u with $b_2 < u < 1$ and an x with $x \nleqslant u$ and $x^* \nleqslant b_2$. Then $x, x^* \nleqslant b_2 = b_1 = a_1$ and we have $(0, x^*) \nleqslant (b_1, b_2)$ and $(x^{**}, x^{**}) \nleqslant (a_1, a_2)$. □

3.5.2 Lemma

(1) *If L is nontrivial, then $\widetilde{L}$ is not subfit.*
(2) *If L is compact, then $\widetilde{L}$ is compact.*

Proof

(1) We have $(0, 1) \nleqslant (0, 0)$ in $\widetilde{L}$. Let (c_1, c_2) be such that $(c_1, c_2) \vee (0, 1) = (1, 1)$. Then in particular $c_1 = 1$ and hence also $c_2 = 1$.
(2) Let L be compact and let $\bigvee_{i \in J}(a_i, b_i) = (1, 1)$. Then, in particular, $\bigvee_{i \in J} a_i = 1$ and hence there are $i_1, \ldots, i_n$ such that $\bigvee_{j=1}^n a_{i_j} = 1$. Since $a_i \leqslant b_i$, we have also $\bigvee_{j=1}^n b_{i_j} = 1$ and finally $\bigvee_{j=1}^n (a_{i_j}, b_{i_j}) = (1, 1)$. □

3.5.3 Corollary *A Hausdorff frame is not necessarily subfit. Even a Hausdorff compact frame is not necessarily subfit, and in particular not necessarily regular.*

(For the latter consider the topology of any compact Hausdorff space without isolated points.)

4 Prime, Semiprime and Maximal Elements

4.1 Recall that an element p of a frame L is *prime* if $a \wedge b \leqslant p$ implies that either $a \leqslant p$ or $b \leqslant p$, and that prime elements can be viewed as spectral points of L (in particular for a sober space X, the primes of $\Omega(X)$ are in a natural one-to-one correspondence with the points of X).

An element p is *semiprime* if $a \wedge b = 0$ implies that either $a \leqslant p$ or $b \leqslant p$. Obviously every prime element is semiprime; it should be noted that unlike the interpretation of primes as points, there is no immediate geometric interpretation of semiprimes (which makes the definition in 4.1.2 below deeper than what meets the eye).

4.1.1 Proposition *A T_0-space X is Hausdorff iff all the semiprime elements $P \in \Omega(X)$ are maximal.*

Proof Let X be Hausdorff and let $P \in \Omega(X)$ be semiprime. Suppose there are distinct x_1, x_2 that are not in P. Choose disjoint open $U_i \ni x_i$. Then at least one of them, say U_1, is contained in P, and hence $x_1 \in P$, a contradiction.

On the other hand let X not be Hausdorff. Then there are distinct x_1, x_2 such that, for $U_i \in \Omega(X)$, $U_i \ni x_i$ implies that $U_1 \cap U_2 \neq \emptyset$. Set

$$P = X \smallsetminus \overline{\{x_1, x_2\}}.$$

Suppose that $U \cap V = \emptyset$ and that $U \nsubseteq P$. Since U is open it means that it contains some of x_1, x_2, say x_1. Then $x_1 \notin V$, but also $x_2 \notin V$ by the assumption on x_1, x_2. Thus, $V \cap \{x_1, x_2\} = \emptyset$ and since it is open, $V \cap \overline{\{x_1, x_2\}} = \emptyset$ and $V \subseteq P$. Hence P is semiprime, and since X is T_0, $P \neq X \smallsetminus \overline{\{x_1\}}$ or $P \neq X \smallsetminus \overline{\{x_2\}}$, that is, it is not maximal. □

4.1.2 This leads to the following definition (Rosický and Šmarda [243]).

A frame L is *point Hausdorff* if

(pH): *every semiprime element in L is maximal.*

Thus, since we agreed that T_0 is automatically assumed, by 4.1.1

(pH) is conservative.

4.1.3 Note Rosický and Šmarda speak simply of Hausdorff frames, but we need to distinguish several concepts and hence have to choose a more specific term. The *"point Hausdorff"* is an allusion to the somewhat related T_1 axiom requiring that every prime element is maximal. It is an ad hoc term and we are fully aware of the fact that it is unjust to the fine observation on semiprimes. Certainly this concept cannot be reduced to the Hausdorffness of the spectrum.

4.2 Proposition *Every Hausdorff frame (from 3.4) is point Hausdorff.*

Proof Suppose (H) holds and p is semiprime in L. If it is not maximal we have an a such that $p < a \neq 1$ and hence by (H) there are $u \nleq a$ and $v \nleq p$ with $u \wedge v = 0$. Now since $v \nleq p$ we have $u \leqslant p < a$, a contradiction. □

4.2.1 The categorical behaviour of (pH) is very satisfactory. In [243, Proposition 2.1] it is proved that the subcategory of point Hausdorff frames is monocoreflective in the category **Frm** of all frames.

4.3 There does not seem to be much about the point Hausdorff frames in the literature. The authors of [243] themselves were not very enthusiastic about it because of the frames with no semiprimes that have the property for too trivial reasons. But this may be a too hasty dismissal. First, one should know more about the frames with no semiprimes (the example given in [243] is Boolean and hence we can easily accept it as Hausdorff; it is not hard to construct a non-subfit L with no semiprimes, but even this, in view of 3.5.3, is not quite so damning). Secondly, there are problems that might be of interest. For instance,

> *what happens if one combines (pH) with subfitness? Are then some of the other Hausdorff axioms implied?*

And are there some interesting results following from (pH) without support by another property?

4.3.1 As for the last question, let us recall a fact following from an axiom even weaker than (pH) [183, 220].

In 1953, Raney [242] observed that the lattice $L = \Omega(X)$ of open sets of a topological space X satisfied in fact a distributivity rule stronger than the frame one, namely, for any system $\{F_i\}_{i\in J}$ of finite $F_i \subseteq L$,

$$\bigvee_{i\in J} \bigwedge F_i = \bigwedge\{\bigvee_{i\in J} \phi(i) \mid \phi \in \prod_{i\in J} F_i\} \tag{RD}$$

(we speak of the *Raney identity*), and asked whether every complete lattice satisfying (RD) (every *quasitopology* in his terminology) is a topology. In the language of point-free topology, we can translate this question to asking whether a frame satisfying (RD) is spatial. The answer in [183] is negative in general, but positive for frames satisfying a very weak condition.

Consider the following relaxation of (pH)

(wpH): *every semiprime element in L is prime*

and speak of the frames satisfying (wpH) as the *weakly point Hausdorff* ones. We have

4.3.2 Theorem *A weakly point Hausdorff frame L satisfying the Raney identity is spatial.*

Proof Because of (wpH) it suffices to prove that each element $a \in L$ is a meet of semiprime elements.

Take an a and consider the system $\mathcal{F} = \{F_i\}_{i\in J}$ of all finite $F_i \subseteq L$ such that $\bigwedge F_i \leqslant a$. Since $\{a\}$ is in $\mathcal{F}$, we have

$$a = \bigvee_{i\in J} \bigwedge F_i = \bigwedge\{\bigvee_{i\in J} \phi(i) \mid \phi \in \prod_{i\in J} F_i\}.$$

We will prove that each $\bigvee_{i\in J} \phi(i)$ with $\phi \in \prod_{i\in J} F_i$ is semiprime. Indeed, if $x_1 \wedge x_2 = 0$, then $\{x_1, x_2\} = F_k$ for some k, and we have an $x_j = \phi(k) \leqslant \bigvee_{i\in J} \phi(i)$. □

5 Strongly Hausdorff Frames

5.1 In classical topology, Hausdorff spaces are obviously precisely the X such that the diagonal $\{(x,x) \mid x \in X\}$ is closed in the product $X \times X$. This led Isbell in [152] to the definition we will discuss from now on. It has a slight disadvantage in not being conservative (which is not very surprising: the product of locales—coproduct of frames—$\Omega(X) \oplus \Omega(X)$ does not quite correspond to the topology $\Omega(X \times X)$ of the product, see the next section), but as we will see further in this chapter this is more than compensated by other merits.

5.2 We will work with the binary coproduct $L \oplus L$ using the terminology and notation from A.8.6. In particular we will have the coproduct injections as

$$\iota_1 = (a \mapsto a \oplus 1)\colon L \to L \oplus L, \quad \iota_2 = (b \mapsto 1 \oplus b)\colon L \to L \oplus L.$$

5.2.1 The diagonal The codiagonal homomorphism Δ^* satisfying $\Delta^*\iota_i = \mathrm{id}$ is given by

$$\Delta^*(U) = \bigvee\{a \wedge b \mid a \oplus b \subseteq U\} = \bigvee\{a \wedge b \mid (a,b) \in U\}$$

(because $a \oplus b = (a \oplus 1) \cap (1 \oplus b)$). Consider the adjunction

$$L \oplus L \underset{\Delta}{\overset{\Delta^*}{\underset{\longleftarrow}{\overset{\longrightarrow}{\perp}}}} L$$

with Δ the associated diagonal localic map. It is easy to check that

$$\Delta(a) = \{(x,y) \mid x \wedge y \leqslant a\}.$$

Hence we have

$$U \subseteq \Delta\Delta^*(U) \quad \text{and} \quad \Delta^*\Delta = \mathrm{id}$$

(the latter because Δ^* is onto). The *diagonal sublocale*

$$\Delta[L]$$

corresponds to the classical diagonal subspace.

5.3 Strongly Hausdorff frames A frame L is *strongly Hausdorff* if the diagonal sublocale $\Delta[L]$ is closed in $L \oplus L$. This property will be referred to as

(sH).

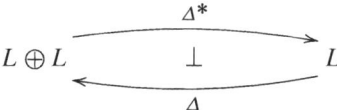

Since the least element of $\Delta[L]$ is

$$d_L = \Delta(0) = \{(x, y) \mid x \wedge y \leqslant 0\} = \downarrow\{(x, x^*) \mid x \in L\},$$

this is the same as saying that

(sH): $\Delta[L] = \uparrow d_L$.

5.3.1 Lemma *Consider the diagram*

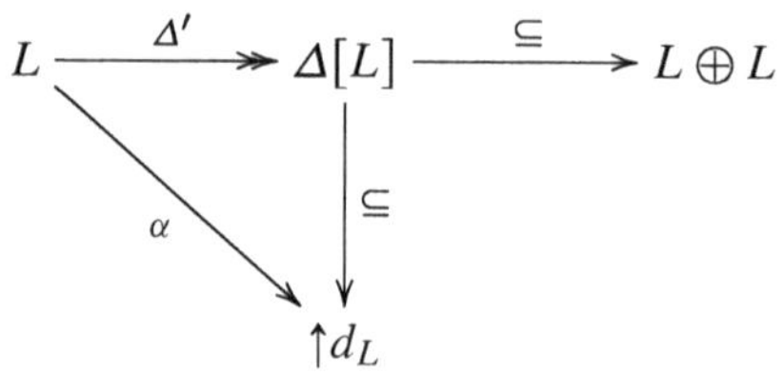

with Δ' and α defined by $\Delta'(x) = \alpha(x) = \Delta(x)$. Then the following statements are equivalent:

(1) $\Delta[L]$ *is closed.*
(2) *For all cp-ideals $U \supseteq d_L$, $\Delta\Delta^*(U) = U$, and the restrictions of Δ, Δ^* are mutually inverse isomorphisms.*
(3) α *is onto and $\alpha\Delta^*(U) = U \vee d_L$ for every $U \in L \oplus L$.*

Proof (1)⇒(2): For $U \supseteq d_L$ choose an a such that $U = \Delta(a)$. Then $\Delta^*(U) = \Delta^*\Delta(a) = a$ and hence $\Delta\Delta^*(U) = \Delta(a) = U$.

(2)⇒(3): We have $\alpha\Delta^*(U) = \Delta\Delta^*(U) = \Delta(\Delta^*(U) \vee 0) = \Delta(\Delta^*(U) \vee \Delta^*(d_L)) = \Delta\Delta^*(U \vee d_L) = U \vee d_L$.

(3)⇒(1): We have $\Delta^*\alpha\Delta^*(U) = \Delta^*(U \vee d_L) = \Delta^*(U)$; since Δ^* is onto, $\Delta^*\alpha = \mathrm{id}$, hence α is one-to-one, and as it is onto, it is an isomorphism. □

5.3.2 Lemma *A frame is strongly Hausdorff iff for every cp-ideal $U \supseteq d_L$ we have the implication*

$$(a \wedge b, a \wedge b) \in U \quad \Rightarrow \quad (a, b) \in U.$$

Proof ⇒: Let L be strongly Hausdorff. Then $U = \Delta\Delta^*(U)$ so that if $(a \wedge b, a \wedge b) \in U$, then $a \wedge b \leqslant \Delta^*(U)$ and hence $(a, b) \in \Delta(a \wedge b) \subseteq \Delta\Delta^*(U) = U$.

⇐: First, observe that if $(a_i, a_i) \in U$ for $i \in J$, then, since U is a down-set, $(a_i \wedge a_j, a_i \wedge a_j) \in U$, and hence by the assumed implication we have

$$\forall i \in J,\ (a_i, a_i) \in U \quad \Rightarrow \quad \forall i, j \in J,\ (a_i, a_j) \in U. \qquad (*)$$

Set $\alpha(a) = (a \oplus a) \vee d_L$. Thus defined $\alpha \colon L \to {\uparrow} d_L$ obviously preserves meets. By $(*)$ it preserves all joins: we have

$$\alpha(\bigvee_{i\in J} a_i) = ((\bigvee_{i\in J} a_i) \oplus (\bigvee_{j\in J} a_j)) \vee d_L =$$
$$\overset{(*)}{=} (\bigvee_{i\in J} (a_i \oplus a_i)) \vee d_L = \bigvee_{i\in J} ((a_i \oplus a_i)) \vee d_L) = \bigvee_{i\in J} \alpha(a_i).$$

We have, using $(*)$ again,

$$\alpha\Delta^*(a \oplus b) = \alpha(a \wedge b) = ((a \wedge b) \oplus (a \wedge b)) \vee d_L = (a \oplus b) \vee d_L$$

and since $L \oplus L$ is join-generated by the $a \oplus b$ we have $\alpha\Delta^*(U) = U \vee d_L$ and the statement follows from (3) in 5.3.1. □

5.3.3 Theorem *A frame L is strongly Hausdorff if and only if*

$$\forall a, b \in L,\ (a \oplus b) \vee d_L = (b \oplus a) \vee d_L \qquad \textbf{(sH)}$$

or, equivalently, if and only if

$$\forall a \in L,\ (a \oplus 1) \vee d_L = (1 \oplus a) \vee d_L. \qquad \textbf{(sH')}$$

Proof First, since a cp-ideal is a down-set, the implication in 5.3.2 is in fact an equivalence and hence the statement can be interpreted as

$$\forall a, b \in L,\ (a \oplus b) \vee d_L = ((a \wedge b) \oplus (a \wedge b)) \vee d_L. \qquad (*)$$

Now the formula (sH) immediately follows from $(*)$, and, on the other hand, if we have (sH) we have

$$(a \oplus b) \vee d_L = ((a \oplus b) \vee d_L) \wedge ((b \oplus a) \vee d_L) =$$
$$= ((a \oplus b) \wedge (b \oplus a)) \vee d_L = ((a \wedge b) \oplus (a \wedge b)) \vee d_L.$$

(sH) is obtained from the formally weaker (sH') using the obvious formula

$$(a \oplus b) \vee d_L = ((a \oplus 1) \vee d_L) \wedge ((1 \oplus b) \vee d_L).$$ □

5.3.4 Proposition *Each sublocale of a strongly Hausdorff frame is strongly Hausdorff.*

Proof Let S be a sublocale of a strongly Hausdorff frame L and let U be a cp-ideal of S such that

$$U \supseteq d_S = \Delta_S(0_S) = \{(x, y) \in S \times S \mid x \wedge y = \nu_S(0)\}.$$

Then $\downarrow U$ (with $\downarrow$ taken in $L \times L$) is a cp-ideal of L: if $(x_i, y) \in \downarrow U$ there are $x'_i \geqslant x_i$, $y' \geqslant y$ with $(x'_i, y') \in U$ and thus

$$(\bigvee_{i \in J}^{S} x'_i, y') \in U \quad \text{and} \quad \bigvee_{i \in J} x_i \leqslant \bigvee_{i \in J} x'_i \leqslant \nu_S(\bigvee_{i \in J} x'_i) = \bigvee_{i \in J}^{S} x'_i,$$

so that $(\bigvee_{i \in J} x_i, y) \in \downarrow U$. Moreover, $\downarrow U \supseteq d_L$, since $x \wedge y = 0$ implies $\nu_S(x) \wedge \nu_S(y) = \nu_S(0)$ and therefore

$$(x, y) \leqslant (\nu_S(x), \nu_S(y)) \in \Delta_S(0_S) \subseteq U.$$

Now let $(a \wedge b, a \wedge b) \in U$ with $a, b \in S$. Then $(a \wedge b, a \wedge b) \in \downarrow U \supseteq d_L$ and applying 5.3.2 to L we get $(a, b) \in \downarrow U$. Finally, since U is a down-set in $S \times S$, $(a, b) \in U$. □

5.4 The next result was proved first by Banaschewski for regular frames [25]. It holds already for strongly Hausdorff ones.

5.4.1 Theorem *Let L be strongly Hausdorff and let $h_1, h_2 \colon L \to M$ be frame homomorphisms. Set*

$$c = \bigvee\{h_1(x) \wedge h_2(y) \mid x \wedge y = 0\}.$$

Then

$$\check{c} = (a \mapsto a \vee c) \colon M \to \uparrow c = \mathfrak{c}(c)$$

is the coequalizer of h_1 and h_2 in **Frm**.

Proof Obviously $c = \Delta^*(h_1 \oplus h_2)(d_L)$. Using (sH') from 5.3.3 we obtain

$$\begin{aligned}\check{c}h_1(a) = h_1(a) \vee c = \Delta^*(h_1 \oplus h_2)((a \oplus 1) \vee d_L) = \\ = \Delta^*(h_1 \oplus h_2)((1 \oplus a) \vee d_L) = h_2(a) \vee c = \check{c}h_2(a).\end{aligned}$$

On the other hand, if $\phi h_1 = \phi h_2 = h$ for a $\phi \colon M \to K$ we have

$$\phi(c) = \bigvee\{h(x \wedge y) \mid x \wedge y = 0\} = h(0) = 0$$

and hence we can define $\overline{\phi} \colon \uparrow c \to K$ by $\overline{\phi}(s) = \phi(s)$ to obtain $\overline{\phi} \cdot \check{c} = \phi$.

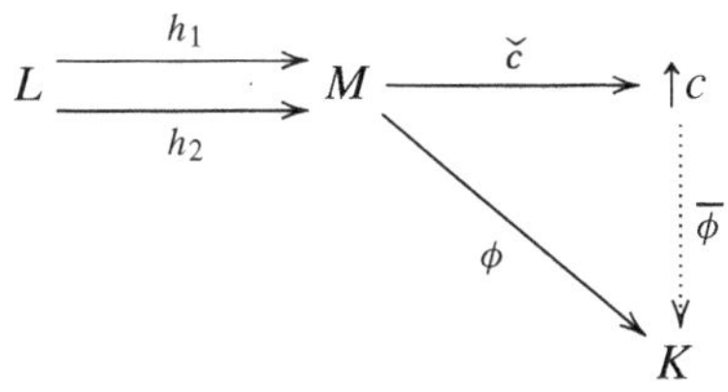

□

The converse is also true and the coequalizer property (which by the proof above amounts only to the identity $\breve{c} \cdot h_1 = \breve{c} \cdot h_2$) characterizes strong Hausdorffness.

5.4.2 Theorem *If $\breve{c} \cdot h_1 = \breve{c} \cdot h_2$ for any frame homomorphisms $h_1, h_2 \colon L \to M$, then L is strongly Hausdorff.*

Proof Let $\breve{c} \cdot h_1 = \breve{c} \cdot h_2$. By the previous proof, we know already that this suffices for $\breve{c}$ to be the coequalizer of h_1 and h_2. Set $h_i = \iota_i$ (the coproduct injections $L \to L \oplus L$). In this case,

$$c = \bigvee\{(x \oplus 1) \wedge (1 \oplus y) \mid x \wedge y = 0\} = \bigvee\{x \oplus y \mid x \wedge y = 0\} = d_L.$$

But the codiagonal $\Delta^* \colon L \oplus L \to L$ is the coequalizer of ι_1 and ι_2. Indeed, if $\phi\,\iota_1 = \phi\,\iota_2 = h$, then

$$\phi(x \oplus y) = \phi(\iota_1(x) \wedge \iota_2(y)) = h(x \wedge y) = h\Delta^*(x \oplus y).$$

Hence $\Delta^* = \breve{d_L}$ is closed. □

5.5 Density Let $f : L \to M$ be a localic map. As in spaces, f is said to be *dense* if the image $f[L]$ is dense in M, that is, if $\overline{f[L]} = M$. This means that for the $h = f^*$ holds thc implication

$$h(x) = 0 \Rightarrow x = 0. \qquad \text{(dense)}$$

(Since $\overline{S} = \uparrow \bigwedge S$ – see A.6.1.1, then $f[L]$ is dense iff $f(0) = 0$, and this is equivalent, by adjunction, to $h(x) \leqslant 0$ iff $x \leqslant f(0) = 0$.)

Note that the implication (dense) is the standard definition of a *dense frame homomorphism*.

5.5.1 Proposition *Let $k \colon M \to L$ be a dense frame homomorphism. Then for any frame homomorphisms $h_1, h_2 \colon N \to M$ with N strongly Hausdorff we have the implication*

$$kh_1 = kh_2 \Rightarrow h_1 = h_2.$$

Thus, in the category of strongly Hausdorff frames, each dense morphism is a monomorphism.

Proof If $kh_1 = kh_2$, then there is a ϕ such that $k = \phi\breve{c}$. Then $k(c) = \phi\breve{c}(c) = 0$ and if k is dense, $c = 0$, $\breve{c}$ is the identity isomorphism, and $h_1 = h_2$. □

6 (sH) Implies (H), and Is Not Conservative

6.1 Proposition *A strongly Hausdorff frame is Hausdorff.*

Proof For $(a, b) \in L \times L$ set

$$I(a, b) = \{(x, y) \mid x \leqslant a \text{ or } y \leqslant b\}.$$

It is straightforward that it is a cp-ideal.

Let L be strongly Hausdorff and let $a, b \in L$ such that $a \neq 1$, $a \nleqslant b$. Then $(a, 1) \in I(a, b)$ and $(1, a) \notin I(a, b)$ and hence by 5.3.3, $d_L \nsubseteq I(a, b)$. Thus, we cannot have generally that

$$x \wedge y = 0 \quad \Rightarrow \quad x \leqslant a \text{ or } y \leqslant b,$$

that is, there are u, v with $u \wedge v = 0$, $u \nleqslant a$ and $v \nleqslant b$. □

6.1.1 Proposition *Let X be a T_0-space and let $\Omega(X)$ be strongly Hausdorff. Then X is Hausdorff.*

Proof Suppose X is not Hausdorff. Then there are $x \neq y$ in X that cannot be separated. Set

$$\mathcal{U} = \{(U, V) \in \Omega(X) \times \Omega(X) \mid x \notin U \text{ or } y \notin V\}.$$

It is easy to check that $\mathcal{U}$ is a cp-ideal, and since x, y cannot be separated by disjoint U, V, $d_{\Omega(X)} \subseteq \mathcal{U}$. Hence if $\Omega(X)$ is strongly Hausdorff there is an open W such that $\mathcal{U} = \{(U, V) \mid U \cap V \subseteq W\}$. In particular $(X \smallsetminus \overline{\{x\}}, X) \in \mathcal{U}$ so that also $(X, X \smallsetminus \overline{\{x\}}) \in \mathcal{U}$ and hence $y \notin (X \smallsetminus \overline{\{x\}})$, that is, $y \in \overline{\{x\}}$; similarly $x \in \overline{\{y\}}$. Hence X is not T_0. □

6.1.2 Note The frame $\Omega(X)$ of a Hausdorff space X is not necessarily strongly Hausdorff. Thus, the property (sH) is not conservative, and in particular, in view of 6.1 (and 3.4.1) it is strictly stronger than (H). This will be discussed in the next chapter.

6.2 Proposition *A strongly Hausdorff frame satisfies T_U.*

Proof Let h_1, h_2 be two frame homomorphisms $h_1, h_2 \colon L \to M$ such that $h_1 \leqslant h_2$. Consider the coproduct $\iota_i \colon L \to L \oplus L$ $(i = 1, 2)$. Let $h = \langle h_1, h_2\rangle \colon L \oplus L \to M$ be the frame homomorphism such that $h\iota_i = h_i$ for each i. For the localic map $f = h_*$ and the diagonal map $\Delta\colon L \to L \oplus L$ we have

$$f[M] \subseteq \uparrow d_L = \Delta[L] \qquad (*)$$

(the equality by (sH)). Indeed, for any $m \in M$ we have $f(m) \geqslant d_L$ iff $m \geqslant h(d_L)$ and thus it suffices to prove that $h(d_L) = 0$:

$$h(d_L) = \bigvee\{h_1(a) \wedge h_2(b) \mid a \wedge b = 0\} \leqslant$$
$$\leqslant \bigvee\{h_2(a) \wedge h_2(b) \mid a \wedge b = 0\} = h_2(0) = 0.$$

Finally, since the diagonal map Δ is one-to-one there is by $(*)$ a localic map $g\colon M \to L$ such that $\Delta g = f$ and we have $g^* = g^*\Delta^*\iota_i = f^*\iota_i = h\iota_i = h_i$. □

6.3 Proposition *A Hausdorff frame, even a subfit one, does not have to satisfy* T_U.

Proof Consider the following example.

Let $\mathbb{R}$ be the standard space of real numbers. Denote by $\mathbb{Q}$, resp. $\mathbb{J}$, the subsets of rationals resp. irrationals. Consider the space X of reals with the topology augmented by adding the set $\mathbb{Q}$; that is, the open sets in X are the

$$U \cup (V \cap \mathbb{Q}) \text{ with } U, V \in \Omega(\mathbb{R}).$$

Consider the following diagram in **Frm** where $\Omega(\mathbb{Q}), \Omega(\mathbb{J})$ replace by isomorphisms the corresponding sublocales of $\Omega(\mathbb{R})$ the meet of which contains the Booleanization $\mathfrak{B}(\Omega(\mathbb{R}))$ (recall A.6.7.1; note that the intersection of these dense sublocales in the coframe of sublocales, containing the smallest dense sublocale $\mathfrak{B}(\Omega(\mathbb{R}))$, does not correspond to the—void—intersection of the associated subspaces).

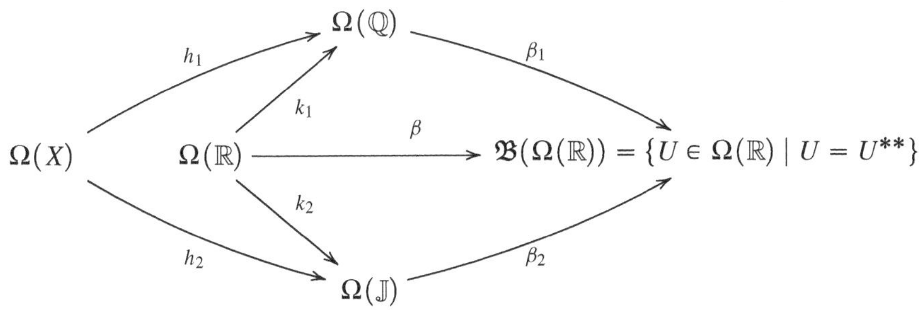

We have $h_1(U) = U \cap \mathbb{Q} = k_1(U)$, $h_2(U) = U \cap \mathbb{J} = k_2(U)$, and for β_i representing the embedding into the intersection, $\beta_1(U \cap \mathbb{Q}) = \beta_2(U \cap \mathbb{J}) = \beta(U)$. Furthermore,

$$\beta_1 h_1(U \cup (V \cap \mathbb{Q})) = \beta_1((U \cup V) \cap \mathbb{Q}) = \beta(U \cup V)$$

and

$$\beta_2 h_2(U \cup (V \cap \mathbb{Q})) = \beta_2(U \cap \mathbb{J}) = \beta(U)$$

so that $\beta_2 h_2 \leqslant \beta_1 h_1$. For $\mathbb{Q} = \emptyset \cup (\mathbb{R} \cap \mathbb{Q})$ we have $\beta_1 h_1(\mathbb{Q}) = \beta(\mathbb{R}) = 1$ while $\beta_2 h_2(\mathbb{Q}) = \beta(\emptyset) = 0$.

Thus X does not satisfy T_U. But it is Hausdorff, and obviously (as a T_1-space) also subfit. □

6.4 Corollary *A subfit Hausdorff frame is not necessarily strongly Hausdorff. Thus, (sH) is strictly stronger than (H).*

(Use 6.2: the X from 6.3—that is, $\Omega(X)$—is not strongly Hausdorff. For the second statement recall 6.1.2.)

6.5 Corollary *The property (sH) is not conservative.*

6.6 Strongly Hausdorff spaces We have seen (recall 6.3) that (sH) is not only strictly stronger than (H) for general frames, but, moreover, that it is so also for spaces: there was an example of a Hausdorff *space* that was not strongly Hausdorff. The reader will naturally ask the following questions.

- How strong the strong Hausdorff property restricted to spaces is? Does it not imply, under this restriction, some stronger property of interest?
- How does one show that a space is strongly Hausdorff and not just plain Hausdorff?

These two questions are interconnected: we need examples of strongly Hausdorff topological spaces to be able to illustrate the phenomena relevant to the first question. Thus in IV.4 we will show that (sH) does not imply subfitness. For this we will have to produce a non-subfit but strongly Hausdorff frame. Of course such a frame cannot be spatial (a Hausdorff space is T_1 and hence more than subfit). But the construction starts with a space, and at an intermediary step we have a strongly Hausdorff space that is not fit. Hence, to the first question, strongly Hausdorff spaces are not necessarily fit (and hence, to the question the reader would probably ask, not necessarily regular[2]). Proving that the space X in the construction is strongly Hausdorff is not quite comfortable: one has to really go into the details of the behaviour of the diagonal in $\Omega(X) \oplus \Omega(X)$. Later, in V.2.4 we will present a simple criterion of strong Hausdorffness (namely being a codomain of an epimorphism starting in a strongly Hausdorff frame) that will allow for a fairly lucid construction of a strongly Hausdorff but not regular frame (in fact, a space).

[2]Of course, the reader will probably ask first whether regularity implies (sH). It does, and the proof is not hard; it will be presented in V.2.2.2. As for the fitness, it is implied by regularity trivially: it suffices to choose the right formulas for the definitions and compare them (see V.5.1).

7 Johnstone and Shu-Hao's Modification of (sH)

7.1 In [171], Johnstone and Shu-Hao suggested a very interesting adjustment of the strong Hausdorff property. One can modify the (co)product in such a way that there is a sort of diagonal the closedness of which results in a conservative property. Surprisingly (another result from [171]) it turned out that this property can be translated into the property (H) discussed in Sect. 3.

We will present the modification trick and the translation.

7.1.1 For a subset $A \subseteq L \times L$ define

$$A^{+} = \{(x_1, x_2) \mid \forall (a_1, a_2) \in A,\ a_1 \leqslant x_1 \text{ or } a_2 \leqslant x_2\} \quad \text{and}$$

$$A^{-} = \{(x_1, x_2) \mid \forall (a_1, a_2) \in A,\ a_1 \geqslant x_1 \text{ or } a_2 \geqslant x_2\}.$$

The assignments are clearly order-reversing and form a Galois connection

$$\mathfrak{P}(L \times L) \to \mathfrak{P}(L \times L)$$

(i.e. $A \subseteq B^{-}$ iff $B \subseteq A^{+}$). The important fact for the construction is that

$$\kappa = (U \mapsto U^{+-})\colon L \oplus L \to L \oplus L$$

is a nucleus creating a sublocale

$$L \otimes L = \{U \in L \oplus L \mid U^{+-} = U\}.$$

Note It turns out that this sublocale is dense in $L \oplus L$ and that it can be viewed as a product in a suitably chosen subcategory of **Loc**. The authors speak on a *weak product*.

7.1.2 The authors define a frame L as Hausdorff (we will show shortly that it does not interfere with the terminology introduced so far) if the preimage of the diagonal $\Delta[L]$ along the embedding $L \otimes L \hookrightarrow L \oplus L$ is closed in $L \otimes L$. More explicitly:

- L is *Hausdorff* if for each cp-ideal U such that $U \supseteq d_L$, *and such that, moreover,* $U = U^{+-}$, we have $U = \Delta(a)$ for some $a \in L$.

Johnstone and Shu-Hao proved that thus modified axiom was already conservative. But not only that. They showed that it could be expressed in a nice first order formula and that it was equivalent with the axiom suggested by Paseka and Šmarda. We have already mentioned it in 3.3 above; now we will reproduce their proof.

7.1.3 Proposition *A frame L is Hausdorff in the sense of 7.1.2 iff it satisfies the requirement* (*H*) *from 3.4.*

Proof For the reader's convenience let us recall the condition (H) explicitly:

(H): *for any* $1 \neq a \nleqslant b$ *there are* u, v *such that* $u \nleqslant a$, $v \nleqslant b$ *and* $u \wedge v = 0$.

I. Let L satisfy the requirement from 7.1.2 and let $1 \neq a \nleqslant b$. Recall the cp-ideal

$$I(a,b) = \{(x,y) \mid x \leqslant a \text{ or } y \leqslant b\}$$

from 6.1. It is easy to check that $I(a,b)^{+-} = I(a,b)$ and hence we can use the reasoning from 6.1 to prove (H).

II. Now let L satisfy (H). Take a cp-ideal U with $U \supseteq d_L$ and $U = U^{+-}$. By the former we see that for $(a,b) \in U^+$ we cannot have $1 \neq a \nleqslant b$, so that either $a = 1$ or $a \leqslant b$, and by the symmetry as in 5.3.3, also either $b = 1$ or $b \leqslant a$. Thus,

$$(a,b) \in U^+ \quad \Rightarrow \quad a = 1 \text{ or } b = 1 \text{ or } a = b.$$

Taking into account that U^+ is obviously an up-set we see that each a with $(a,a) \in U^+$ is maximal (if $a < x < 1$ consider $(a,a) \leqslant (a,x)$) and hence

$$U^+ = (L \times \{1\}) \cup (\{1\} \times L) \cup \{(a,a) \mid a \in M_U\},$$

where M_U is a subset of the set of maximal elements of L.

Since a maximal element is prime we have $x \leqslant a$ or $y \leqslant a$ iff $x \wedge y \leqslant a$ and hence

$$U^{+-} = \{(x,y) \mid x \wedge y \leqslant a \text{ for all } (a,a) \in U^+\} = \{(x,y) \mid x \wedge y \leqslant \bigwedge M_U\},$$

that is, $U = U^{+-} = \Delta(\bigwedge M_U)$. □

8 A Few Technical Facts

Recall 5.5.1. It follows from it that, similarly like in spaces,

> *if L is strongly Hausdorff and if $f_1, f_2\colon M \to L$ coincide on a dense sublocale of M then they are equal*

(and in fact, (sH) is characterized by this property, see IV.2.2 below).

In the following section we will show that the strong Hausdorff property behaves very satisfactorily in further respects:

- compact strongly Hausdorff frames are regular,[3] and
- compact sublocales of strongly Hausdorff frames are closed.

[3] And (sH) alone does not imply regularity as we will learn in Chap. V.

Some of it needs a technical setup. In this section we will prepare a few useful tools.

8.1 Lemma *Let S and T be sublocales of L. If, for any $a, b \in L$,*

$$T \cap \mathfrak{o}(a) \subseteq \mathfrak{o}(b) \;\Rightarrow\; S \cap \mathfrak{o}(a) \subseteq \mathfrak{o}(b),$$

then $S \subseteq T$.

Proof Recall A.6.4: each sublocale S can be written as an intersection $\bigcap_{i\in J}(\mathfrak{o}(a_i) \vee \mathfrak{c}(b_i))$. Consequently, for each j,

$$T \cap \mathfrak{o}(b_j) = \bigcap_{i\in J} \big((\mathfrak{o}(a_i) \vee \mathfrak{c}(b_i)) \cap \mathfrak{o}(b_j)\big) \subseteq (\mathfrak{o}(a_j) \vee \mathfrak{c}(b_j)) \cap \mathfrak{o}(b_j) \subseteq \mathfrak{o}(a_j).$$

Hence also $S \cap \mathfrak{o}(b_j) \subseteq \mathfrak{o}(a_j)$ for every j and then

$$S = (S \cap \mathfrak{c}(b_j)) \vee (S \cap \mathfrak{o}(b_j)) \subseteq (S \cap \mathfrak{c}(b_j)) \vee \mathfrak{o}(a_j) \subseteq \mathfrak{c}(b_j) \vee \mathfrak{o}(a_j)$$

that is, $S \subseteq \bigcap_{j\in J}(\mathfrak{o}(a_j) \vee \mathfrak{c}(b_j)) = T$. □

8.1.1 In particular, in order to show that a sublocale S is closed it suffices to check that

$$S \cap \mathfrak{o}(a) \subseteq \mathfrak{o}(b) \;\Rightarrow\; \overline{S} \cap \mathfrak{o}(a) \subseteq \mathfrak{o}(b)$$

holds for any a, b.

8.2 Proposition *Let $f\colon L \to M$ be a localic map. For any sublocale S of L and any $b \in M$,*

$$f[S \cap f_{-1}[\mathfrak{o}(b)]] = f[S] \cap \mathfrak{o}(b).$$

Proof Since for a complemented x in a co-Heyting lattice we have obviously $y \smallsetminus x = y \wedge x^{\#}$ (recall A.4.1.2), and $f[-]$ is a colocalic map adjoint to $f_{-1}[-]$ (see A.7.4.4), the formula is an immediate consequence of A.4.4 and A.7.3:

$$f[S \cap f_{-1}[\mathfrak{o}(b)]] = f[S \smallsetminus \mathfrak{c}(f^*(b))] = f[S \smallsetminus f_{-1}[\mathfrak{c}(b)]] = f[S] \smallsetminus \mathfrak{c}(b). \;\square$$

8.3 Stable down-sets and cp-ideals A down-set $R \in \mathfrak{D}(L \times M)$ is *finitely left-stable*, resp. *directedly left-stable*, resp. *left-stable*, if each left class

$$Ry = \{x \in L \mid (x, y) \in R\}$$

is closed under finite, resp. directed, resp. all suprema; similarly *finitely, directedly right-stable* and *right-stable* sets. Note that finitely and directedly stable means stable, and that cp-ideals are the $R \in \mathfrak{D}(L \times M)$ that are both left- and right-stable.

Proposition *Let $R, S \in \mathfrak{D}(L \times M)$ be right-stable, let R be finitely and S directedly left-stable, and let $R \subseteq S$. Then there exists a cp-ideal U such that*

$$R \subseteq U \subseteq S.$$

Proof Define $S' \subseteq L \times M$ by setting

$$(x, y) \in S' \quad \equiv_{\mathrm{df}} \quad \forall a\, \forall b \leqslant y,\ \ ((a, b) \in R \Rightarrow (x \vee a, b) \in S).$$

Obviously S' is a down-set. Further:

1. *S' is right-stable.* If $(x, y_i) \in S'$ and $b \leqslant \bigvee_i y_i$, then we have for $(a, b) \in R$ and each i, $(x \vee a, b \wedge y_i) \in S$ and hence $(x \vee a, b) \in S$ (since S is right-stable). This includes also void suprema.
2. *S' is directedly left-stable.* If $(x_i, y) \in S'$ and $x = \bigvee_i x_i$ is a directed supremum, then for $b \leqslant y$, $(x_i \vee a, b) \in S$ and hence $(x \vee a, b) \in S$ since $x \vee a = \bigvee_i (x_i \vee a)$ is a directed supremum.
3. $R \subseteq S'$: Consider $(x, y) \in R$. If $(a, b) \in R$ and $b \leqslant y$, then $(x \vee a, b) \in R \subseteq S$ since R is finitely right-stable (and a down-set).
4. $S' \subseteq S$: If $(x, y) \in S'$ take $(0, y)$. As R is finitely right-stable, $(0, y)$ is in R and we obtain $(x \vee 0, y) \in S$.
5. *If $(x, y) \in S'$ and $(x', y) \in R$, then $(x \vee x', y) \in S'$*: for any (a, b) with $b \leqslant y$ we have $(x' \vee a, b) \in R$ and hence $((x \vee x') \vee a, b) = (x \vee (x' \vee a), b) \in S$.

   Now define

$$(x, y) \in U \quad \equiv_{\mathrm{df}} \quad \forall a\, \forall b \leqslant y,\ \ ((a, b) \in S' \Rightarrow (x \vee a, b) \in S').$$

Using 5 for the first inclusion and the same reasoning as above for the others we learn that

$$R \subseteq U \subseteq S' \subseteq S$$

and that U is right-stable and directedly left-stable. Further, as $R \subseteq U$, $(0, y) \in U$ for all y, and finally, if $(x, y), (x', y) \in U$, then $(x \vee x', y) \in U$. Thus, U is also left-stable, that is, a cp-ideal. □

8.4 Efficient down-sets A $T \in \mathfrak{D}(L \times M)$ is said to be *efficient* if $(0, 1) \in T$ and $(x, y), (x', y') \in T$ implies $(x \vee x', y \wedge y') \in T$.

(Note that a finitely left-stable T is efficient; but an efficient T is not necessarily finitely left-stable.)

Lemma *Let L be compact. If $T \in \mathfrak{D}(L \times M)$ is efficient and $(1, y) \in \bigvee\{a \oplus b \mid (a, b) \in T\}$, then $y \leqslant \bigvee\{z \mid (1, z) \in T\}$.*

Proof Set $j(x) = \bigvee\{z \mid (x, z) \in T\}$. Since T is a down-set we have

$$x \leqslant x' \Rightarrow \{z \mid (x', z) \in T\} \subseteq \{z \mid (x, z) \in T\} \text{ and hence } j(x') \leqslant j(x). \qquad (*)$$

Further we have,

$$j(x \vee y) = j(x) \wedge j(y).$$

(By $(*)$, $j(x) \wedge j(y) \geqslant j(x \vee y)$. On the other hand,

$$\begin{aligned} j(x) \wedge j(y) &= \bigvee\{z \mid (x, z) \in T\} \wedge \bigvee\{z \mid (y, z) \in T\} \\ &= \bigvee\{z_1 \wedge z_2 \mid (x, z_1), (y, z_2) \in T\} \\ &\leqslant \bigvee\{z_1 \wedge z_2 \mid (x \vee y, z_1 \wedge z_2) \in T\} \\ &\leqslant \bigvee\{z \mid (x \vee y, z) \in T\} = j(x \vee y).) \end{aligned}$$

Now define

$$\begin{aligned} (x, y) \in R \quad &\equiv_{\mathrm{df}} \quad y \leqslant j(x) \quad \text{and} \\ (x, y) \in S \quad &\equiv_{\mathrm{df}} \quad x \neq 1 \text{ or } (1, y) \in R. \end{aligned}$$

Obviously $R, S \in \mathfrak{D}(L \times M)$, both right-stable, and $(0, 1) \in R \subseteq S$. If (x, y) and (x', y) are in R, then $y \leqslant j(x) \wedge j(x') = j(x \vee x')$; hence $(x \vee x', y) \in R$, and R is finitely left-stable.

Finally, if $(x_i, y) \in S$ and $\bigvee_i x_i = x$ is a directed supremum, then either $x \neq 1$ and $(x, y) \in S$ trivially, or $x = 1$ and by compactness $x = x_i$ for some i and $(x, y) \in S$ again. Hence S is directedly left-stable and we can apply 8.3 to obtain a cp-ideal U with $R \subseteq U \subseteq S$. Since obviously $T \subseteq R$ we conclude that

$$(1, y) \in \bigvee\{a \oplus b \mid (a, b) \in T\} \leqslant \bigvee\{a \oplus b \mid (a, b) \in U\} = \bigcup\{a \oplus b \mid (a, b) \in U\}$$

and $(1, y) \in U \subseteq S$, hence $(1, y) \in R$, and $y \leqslant \bigvee\{z \mid (1, z) \in T\}$. □

8.5 Proposition *Let p_M be the projection $L \oplus M \to M$ (that is, the right adjoint to the injection $\iota_M \colon (a \mapsto 1 \oplus a) \colon M \to L \oplus M$), with L a compact locale. For any $U \in L \oplus M$ and any $a, b \in M$,*

$$(p_M)_{-1}[\mathfrak{o}(a)] \subseteq \mathfrak{o}(U) \vee (p_M)_{-1}[\mathfrak{o}(b)] \;\Rightarrow\; \mathfrak{o}(a) \subseteq \overline{(p_M[\mathfrak{c}(U)])}^{\#} \vee \mathfrak{o}(b).$$

Proof The left-hand side means that $\mathfrak{o}(\iota_M(a)) \subseteq \mathfrak{o}(U) \vee \mathfrak{o}(\iota_M(b))$, that is,

$$\mathfrak{o}(1 \oplus a) \subseteq \mathfrak{o}(U) \vee \mathfrak{o}(1 \oplus b) = \mathfrak{o}(U \vee (1 \oplus b))$$

and hence it amounts to

$$1 \oplus a \subseteq U \vee (1 \oplus b) = \bigvee\{x \oplus y \mid (x, y) \in U \text{ or } y \leqslant b\}.$$

Now, consider a relation T on $L \times M$ defined by

$$(x, y) \in T \equiv_{\text{df}} \quad (x, y) \in U \text{ or } y \leqslant b.$$

T is obviously a down-set containing $(0, 1)$. Further, if $(x, y) \in T$ and $(x', y') \in T$, then either $(x, y) \in U$ and $(x', y') \in U$ and then $(x \vee x', y \wedge y') \in U$, or $y \wedge y' \leqslant b$; in either case, $(x \vee x', y \wedge y') \in T$. Thus, T is efficient, and we can apply 8.4 to obtain $a \leqslant \bigvee\{z \mid (1, z) \in T\}$. Hence

$$\mathfrak{o}(a) = \bigvee\{\mathfrak{o}(z) \mid (1, z) \in T\}$$

and thus it suffices to show that $\mathfrak{o}(z) \subseteq \overline{(p_M[\mathfrak{c}(U)])}^{\#}$ for each $(1, z) \in U$.

Let $(1, z) \in U$. Then $z \leqslant p_M(U)$, that is, $\mathfrak{o}(z) \subseteq \mathfrak{o}(p_M(U))$, so that

$$\mathfrak{o}(z) \cap p_M[\mathfrak{c}(U)] \subseteq \mathfrak{o}(p_M(U)) \cap p_M[\mathfrak{c}(U)] \subseteq \mathfrak{o}(p_M(U)) \cap \mathfrak{c}(p_M(U)) = \mathsf{O}.$$

Hence $\mathfrak{o}(z) \subseteq \bigvee\{\mathfrak{o}(x) \mid \mathfrak{o}(x) \cap p_M[\mathfrak{c}(U)] = \mathsf{O}\} = \overline{(p_M[\mathfrak{c}(U)])}^{\#}$. □

9 Two Important Properties of Strongly Hausdorff Frames

9.1 Strong Hausdorff property and regularity In Chap. V (Theorem 2.2.2) we will show that, similarly like in spaces, a regular frame is strongly Hausdorff. Now we will prove, again in parallel with the classical facts, that a compact strongly Hausdorff frame is regular. Recall from 3.5.3 that the strong Hausdorff property is essential: a (plain) Hausdorff compact frame is not necessarily regular (indeed not even subfit).

9.1.1 Theorem *A compact strongly Hausdorff locale is regular.*

Proof Let $a \in L$. By 5.3.3

$$(1 \oplus a) \leqslant (a \oplus 1) \vee d_L = \bigvee\{x \oplus y \mid x \leqslant a \text{ or } x \wedge y = 0\}.$$

Define $T \subseteq L \times L$ by

$$(x, y) \in T \equiv_{\text{df}} \quad x \leqslant y^* \vee a.$$

Obviously T is a down-set and if $(x, y), (z, w) \in T$ we have $x \leqslant y^* \vee a$ and $z \leqslant w^* \vee a$, hence $x \vee z \leqslant (y^* \vee w^*) \vee a \leqslant (y \wedge w)^* \vee a$, so that $(x \vee z, y \wedge w) \in T$ and since trivially $(0, 1) \in T$, it is efficient (it is not finitely left-stable, however).

Note that $(1, y) \in T$ iff $y \prec a$, and that $(x, y) \in T$ whenever $x \leqslant a$ or $x \wedge y = 0$. Hence,

$$1 \oplus a \leqslant \bigvee\{x \oplus y \mid (x, y) \in T\}.$$

Since T is efficient we may then conclude that

$$a \leqslant \bigvee\{y \mid (1, y) \in T\} = \bigvee\{y \mid y \prec a\}. \qquad \square$$

9.2 Compact sublocales of strongly Hausdorff frames First, let us prove the following

Proposition *For a compact frame L, the projection*

$$p_M : L \oplus M \to M$$

is closed.

Proof Let $\mathfrak{c}(U)$ be a closed sublocale of $L \oplus M$. By 8.1.1 it suffices to check that, for any $a, b \in M$,

$$p_M[\mathfrak{c}(U)] \cap \mathfrak{o}(a) \subseteq \mathfrak{o}(b) \;\Rightarrow\; \overline{p_M[\mathfrak{c}(U)]} \cap \mathfrak{o}(a) \subseteq \mathfrak{o}(b),$$

that is,

$$p_M[\mathfrak{c}(U)] \cap \mathfrak{o}(a) \subseteq \mathfrak{o}(b) \;\Rightarrow\; \mathfrak{o}(a) \subseteq (\overline{p_M[\mathfrak{c}(U)]})^{\#} \vee \mathfrak{o}(b).$$

First, we show that

$$p_M[\mathfrak{c}(U)] \cap \mathfrak{o}(a) \subseteq \mathfrak{o}(b) \Leftrightarrow \mathfrak{c}(U) \cap (p_M)_{-1}[\mathfrak{o}(a)] \subseteq (p_M)_{-1}[\mathfrak{o}(b)].$$

$\Rightarrow$: Since $(p_M)_{-1}[-]$ is right adjoint to $p_M[-]$, $p_M[\mathfrak{c}(U)] \cap \mathfrak{o}(a) \subseteq \mathfrak{o}(b)$ implies

$$\begin{aligned}\mathfrak{c}(U) \cap (p_M)_{-1}[\mathfrak{o}(a)] &\subseteq (p_M)_{-1} p_M[\mathfrak{c}(U)] \cap (p_M)_{-1}[\mathfrak{o}(a)] \\ &= (p_M)_{-1}[p_M[\mathfrak{c}(U)] \cap \mathfrak{o}(a)] \subseteq (p_M)_{-1}[\mathfrak{o}(b)].\end{aligned}$$

$\Leftarrow$: By 8.2, if $\mathfrak{c}(U) \cap (p_M)_{-1}[\mathfrak{o}(a)] \subseteq (p_M)_{-1}[\mathfrak{o}(b)]$, then

$$p_M[\mathfrak{c}(U)] \cap \mathfrak{o}(a) = p_M[\mathfrak{c}(U) \cap (p_M)_{-1}[\mathfrak{o}(a)]] \subseteq p_M(p_M)_{-1}[\mathfrak{o}(b)] \subseteq \mathfrak{o}(b)$$

(since p_M is a right adjoint and hence preserves meets, and since it is onto and hence $p_M(p_M)_{-1}[\mathfrak{o}(a)] = \mathfrak{o}(a)$). Moreover,

$$\mathfrak{c}(U) \cap (p_M)_{-1}[\mathfrak{o}(a)] \subseteq (p_M)_{-1}[\mathfrak{o}(b)]$$

iff

$$(p_M)_{-1}[\mathfrak{o}(a)] \subseteq \mathfrak{o}(U) \vee (p_M)_{-1}[\mathfrak{o}(b)].$$

Finally, by 8.5,

$$(p_M)_{-1}[\mathfrak{o}(a)] \subseteq \mathfrak{o}(U) \vee (p_M)_{-1}[\mathfrak{o}(b)] \ \Rightarrow\ \mathfrak{o}(a) \subseteq (\overline{p_M[\mathfrak{c}(U)]})^{\#} \vee \mathfrak{o}(b). \quad \square$$

9.2.1 Theorem *Each compact sublocale of a strongly Hausdorff locale is closed.*

Proof Let $j_L\colon L \to M$ be the embedding of a compact sublocale into a strongly Hausdorff locale M. Consider the embedding $\langle \mathrm{id}_L, j_L\rangle$ given by the universal property of products:

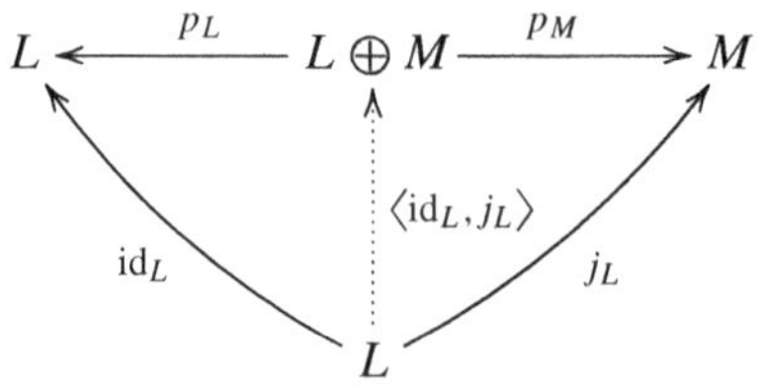

Since

$$(j_L \oplus \mathrm{id}_M)\langle \mathrm{id}_L, j_L\rangle[L] = \langle j_L, j_L\rangle[L] = \Delta\, j_L[L] \subseteq \Delta[M]$$

and moreover (as can be easily checked) $\langle \mathrm{id}_L, j_L\rangle[L]$ is the largest sublocale of $L \oplus M$ contained in $(j_L \oplus \mathrm{id}_M)^{-1}[\Delta[M]]$, then

$$\langle \mathrm{id}_L, j_L\rangle[L] = (j_L \oplus \mathrm{id}_M)_{-1}[\Delta[M]].$$

Hence, by the strong Hausdorff property of M, $\langle \mathrm{id}_L, j_L\rangle[L]$ is a closed sublocale of $L \oplus M$ and, by 9.2.1, $p_M[\langle \mathrm{id}_L, j_L\rangle[L]] = j_L[L]$ is a closed sublocale of M. $\square$

10 Regular Monomorphisms and Epimorphisms in Strongly Hausdorff Frames

Representing subobjects of locales is very natural. What one needs are extremal monomorphisms in the category **Loc**, that is, the adjoints of extremal epimorphisms in **Frm**. Now as it happens, this is very simple: the extremal epimorphisms in **Frm** are the onto frame homomorphisms. The structure of plain epimorphisms is in general very complicated (see, e.g., [161, 196, 220]) but we will be able to show that in the strongly Hausdorff case it is more transparent.

With representing quotient locales (nice epimorphisms in **Loc**, nice monomorphisms in **Frm**) the situation is reverse (and even more complicated than that). While plain monomorphisms in **Frm** are simply the one-to-one homomorphisms, the structure of special (extremal, regular) monomorphisms is very complex. In case of strongly Hausdorff frames we can, however, characterize the *regular* ones (those that are able to equalize pairs of homomorphisms).

10.1 For a subframe L of a frame M (we are now discussing subframes, not sublocales, and wish to get some insight into *quotients* of locales) set

$$d_L^M = \bigvee\{x \oplus y \mid x, y \in L,\ \ x \wedge y = 0\} = (j \oplus j)(d_L),$$

where $j\colon L \subseteq M$ is the embedding homomorphism.

10.1.1 Observations

1. *If L is strongly Hausdorff, then*

$$\breve{d}_L^M = (U \mapsto U \vee d_L^M)$$

   is the coequalizer of $\iota_1 j$ and $\iota_2 j$.
2. *If $L \subseteq K \subseteq M$, then $d_L^M \leqslant d_L^K$.*

(The first is in 5.4.1, the second is obvious.)

10.2 Regular monomorphisms in Frm For a subframe L of M set

$$\gamma(L) = \{a \in M \mid (a \oplus 1) \vee d_L^M = (1 \oplus a) \vee d_L^M\}.$$

Obviously,

the embedding $\gamma(L) \subseteq M$ is the equalizer of $\breve{d}_L^M \iota_1$ and $\breve{d}_L^M \iota_2$.

10.2.1 Proposition *We have the following facts.*

(1) $\gamma(\mathbf{2}) = \mathbf{2}$ *(where* **2** *denotes the 2-chain* $\{0 < 1\}$*).*
(2) $L \subseteq K \ \Rightarrow\ \gamma(L) \subseteq \gamma(K)$.
(3) *If L is strongly Hausdorff, then $L \subseteq \gamma(L)$.*
(4) *If L is strongly Hausdorff, then $\gamma(\gamma(L)) = \gamma(L)$.*

Proof (1) and (2) are straightforward and (3) follows from 5.3.3.

(4): Put $K = \gamma(L)$. We have to prove that $d_K^M \leqslant d_L^M$. Let $x, y \in K$ and $x \wedge y = 0$. Then

$$(x \oplus y) \vee d_L^M = ((x \oplus 1) \wedge (1 \oplus y)) \vee d_L^M =$$
$$= ((x \oplus 1) \wedge (y \oplus 1)) \vee d_L^M = ((x \wedge y) \oplus 1) \vee d_L^M = d_L^M$$

and hence $x \oplus y \leqslant d_L^M$. □

10.2.2 Theorem *Let L be strongly Hausdorff. Then a one-to-one frame homomorphism $h\colon L \to M$ is a regular monomorphism iff $\gamma(h[L]) = h[L]$.*

Proof It is a well-known (and easy) fact that a monomorphism $h\colon L \to M$ is regular iff it is the equalizer of the f, g from the pushout

$$\begin{array}{ccc} L & \xrightarrow{\ h\ } & M \\ {\scriptstyle h}\big\downarrow & & \big\downarrow{\scriptstyle f} \\ M & \xrightarrow[\ g\ ]{} & P \end{array}$$

Now the pushout can be constructed considering $f = k\iota_1$ and $g = k\iota_2$ with k the coequalizer of $\iota_1 h$ and $\iota_2 h$. By 10.1.1, $k = \check{d}^M_{h[L]}$ and by 10.2 the equalizer of $\check{d}^M_{h[L]}\iota_1$ and $\check{d}^M_{h[L]}\iota_2$ is the embedding $\gamma(h[L]) \subseteq M$. □

10.2.3 Note In [233] it was proved that even under the restriction to strongly Hausdorff frames, the regular monomorphisms do not represent "quotient locales" quite satisfactorily. But, nevertheless, they constitute a step forward in the problem, at least for $\Omega(X)$ of not quite exceptional spaces. For more see [231, 238].

10.3 Epimorphisms in strongly Hausdorff frames

10.3.1 Lemma *Let L be a subframe of M and let $f, g\colon M \to N$ coincide on L. Then they coincide on $\gamma(L)$.*

Proof We have

$$\begin{aligned} \Delta_N^*(f \oplus g)(d_L^M) &= \bigvee\{f(x) \wedge g(y) \mid x, y \in L, x \wedge y = 0\} \\ &= \bigvee\{f(x \wedge y) \mid x, y \in L,\ x \wedge y = 0\} = 0, \end{aligned}$$

$$\Delta_N^*(f \oplus g)(a \oplus 1) = f(a) \quad \text{and} \quad \Delta_N^*(f \oplus g)(1 \oplus a) = g(a).$$

If $(a \oplus 1) \vee d_L^M = (1 \oplus a) \vee d_L^M$, then

$$f(a) = \Delta_N^*(f \oplus g)(a \oplus 1) = \Delta_N^*(f \oplus g)(1 \oplus a) = g(a).$$ □

10.3.2 Theorem *Let L be a strongly Hausdorff frame. Then a morphism $h\colon L \to M$ is an epimorphism iff $\gamma(h[L]) = M$.*

Proof If $\gamma(h[L]) = M$, then h is an epimorphism by 10.3.1. On the other hand, the embedding $j\colon \gamma(h[L]) \subseteq M$ is the equalizer of $\check{d}^M_{h[L]}\iota_1$ and $\check{d}^M_{h[L]}\iota_2$ and hence, if j is not onto,

$$\check{d}^M_{h[L]}\iota_1 \neq \check{d}^M_{h[L]}\iota_2.$$

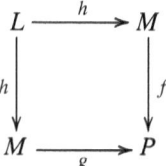

But

$$\breve{d}^M_{h[L]}\iota_1 \cdot h = \breve{d}^M_{h[L]}\iota_2 \cdot h.$$

Hence h is not an epimorphism. □

11 (Co)reflection

In this section we will prove that the category of strongly Hausdorff frames is coreflective in the category of frames.

11.1 Lemma *If $K_i \subseteq L$ $(i \in J)$ are strongly Hausdorff subframes of L, then the subframe K generated by them is also strongly Hausdorff.*

Proof We will use 5.4.2. Let $h_1, h_2 \colon K \to N$ be frame homomorphisms and set

$$c = \bigvee\{h_1(x) \wedge h_2(y) \mid x \wedge y = 0\}$$

and

$$c_i = \bigvee\{h_1(x) \wedge h_2(y) \mid x, y \in K_i,\ \ x \wedge y = 0\}.$$

Then $c_i \leqslant c$. Consider the embeddings $m_i \colon K_i \to K$ and

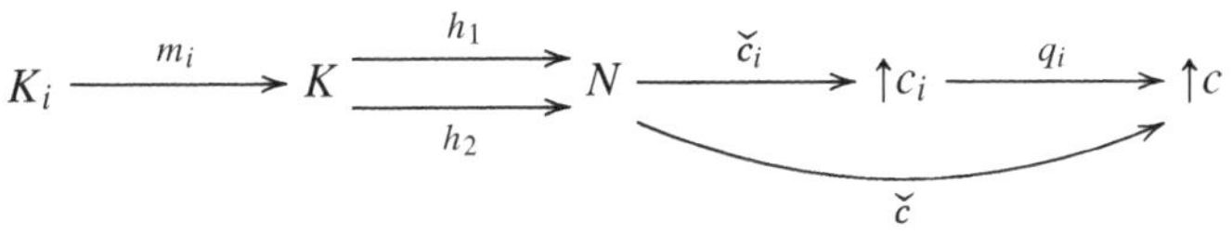

where $\breve{c}_i \colon a \mapsto a \vee c_i$ and $q_i \colon a \mapsto a \vee c$, hence $q_i \cdot \breve{c}_i = \breve{c}$. Then $\breve{c}h_1 m_i = \breve{c}h_2 m_i$ for every i, hence $\breve{c} \cdot h_1 = \breve{c} \cdot h_2$ (since K is generated by the K_i). □

11.2 For any subframe K of L define a map $p_K \colon L \to K$ by setting

$$p_K(x) = \bigvee\{y \in K \mid y \leqslant x\}.$$

It is easy to check that this is a right adjoint to the embedding homomorphism $K \subseteq L$; hence it is a localic map.

Lemma *Let $f \colon L \to M$ be a localic map with adjoint frame homomorphism f^* and let K be a subframe of L. If $f^*[M] \subseteq K$, then the restriction $f|_K \colon K \to M$ is a localic map such that $f|_K \cdot p_K = f$.*

Proof By the assumption, f^* factorizes through the embedding $K \subseteq L$, and $f|_K$ is the right adjoint of f^*. □

11.3 For a frame L define $H_s(L)$ as the largest strongly Hausdorff subframe of L (which exists by 11.1).

Theorem *Define* $\sigma_L : L \to H_s(L)$ *by setting*

$$\sigma_L(a) = p_{H_s(L)}(a) = \bigvee\{x \in H_s(L) \mid x \leqslant a\}.$$

Then the system $(\sigma_L)_L$ *constitutes an epireflection of the category of locales onto the subcategory of the strongly Hausdorff ones.*

Proof The maps σ_L are localic maps: they are adjoints to the frame embeddings $H_s(L) \subseteq L$. Let M be a strongly Hausdorff locale and let $f : L \to M$ be a localic map. By 5.3.4, $f^*[M]$ is strongly Hausdorff. Hence $f^*[M] \subseteq H_s(L)$ and by 11.2 there is a localic map $f|_{H_s(L)} : H_s(L) \to M$ such that

$$f|_{H_s(L)} \cdot \sigma_L = f|_{H_s(L)} \cdot p_{H_s(L)} = f. \qquad \square$$

Chapter IV
Summarizing Low Separation

So far (with just a few exceptions) we have discussed the "low separation axioms": those that can be used to replace the classical T_1, and several conditions mimicking the Hausdorff property. Before turning to the stronger ones we will now summarize relations between them and add some facts that will help to understand the role they play and can play in further applications.

1 Axioms Akin to Classical T_1

1.1 We have in mind the (point-free) T_1, T_U, and subfitness (abbreviated (sfit)).

T_1 from II.1.1 is not very satisfactory, stating only that the spectrum is T_1 in the classical sense. Just in itself, hence, it has not much importance in the point-free context. But the relations to the other axioms help to clarify some phenomena.

T_U is much more interesting, and we venture to say that it is not yet quite well understood. Understood formally just in spaces, the condition reduces simply to symmetry, a very weak property. But in the point-free generality the situation changes. We will see that it is stronger than the (point-free) T_1.

Subfitness is particularly interesting. The reader has already seen a number of important consequences in Chap. II. There will be more in the sequel.

1.2 Proposition *T_1 and subfitness are independent; that is, none of them implies the other.*

Proof

(sfit) does not imply T_1: Take an infinite set and the topology L consisting of $\emptyset$ and all complements of finite subsets of X. Then L is obviously subfit but it is not T_1 because $\emptyset$ is prime in L but not maximal.

J. Picado, A. Pultr, *Separation in Point-Free Topology*,
https://doi.org/10.1007/978-3-030-53479-0_4

T_1 does not imply (sfit): Consider a Boolean algebra B without atoms (for instance, the set of regular open[1] subsets of the unit interval) and set

$$L = \{(a, b) \mid a, b \in B,\ a \leqslant b\} \text{ with the natural coordinatewise order.}$$

Then L has no primes (if (p, q) were a prime in L, then p would be equal to q and it would be a prime in B) and hence satisfies T_1. But it is not subfit: $(0, 1) \nleqslant (0, 0)$ but if $(0, 1) \vee (c_1, c_2) = (1, 1)$ with $c_1 \leqslant c_2$, then $(c_1, c_2) = (1, 1)$. □

1.2.1 Notes

1. The reader may wonder about the first part: the cofinal space (X, L) is T_1 and hence subfit. But the spectrum of L is the sobrification of (X, L) which has one more point and is not a T_1-space.
2. In the second part of the proof we saw that T_1 does not imply even weak subfitness.
3. In spaces, however, T_1 does imply subfitness.

1.3 Proposition *T_U implies T_1.*

Proof Recall that an element $p \in L$ is prime iff $P(p) = \{p, 1\}$ is a sublocale (A6.6). Thus the point is in proving that in a T_U-frame all the one-point sublocales $P(p)$ are closed.

Define a map $g\colon \overline{P(p)} = {\uparrow}p \to L$ by setting

$$g(1) = 1 \quad \text{and} \quad g(a) = p \text{ for } a \neq 1.$$

We will prove that it is a localic map. Since it obviously preserves meets and since

$$g(x) = 1 \Rightarrow x = 1$$

it remains to prove that (recall A.4.4) $g(g^*(x) \to a) = x \to g(a)$. We have

$$x \to g(a) = \begin{cases} x \to 1 = 1 & \text{if } a = 1 \\ x \to p & \text{if } a \neq 1, \end{cases}$$

that is,

$$x \to g(a) = \begin{cases} 1 & \text{if } a = 1 \text{ or } x \leqslant p \\ p & \text{otherwise} \end{cases}$$

[1] *Regular* open in the sense that $\mathrm{int}(\overline{U}) = U$.

and hence it suffices to check that

$$g^*(x) \to a = 1 \quad \text{iff} \quad a = 1 \text{ or } x \leqslant p$$

which is straightforward because

$$g^*(x) = \begin{cases} p & \text{if } x \leqslant p \\ 1 & \text{otherwise.} \end{cases}$$

Now consider the embedding of the closure $j\colon \overline{P(p)} \hookrightarrow L$. We have $g \leqslant j$ and hence if L is T_U, $g = j$. Thus, $\overline{P(p)} = P(p)$. □

1.4 In dealing with T_U we may use equivalently the localic or frame language, because the order of localic maps reflects contravariantly in the order of the associated frame homomorphisms.

1.4.1 Proposition *The following statements about localic maps $f, g\colon L \to M$ are equivalent.*

(1) $g \leqslant f$.
(2) *For the adjoint frame homomorphisms, $f^* \leqslant g^*$.*
(3) *For each $b \in M$, $f_{-1}[\mathfrak{o}(b)] \subseteq g_{-1}[\mathfrak{o}(b)]$.*
(4) *For each $b \in M$ and $a \in L$, $b \to g(f^*(b) \to a) = g(f^*(b) \to a)$.*

Proof (1)⇔(2): If $g \leqslant f$ and $g^*(y) \leqslant x$, then $y \leqslant g(x) \leqslant f(x)$ and hence $f^*(y) \leqslant x$; thus, $f^*(y) \leqslant g^*(y)$. Similarly in the other direction.

(2)⇔(3): We have $f_{-1}[\mathfrak{o}(b)] = \mathfrak{o}(f^*(b))$ by A.7.3 and $\mathfrak{o}(x) \subseteq \mathfrak{o}(y)$ iff $x \leqslant y$.

(3)⇔(4): From the adjunction $g[-] \dashv g_{-1}[-]$ (recall A.7.2) we get for each $b \in M$

$$\begin{aligned} f_{-1}[\mathfrak{o}(b)] \subseteq g_{-1}[\mathfrak{o}(b)] &\Leftrightarrow g f_{-1}[\mathfrak{o}(b)] \subseteq \mathfrak{o}(b) \Leftrightarrow g[\mathfrak{o}(f^*(b))] \subseteq \mathfrak{o}(b) \Leftrightarrow \\ &\Leftrightarrow \forall a \in L,\ g(f^*(b) \to a) \in \mathfrak{o}(b) \Leftrightarrow \\ &\Leftrightarrow \forall a \in L,\ b \to g(f^*(b) \to a) = g(f^*(b) \to a). \end{aligned}$$ □

Notes

1. It follows immediately that the property T_U is *hereditary* (that is, any sublocale of a T_U-locale is T_U) as well as *productive* (that is, the product of any family of T_U-locales is T_U). Hence, one has that

   the subcategory of T_U-locales is reflective in the category ***Loc****.*

2. Furthermore, if L is T_U, then for every sublocale S of L and any localic map $g\colon S \to L$,

$$(\forall s \in S,\ g(s) \geqslant s) \Rightarrow (\forall s \in S,\ g(s) = s).$$

In particular, the 3-chain $\mathbf{3} = \{0 < a < 1\}$ cannot be a sublocale of L since $g\colon \mathbf{3} \to L$ defined by $g(0) = g(a) = 0$ and $g(1) = 1$ is a localic map.[2]

1.5 Proposition *Subfitness does not imply T_U, but fitness does.*

Proof The example in III.6.3 is classically T_1 and hence subfit, and it is not T_U.

Now let L be fit and let $f, g\colon L \to M$ be localic maps with $g \leqslant f$. By 1.4.1, $f_{-1}[\mathfrak{o}(b)] \subseteq g_{-1}[\mathfrak{o}(b)]$ for any $b \in M$.

Preimage maps $f_{-1}[-]$ preserve (existing) complements. Hence $g_{-1}[L \smallsetminus \mathfrak{o}(b)]$ is the complement of $g_{-1}[\mathfrak{o}(b)]$. By fitness, $L \smallsetminus \mathfrak{o}(b) = \bigcap_{i \in J} \mathfrak{o}(b_i)$. We have $f_{-1}[\mathfrak{o}(b)] \subseteq g_{-1}[\mathfrak{o}(b)]$ but also $f_{-1}[\mathfrak{o}(b_i)] \subseteq g_{-1}[\mathfrak{o}(b_i)]$ and hence

$$M \smallsetminus g_{-1}[\mathfrak{o}(b)] = g_{-1}[\bigcap_{i \in J} \mathfrak{o}(b_i)] = \bigcap_{i \in J} g_{-1}[\mathfrak{o}(b_i)] \supseteq$$

$$\supseteq \bigcap_{i \in J} f_{-1}[\mathfrak{o}(b_i)] = f_{-1}[\bigcap_{i \in J} \mathfrak{o}(b_i)] = M \smallsetminus f_{-1}[\mathfrak{o}(b)]$$

so that also $g_{-1}[\mathfrak{o}(b)] \subseteq f_{-1}[\mathfrak{o}(b)]$ and, finally, $g_{-1}[\mathfrak{o}(b)] = f_{-1}[\mathfrak{o}(b)]$. □

1.5.1 Note Thus, in turn, fitness implies T_1. It might be of interest to see how this follows directly. Suppose L is fit, p prime and $p < x$. Then by fitness there is a c such that $x \vee c = 1$ and $c \to p \nleqslant p$. Since $c \wedge (c \to p) \leqslant p$ necessarily $c \leqslant p$ and hence $x = x \vee p \geqslant x \vee c = 1$.

1.6 The situation with the axioms we have discussed is depicted in the following diagram where none of the implications can be reversed, and the wiggles indicate independence. We put there fitness as well; it belongs to the stronger separation conditions, but its kinship with subfitness is certainly a sufficient reason for being included.

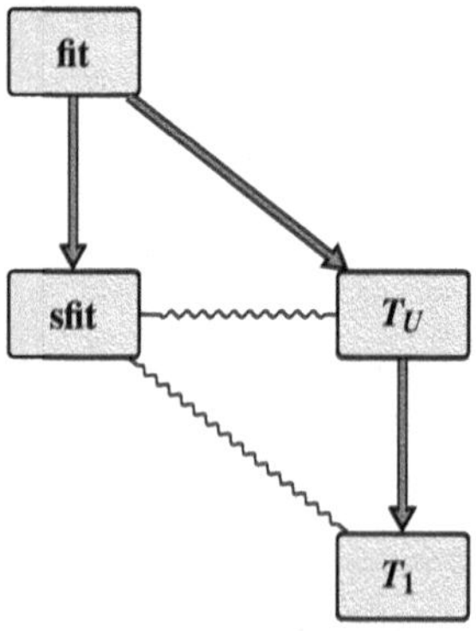

[2]The condition "**3** is not a sublocale of L" is the T_1-definition for locales proposed by Han and Hong in [135]. Hence, also this variant of T_1 is weaker than T_U.

There are two facts in the scheme we have not proved yet. It remains to be shown that

- T_1 *does not imply* T_U and that
- T_U *does not imply* (*sfit*).

The former follows from III.6.3: the space in the example is T_1 (both in the classical and in the point-free sense). The latter is harder and will follow from the quite involved example in Sect. 4 below that (sH) does not imply (sfit) (since we already know from III.6.2 that (sH) implies T_U).

2 Some Statements Equivalent to (sH)

Before summarizing the Hausdorff type conditions let us introduce some equivalents of the strongest of them.

2.1 Dowker and Strauss' formula In [85], Dowker and Strauss considered a condition stronger than T_U and then proved that it is equivalent with (sH). Here it is.

(sH"): *For any frame homomorphisms* $h_1, h_2\colon L \to M$, *if* $h_1(x) \wedge h_2(y) = 0$ *whenever* $x \wedge y = 0$, *then* $h_1 = h_2$.

Note Of course, (sH") can be written more concisely, say as

$$(\forall x \in L,\ h_1(x^*) \leqslant h_2(x)^*) \ \Rightarrow\ h_1 = h_2,$$

but the explicit formula is more expedient in proofs.

2.2 Theorem *The following statements about a frame L are equivalent.*

(1) *L satisfies (sH").*
(2) *For any two frame homomorphisms* $h_1, h_2\colon L \to M$, *the coequalizer S is closed in M.*
(3) *L is strongly Hausdorff.*

Proof (1)⇒(2): Set

$$c = \bigvee\{h_1(x) \wedge h_2(y) \mid x \wedge y = 0\}$$

and define $\breve{c}\colon L \to {\uparrow}c$ by $\breve{c}(a) = c \vee a$. Then

$$x \wedge y = 0 \quad \Rightarrow \quad (c \vee h_1(x)) \wedge (c \vee h_2(y)) = c = 0_{\uparrow c}$$

and hence, by (sH"), $\breve{c}h_1 = \breve{c}h_2$.

Now let $g: M \to N$ be such that $gh_1 = gh_2$. Then

$$g(c) = \bigvee\{gh_1(x) \wedge gh_2(y) \mid x \wedge y = 0\} = \bigvee\{gh_1(x \wedge y) \mid x \wedge y = 0\} = 0$$

and hence we can define $\overline{g}: \uparrow c \to N$ by setting $\overline{g}(a) = g(a)$ to obtain $\overline{g}\,\check{c} = g$. Thus, $\check{c}$ is the coequalizer of h_1, h_2.

(2)⇒(3): The codiagonal

$$\Delta^*: L \oplus L \to L$$

is the coequalizer of ι_1 and ι_2, as we have observed in the proof of III.5.4.2.

(3)⇒(1): Use the notation from III.5.2–5.3. Hence,

$$d_L = \{(x, y) \mid x \wedge y = 0\} = \bigvee\{x \oplus y \mid x \wedge y = 0\}$$

and $\alpha: L \to \uparrow d_L$ is such that $\alpha\Delta^*(U) = U \vee d_L = \check{d}_L(U)$.

Consider $h_i: L \to M$ $(i = 1, 2)$ such that $h_1(x) \wedge h_2(y) = 0$ whenever $x \wedge y = 0$, and take the $h: L \oplus L \to M$ defined by $h\,\iota_i = h_i$. If $(x, y) \in d_L$, that is, $x \oplus y \subseteq d_L$, then $x \wedge y = 0$ and therefore

$$h(x \oplus y) = h(\iota_1(x) \wedge \iota_2(y)) = h_1(x) \wedge h_2(y) = 0.$$

Hence $h(d_L) = 0$ and if we define $h': \uparrow d_L \to M$ by $h'(U) = h(U)$, we have $h'\,\check{d}_L = h$.

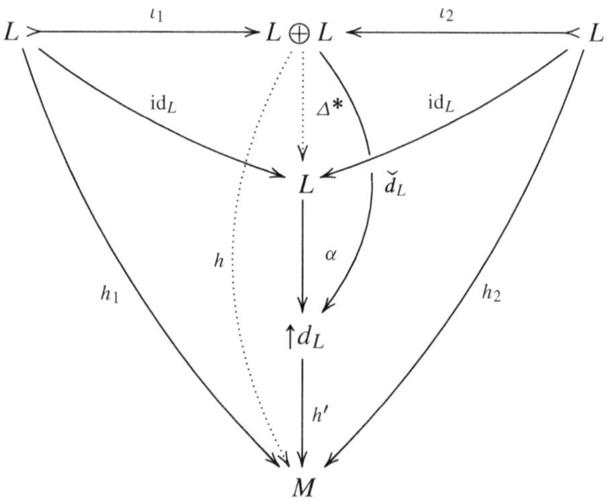

By III.5.3.1, $\alpha\,\Delta^* = \check{d}_L$, hence

$$h_i = h \cdot \iota_i = h' \cdot \check{d}_L \cdot \iota_i = h' \cdot \alpha \cdot \Delta^* \cdot \iota_i = h' \cdot \alpha,$$

and $h_1 = h_2$. □

Note In particular compare the equivalence (2)⇔(3) (which extends Theorem III.5.4.1) with the situation in spaces.

2.3 Viewing a frame as a generalized space we see that while the join of elements is of an utmost importance, in many questions the role of the finite infimum is just in the decision whether two nonzero places meet or not. This led to the concept of a *weak homomorphism* [54] as a mapping $h\colon L \to M$ preserving all joins and such that

$$h(1) = 1 \quad \text{and} \quad a \wedge b = 0 \Rightarrow h(a) \wedge h(b) = 0.$$

It turned out that, e.g. for regular L each weak homomorphism $h\colon L \to M$ is indeed a homomorphism which led to the question how much one had to assume for such a statement, and ultimately to the formulation of the separation axiom (on L)

(W): *every weak homomorphism $h\colon L \to M$ is a (frame) homomorphism.*

2.4 Lemma *A frame L is strongly Hausdorff iff*

$$\forall a, b \in L,\ (a \oplus b) \vee d_L = ((a \wedge b) \oplus 1) \vee d_L.$$

Proof The formula obviously implies the (sH) from III.5.3.3. On the other hand, if (sH) holds, we have

$$\begin{aligned}(a \oplus b) \vee d_L &= ((a \oplus 1) \wedge (1 \oplus b)) \vee d_L = \\ &= ((a \oplus 1) \wedge (b \oplus 1)) \vee d_L = ((a \wedge b) \oplus 1) \vee d_L.\end{aligned}$$ □

2.5 Theorem *A frame is strongly Hausdorff iff it satisfies* (W) *and* T_U.

Proof ⇒: By III.6.2 a strongly Hausdorff frame satisfies T_U. Hence, we have only to prove that it satisfies (W).

Let $h\colon L \to M$ be a weak homomorphism. We will use the coproduct $\oplus$ in its facet as a tensor product in the category of complete lattices with the join-preserving mappings (see Sect. A.8.4 of the Appendix). In particular, we have a join-preserving mapping

$$h \oplus h\colon L \oplus L \to M \oplus M$$

such that $(h \oplus h)(a \oplus b) = h(a) \oplus h(b)$. Since $x \wedge y = 0$ implies $h(x) \wedge h(y) = 0$ we have

$$(h \oplus h)(d_L) \leqslant d_M. \tag{$*$}$$

By Lemma 2.4, $(h \oplus h)(((a \wedge b) \oplus 1) \vee d_L) = (h \oplus h)((a \oplus b) \vee d_L)$ and hence

$$(h(a \wedge b) \oplus 1) \vee (h \oplus h)(d_L) = (h \oplus h)((a \wedge b) \oplus 1) \vee (h \oplus h)(d_L) = \\ = (h \oplus h)(a \oplus b) \vee (h \oplus h)(d_L) = (h(a) \oplus h(b)) \vee (h \oplus h)(d_L),$$

and joining both sides with d_M and using $(*)$ we see that

$$(h(a \wedge b) \oplus 1) \vee d_M = (h(a) \oplus h(b)) \vee d_M.$$

Now applying the diagonal ∇_M on this equality we conclude that

$$h(a \wedge b) = h(a) \wedge h(b).$$

$\Leftarrow$: Let $f, g \colon L \to M$ be homomorphisms such that $f(x) \wedge g(y) = 0$ whenever $x \wedge y = 0$. Setting $h(x) = f(x) \vee g(x)$ we easily check that we thus have a weak homomorphism $h \colon L \to M$. Hence if L satisfies (W), this h is a frame homomorphism, and we have $f, g \leqslant h$. Thus, if L also satisfies (T_U),

$$f = g = h.$$

Now consider, in particular, the $f, g \colon L \to {\uparrow} d_L$ given by

$$f(a) = (a \oplus 1) \vee d_L \quad \text{and} \quad g(a) = (1 \oplus a) \vee d_L.$$

Then the resulting

$$(a \oplus 1) \vee d_L = (1 \oplus a) \vee d_L$$

is the (sH') from III.5.3.3. □

2.6 Note The T_U in previous theorem is essential. Consider the 3-chain $\mathbf{3} = \{0 < a < 1\}$. For any weak homomorphism $h \colon \mathbf{3} \to M$ we have

$$h(a \wedge b) = \begin{cases} h(a) = h(a) \wedge h(b) & \text{for } b = a, 1 \\ h(0) = 0 = h(a) \wedge h(b) & \text{for } b = 0. \end{cases}$$

Hence **3** satisfies (W) (and the same holds for any linearly ordered frame).

3 The Hausdorff Type Axioms: Merits of the Strong One

3.1 First, we have the implications

$$\text{(sH)} \;\Rightarrow\; \text{(H)} \;\Rightarrow\; \text{(wH)},$$

all of them irreversible (see Chap. III). Furthermore, we have

$$\text{(H)} \Rightarrow \text{(pH)} \Rightarrow \text{(wpH)} \quad \text{and} \quad \text{(wH)} \Rightarrow \text{(wpH)} \quad \text{(see [183])}$$

about which we do not really know enough (III.4.3).

3.2 The conditions applied in spaces The first question one naturally asks is whether this or other of the point-free axioms when applied for spaces is *conservative*. That is, when applied for a space yields precisely the classical homonymous condition it should mimic. Thus we ask whether spaces X with $\Omega(X)$ satisfying (wH), (H), (sH) or (pH) are precisely the Hausdorff ones.

We know from Chap. III (3.4.1 and 4.1.2) that

3.2.1 *(H) and (pH) are conservative*

while

3.2.2 *(wH) and (sH) are not* (III.2.3, III.6.5).

Regarding (wH), this was mended already in the original paper [81] by considering

$$\text{(wH) \& (sfit)}$$

instead (by the way, adding (sfit) makes all kind of variants of (wH) and also (H) coincide—III.2.3.1 and III.3.3.2; it is an open problem what adding subfitness does to (pH)).

On the other hand, (sH) is for spaces strictly stronger than the classical T_2.

3.3 Relations to the axioms from 3.1 One should expect the axioms mimicking T_2 to be stronger than those replacing T_1, but it is not quite so.

First, none of them implies subfitness. This is a rather involved fact to be proved in the next section. Further, (wH) does not imply even T_1, for trivial reasons (chains are (wH)).

The situation with the others is more satisfactory. We have that (see III.6.2)

3.3.1 *(sH) implies T_U (and hence also T_1)*, and

3.3.2 *(H) implies T_1 (although it does not imply T_U).*

Proof Suppose L satisfies (H). Let p be prime and let there be an a such that $p < a \neq 1$. Then there are u, v such that $u \nleqslant a$, $v \nleqslant p$ and $u \wedge v = 0$. Since p is prime and $v \nleqslant p$ we have to have $u \leqslant p$; but then $u \leqslant a$, a contradiction. □

3.4 Merits of the strong Hausdorff property The condition (sH) is not conservative which is somewhat awkward. But, still, it can be argued that, when we compare its behaviour with that of classical Hausdorff spaces, it is the best one.

3.4.1 We have the basic fact that

the equalizer of two localic maps $L \to M$ is closed

(and hence *epimorphisms are dense maps*). Since this is a necessary condition for (sH) (see III.5.5.1)

it does not hold for any other of the discussed axioms.

3.4.2 We have that

compact sublocales of a (sH) frame are closed (III.9.2.1).

Here we do not know whether we might not assume substantially less, but considering how one had to use the specific (sH) features in the proof it is unlikely.

3.4.3 In classical spaces, compact Hausdorff spaces have automatically stronger separation properties. And so it is also with (sH) frames. We have seen in III.9.1.1 that

a compact (sH) frame is regular,

(which we will see, again, to lead to higher separation as well). On the other hand, compact (H) frames are not necessarily regular (III.3.5.3).

4 Strong Hausdorff Does Not Imply Subfit

The proof to be presented is based on the constructions from the appendix of [28].

4.1 Barely dense elements An element a of a frame L is *barely dense* if it is dense in L (i.e., $a^* = 0$) but not in $\uparrow b$ with $0 < b \leqslant a$ (i.e., whenever $0 < b \leqslant a$, there is $c > b$ such that $a \wedge c = b$).

For $x \in \downarrow a$ set

$$x^\sim = \bigvee\{y \in L \mid x = y \wedge a\}.$$

If a is barely dense, then $x^\sim > x$ for any nonzero $x \in \downarrow a$; that is,

$$\forall x \in \downarrow a,\ x^\sim = x \ \Rightarrow\ x = 0. \tag{$*$}$$

4.1.1 Denote the pseudocomplements in $\downarrow a$ by x^{*_a}. Clearly,

$$x^{*_a} = x^* \wedge a \ \text{ and } \ x^{*_a *_a} = (x^* \wedge a)^* \wedge a.$$

Lemma *Let a be a barely dense element such that*

$$c \vee a = 1 \quad \textit{and} \quad (a \vee (c \wedge a)^{*_a *_a})^{*_a *_a} = a.$$

Then $(c \wedge a)^{*_a *_a} = a$.

Proof Consider an arbitrary $d \in {\downarrow}a$ satisfying $d \leqslant (c \wedge a)^{*_a}$, that is, $(c \wedge a) \wedge d = 0$. Then $a \wedge (c \wedge a)^{\sim} \wedge d^{\sim} = c \wedge a \wedge d = 0$ and $(c \wedge a)^{\sim} \wedge d^{\sim} = 0$ (since $a^* = 0$). Hence also $c \wedge d^{\sim} = 0$ (because $c \leqslant (c \wedge a)^{\sim}$), and thus $d^{\sim} \leqslant a$ (since $c \vee a = 1$). It follows that $d^{\sim} = d$, which implies $d = 0$, by $(*)$. This means that $c \wedge a$ is dense in ${\downarrow}a$ and, consequently, $(c \wedge a)^{*_a *_a} = a$. □

4.2 A construction producing non-subfit frames Recall the Booleanization $\mathfrak{B}(L)$ of a frame L (A.6.7.1). For each $a \in L$, consider the frame $({\uparrow}a) \times \mathfrak{B}({\downarrow}a)$ and the frame homomorphism

$$h = h_a \colon L \longrightarrow ({\uparrow}a) \times ({\downarrow}a) \longrightarrow ({\uparrow}a) \times \mathfrak{B}({\downarrow}a)$$

$$x \mapsto (x \vee a, x \wedge a) \mapsto (x \vee a, (x \wedge a)^{*_a *_a}).$$

Set $P_a^L = h_a[L]$. This is a subframe of $({\uparrow}a) \times \mathfrak{B}({\downarrow}a)$ and $h_*[P_a^L]$ is a sublocale of L isomorphic to P_a^L.

4.2.1 Proposition *For any barely dense element a, the frame P_a^L is not subfit.*

Proof We will show that the subfitness condition does not hold for the pair $(a, a) \nleqslant (a, 0)$. For this, consider any $c \in L$ such that

$$(a, a) \vee (c \vee a, (c \wedge a)^{*_a *_a}) = (1, a). \tag{$*$}$$

We need to show that $(a, 0) \vee (c \vee a, (c \wedge a)^{*_a *_a}) = (c \vee a, (c \wedge a)^{*_a *_a})$ is equal to $(1, a)$. By $(*)$, $c \vee a = 1$, so that it amounts to showing that for each barely dense element a, $c \vee a = 1$ and $(a \vee (c \wedge a)^{*_a *_a})^{*_a *_a} = a$ imply $(c \wedge a)^{*_a *_a} = a$, which is true by 4.1.1. □

4.2.2 Remark Since fitness is an hereditary property it follows, in particular, that a frame with a barely dense element is not fit.

4.3 Let X be dense in $Y \supsetneq X$. Define a new space

$$E = E_{XY}$$

carried by the set Y with the topology changed to

$$\Omega(E) = \{(U \cap X) \cup M \mid U \in \Omega(Y),\ M \subseteq (Y \smallsetminus X) \cap U\}$$

(note that $\Omega(X) \subseteq \Omega(E)$ and X is a subspace of E).

4.3.1 Proposition

(a) *X is a dense open subspace of E_{XY} and the remainder $E \smallsetminus X$ is discrete.*[3]
(b) *If Y is Hausdorff, then also E_{XY} is Hausdorff.*

Proof

(a) If $V = (U \cap X) \cup M$ is nonempty, then $U \in \Omega(Y)$ is nonempty and since X is dense in Y, $V \cap X = U \cap X$ is nonempty. If M is any subset of $Y \smallsetminus X$, then $X \cup M = (Y \cap X) \cup M$ is in $\Omega(E)$ and $M = ((Y \smallsetminus X) \cup M) \cap M$ is open in $E \smallsetminus X$.
(b) If $x, y \in E$ are distinct, then there are disjoint $U \ni x$, $V \ni y$ in $\Omega(Y)$ and we have disjoint $(U \cap X) \cup \{x\}$, $(V \cap X) \cup \{y\}$ in E (note that this includes the cases with x or y in X). □

4.3.2 Lemma *For any frame L and any $a \in L$, the frame homomorphisms*

$$\alpha_1 = (x \oplus y \mapsto (x \wedge a) \oplus (y \wedge a)) \colon L \oplus L \to (\downarrow a) \oplus (\downarrow a),$$

$$\alpha_2 = (x \oplus y \mapsto (x \wedge a) \oplus (y \vee a)) \colon L \oplus L \to (\downarrow a) \oplus (\uparrow a),$$

$$\alpha_3 = (x \oplus y \mapsto (x \vee a) \oplus (y \wedge a)) \colon L \oplus L \to (\uparrow a) \oplus (\downarrow a),$$

$$\alpha_4 = (x \oplus y \mapsto (x \vee a) \oplus (y \vee a)) \colon L \oplus L \to (\uparrow a) \oplus (\uparrow a)$$

are jointly monic (that is, if $\alpha_i(U) = \alpha_i(V)$ for all i, then $U = V$).

Proof It easily follows from the standard fact that in a distributive lattice $x \wedge a = y \wedge a$ and $x \vee a = y \vee a$ only if $x = y$. □

4.3.3 Theorem *If X is strongly Hausdorff (in particular, if it is regular) and if Y is Hausdorff, then $E = E_{XY}$ is strongly Hausdorff.*

Proof Recall the characterization (sH') from III.5.3.3. We will prove that for

$$d = d_{\Omega(E)} = \bigvee \{A \oplus B \mid A, B \in \Omega(X), A \cap B = \emptyset\}$$

[3] A space E with these properties is called a *simple extension* of X. A standard construction of a simple extension of a space X (due to Katětov [176]) is to present a family $(\mathcal{F}_r)_{r \in R}$ of filters in $\Omega(X)$ indexed by a set R disjoint from X; one takes the topology determined by the neighbourhood bases

$$\{U \in \Omega(X) \mid x \in U\} \ \text{ for } x \in X, \text{ and}$$

$$\{W \cup \{r\} \mid W \in \mathcal{F}_r\} \ \text{ for } r \in R.$$

In our case we have taken $R = Y \smallsetminus X$ and $\mathcal{F}_r = \{U \cap \{r\} \mid U \in \Omega(Y), r \in U\}$.

we have

$$(U \oplus E) \vee d = (E \oplus U) \vee d. \tag{$*$}$$

Set $R = Y \smallsetminus X = E \smallsetminus X$. By 4.3.2, and considering the obvious isomorphisms

$$(U \mapsto U \cap R)\colon \uparrow X \cong \Omega(R) = \mathfrak{P}(R) \text{ and}$$

$$(U \oplus V \mapsto U \times V)\colon \Omega(R) \oplus \Omega(R) \cong \Omega(R \times R)$$

(note that R is discrete), it suffices to prove that the terms of $(*)$ are equalized by the maps

$$\alpha = (U \oplus V \mapsto (U \cap X) \oplus (V \cap X))\colon \Omega(E) \oplus \Omega(E) \to \Omega(X) \oplus \Omega(X),$$

$$\beta = (U \oplus V \mapsto (U \cap X) \oplus (V \cap R))\colon \Omega(E) \oplus \Omega(E) \to \Omega(X) \oplus \Omega(R),$$

$$\gamma = (U \oplus V \mapsto (U \cap R) \oplus (V \cap X))\colon \Omega(E) \oplus \Omega(E) \to \Omega(R) \oplus \Omega(X),$$

$$\delta = (U \oplus V \mapsto (U \cap R) \times (V \cap R))\colon \Omega(E) \oplus \Omega(E) \to \Omega(R \times R).$$

First, let us compute the images of d. We have

$$\begin{aligned} \alpha(d) &= \bigvee\{(U \cap X) \oplus (V \cap X) \mid U \cap V = \emptyset\} = \\ &= \bigvee\{(U \cap X) \oplus (V \cap X) \mid (U \cap X) \cap (V \cap X) = \emptyset\} = \\ &= \bigvee\{A \oplus B \mid A, B \in \Omega(X), A \cap B = \emptyset\} = d_{\Omega(X)}. \end{aligned}$$

Next, since E is Hausdorff (recall 4.3.1(b)) we have for any $x \in X$ and $y \in Y$ disjoint open U_x, V_y in E such that $x \in U_x$ and $y \in V_y$, and see that

$$\beta(d) \supseteq \bigvee\{(U_x \cap X) \oplus (V_y \cap R) \mid (x, y) \in X \times R\} = X \oplus R = 1_{\Omega(X)\oplus\Omega(R)}$$

and for the same reason, $\gamma(d) = R \oplus X$. Similarly, using for distinct $x, y \in R$ separating U_x, V_y we see that

$$\delta(d) \supseteq \bigvee\{(U_x \cap R) \oplus (V_y \cap R) \mid x \neq y\} \supseteq (R \times R) \smallsetminus \Delta_R \supseteq \delta(d)$$

(where Δ_R is the diagonal $\{(x, x) \mid x \in R\}$ of R).

Since $\Omega(X)$ is strongly Hausdorff, we can then conclude that

$$\begin{aligned} \alpha((U \oplus E) \vee d) &= ((U \cap X) \oplus X) \vee d_{\Omega(X)} = \\ &= (X \oplus (U \cap X)) \vee d_{\Omega(X)} = \alpha((E \oplus U) \vee d). \end{aligned}$$

Trivially β resp. γ equalizes the two sides making them both $X \oplus R$ resp. $R \oplus X$, and finally, for $U = X \cup M$,

$$\delta((U \oplus E) \vee d) = (M \times R) \cup ((X \times X) \smallsetminus \Delta_X) = (M \times M) \cup ((X \times X) \smallsetminus \Delta_X)$$

same as $\delta((E \oplus U) \vee d)$. □

4.4 Proposition *Let both X and $Y \smallsetminus X$ be dense in Y. Then X is barely dense in $\Omega(E_{XY})$.*

Proof Let $\emptyset \neq U \subseteq X$. We have to prove that, in $\Omega(E)$, X is not dense in $\uparrow U$. Thus (recall (∗)), we need a $V \in \Omega(E)$ such that $V \supsetneq U$, and $X \cup V = U$. Since X is a subspace of Y there is an open W in Y such that $U = W \cap X$. Since $Y \smallsetminus X$ is dense in Y, $W \cap (Y \smallsetminus X) \neq \emptyset$. Choose some $x \in W \cap (Y \smallsetminus X)$ and set $V = U \cup \{x\}$. □

4.4.1 Theorem *Let Y be regular and let both X and $Y \smallsetminus X$ be dense in Y. Set $E = E_{XY}$. Then:*

(a) $P_X^{\Omega(E)}$ *(recall 4.2) is strongly Hausdorff but not subfit.*
(b) *E is strongly Hausdorff but not fit.*

Proof

(a) By 4.3.3, E is strongly Hausdorff, and since $P_X^{\Omega(E)}$ is isomorphic to a sublocale of E, also this frame is strongly Hausdorff. By 4.4 and 4.2.1 it is not subfit.
(b) Since all sublocales of a fit frame are subfit, E is not fit.

□

5 Fit Does Not Imply Hausdorff

Here we will be able to produce a spatial example. It is inspired by an exercise[4] from [64].

5.1 Construction Let $\mathbb{N}$ be the set of natural numbers and let u, v be distinct elements, $u, v \notin \mathbb{N}$. For $t = u, v$ and $m \in \mathbb{N}$ set

$$B(t, m) = \{t\} \cup \{n \in \mathbb{N} \mid n \geqslant m\}$$

and consider the system

$$\mathcal{B} = \{U \mid U \subseteq \mathbb{N}\} \cup \{B(t, m) \mid t = u, v;\ m \in \mathbb{N}\}.$$

Obviously $\mathcal{B}$ is closed under intersections and hence constitutes a basis of a topology L on $X = \mathbb{N} \cup \{u, v\}$.

[4]Exercise 5(b), p. 142. See also Example 99 of [265].

5.2 Proposition *(X, L) is a T_1-space which is not Hausdorff in the classical sense; hence L does not satisfy (H).*

Proof For $m \in \mathbb{N}$ we have $X \smallsetminus \{m\} = \{n \mid n < m\} \cup B(u, m) \cup B(v, m)$ open, and for, say, u we have $X \smallsetminus \{u\} = B(v, 0)$ open.

Now if U, V are open sets such that $u \in U$ and $v \in V$, we have $B(u, m) \subseteq U$ and $B(v, n) \subseteq V$ and hence $U \cap V \supseteq \{k \mid k \geqslant \max(m, n)\}$. Hence (X, L) is not classically Hausdorff, and hence L does not satisfy (H) by III.3.4.1. □

5.3 Proposition *The frame L is fit.*

Proof We will use the formula from II.4.4.1. Thus, if $A \nsubseteq B$ in L, we should produce a $C \in L$ such that $A \cup C = X$ and $C \to B \neq B$. Since generally $B \subseteq C \to B$ the latter amounts to $C \to B \nsubseteq B$ so that we have to produce $C, D \in L$ such that

$$A \cup C = X,\ C \cap D \subseteq B \text{ and } D \nsubseteq B.$$

If $A \cap \mathbb{N} \nsubseteq B$, consider an $m \in \mathbb{N}$ with $m \in A \smallsetminus B$ and $C = X \smallsetminus \{m\}$, $D = \{m\}$. Hence we are left with the case of A, B such that

$$A \nsubseteq B \quad \text{while} \quad A \cap \mathbb{N} \subseteq B$$

so that, say, $u \in A$ and $u \notin B$. Then $B(u, m) \subseteq A$ for some m and we can take $C = X \smallsetminus \{u\}$ and $D = B(u, m)$ (indeed $D \cap C = \{n \mid n \geqslant m\} = A \cap \mathbb{N} \subseteq B$ by the assumption). □

5.4 Note Here we have a spatial example. It should be noted, concerning the topic of the previous section, that Isbell presented in [152] another *spatial* example showing that fitness does not imply (sH). The procedures from the previous section can produce a similar counterexample.

6 The Tangle of Implications

We summarize the relations between the so far discussed axioms and conditions in the following diagram. None of the implications indicated is reversible, and the wiggles indicate that the conditions in question are independent.

At the top we have added the regularity, stronger than any of them. It will be discussed in the following chapter. In particular, we will learn (in Sects. 2.2 and 2.4) that regularity implies (sH) and that (sH) does not imply regularity. The irreversible implication (reg) ⇒ (fit) is standard (and will be also mentioned later). Then, the implication (reg) ⇒ (H) & (sfit) follows from the just mentioned ones (and from the implication (sH) ⇒ (H) proved in III.6.1) and its irreversibility is obvious already in spaces.

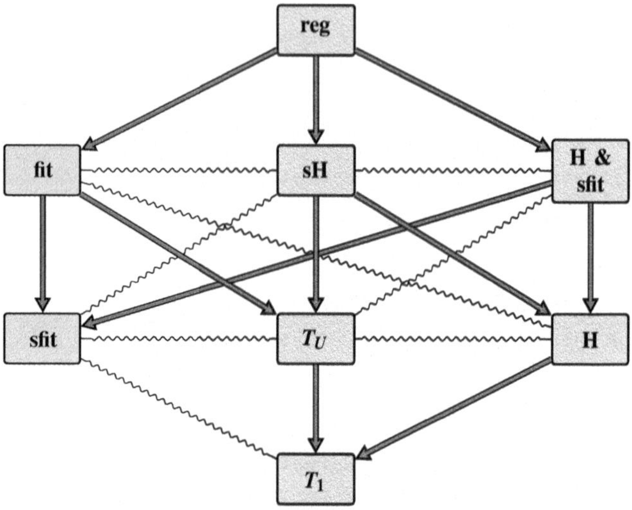

Now let us proceed systematically throughout the diagram.

- *The independence of (H) & (sfit) and (sH)*: (sH) does not imply (sfit) by Sect. 4, and the Example in III.6.3 shows that (H) & (sfit) does not imply T_U while (sH) does (III.6.2).
- *The independence of (H) & (sfit) and T_U*: (H) & (sfit) does not imply T_U by III.6.3. That T_U does not imply (sfit) is the exception left out in 1.6. Now we know, because of Sect. 4 and III.6.2: $T_U \Rightarrow$ (sfit) would yield (sH) $\Rightarrow$ (sfit).
- *The independence of (sH) and (fit) (and of (sH) and (sfit))*: In one direction by Sect. 4, then by Sect. 5 (fit) does not imply even (H).
- (sH) implies (H) by III.6.1 and this implication cannot be reversed by III.6.1.2.
- The part between (fit) and T_1 was already discussed in 1.6 with the exception that T_U does not imply (sfit) that has been already dealt with a few lines above.
- (sH) implies T_U by III.6.2 and T_U cannot imply (sH) because (fit) implies T_U and does not imply (sH).
- *The independence of (H) and (fit)*: (fit) does not imply (H) by Sect. 5, and (H) does not imply (fit) because it does not imply T_U.
- *The independence of (H) and T_U*: in the one direction it was just mentioned, and T_U does not imply (H) because it is implied by (fit) and (fit) does not imply (H).
- Finally, (H) implies T_1 is in 3.3.2, and the reverse does not hold even in spaces.

In the following diagrams we summarize the references for the proofs of each implication.

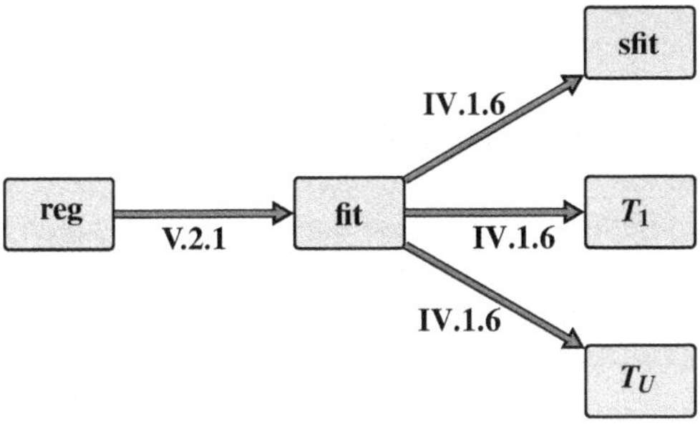

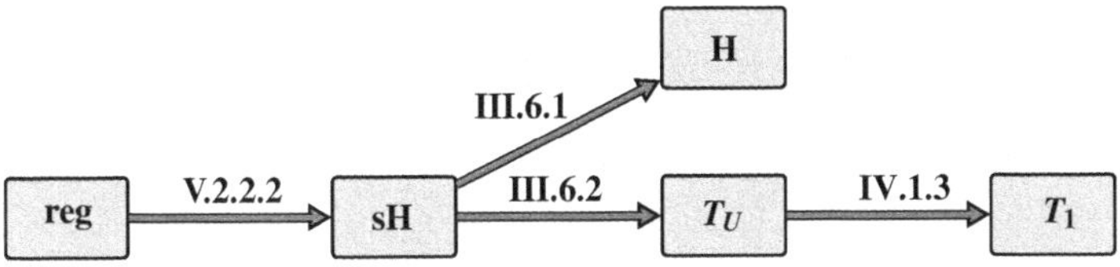

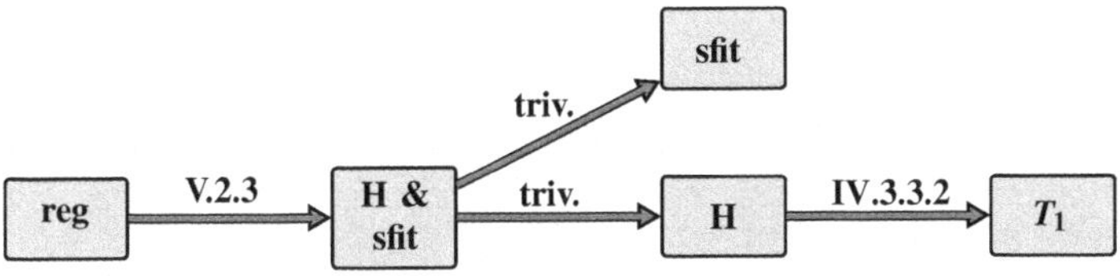

7 Concluding Remarks

7.1 The reader might wonder why we have, in the preceding summary, neglected the weak Hausdorff axioms ((wH), (wH') and (wH") from III.2) and the point Hausdorff property (pH) from III.4.

The weak Hausdorff axioms have, first of all, a historical importance: they appeared in the first comprehensive article [81] about separation in the point-free context. The authors of [81] have, right away, recognized that the formulas are too weak, and supported them with subfitness to obtain a nice conservative condition. But (wH) & (sfit) is already equivalent with (H) & (sfit) (III.3.3.2). It is, however, interesting to observe how a small formal change (replacing the symmetric "$a \nleq b$ and $b \nleq a$" by "$1 \neq a \nleq b$") transforms a weak condition (wH') to an expedient (H) (recall III.3), conservative without any support.

The situation with (pH) is different. We know it is conservative, that it is implied by (H), and that it has a weaker variant with an interesting consequence (III.4.3). But this is about all. Rather, there are interesting problems:

- The authors of [243], where (pH) was formulated, were a little sceptical about it. They assumed that there might be very badly behaved frames satisfying (pH) for the trivial reason of lack of weakly prime elements. One can share this intuition, but it would be nice to see further examples, preferably more than one, and of different characters.
- Since (pH) is implied by (H) and one does not see in which sense it is weaker, and since (wH) implies (H) under subfitness, the question naturally arises how strong is (pH) & (sfit). In particular, does it imply (H)?
- And it would be nice to know about interesting consequences of plain (pH).

7.2 Thus, after all, we are left with two reasonably well understood variants of a Hausdorff condition,

- the property (H), conservative and fairly simple, but not following some natural requirements typical for the classical Hausdorff property, and
- (sH), behaving like the classical Hausdorff property, but not quite so simple, and not conservative

(of course possibly combined with subfitness, which, however does not change anything about their relation), and an interesting conservative (pH) which is still somewhat of a riddle.

Chapter V
Regularity and Fitness

Among the conditions of separation type, the axiom of regularity is very special. From the earliest stages of point-free topology there was natural interest in conditions that would capture classical separation phenomena as convincingly as possible in the new, more general, context. Classical separation axioms are typically formulated in the language of points and point-dependent notions; hence, one looked for equivalent formulations, or imitated the geometric intuition to obtain suitable replacements or at least analogies (see, e.g., [81, 152, 153, 161, 243, 254]). In this company, regularity stands out. As we have already seen in Chap. I, it can be translated very easily, and the obtained formula has a natural appeal even in classical spaces (in fact, in an obviously equivalent form it is used classically anyway). There is no reasonable doubt that this formula makes a fully satisfactory point-free extension. It can be used without problems for proving useful facts parallel with the classical ones; moreover, it is algebraically versatile and easy to work with.[1]

The reader already knows that there are good and interesting replacements of the Hausdorff axiom. Nevertheless, none of them is without disadvantages and, besides, it took some time till the techniques developed to easier dealing with them. Therefore it was only too natural to use the stronger regularity instead (thus, while in classical theory compactness gains important new features starting with compact Hausdorff spaces, here one settled with compact regular frames, etc.). It paid well: one could easily prove the density theorem, closedness of compact sublocales or normality of compact regular frames. Such facts are classical Hausdorff phenomena, and we have shown in previous chapters that they can be proved as such also in the point-free context. Nevertheless, we will now present also the proofs based on regularity. We believe that it is in order not only for historical reasons, but also

[1]This holds, to some extent, also for the closely related complete regularity. It needs more explanation, but the natural formulation is equally compelling. Now, however, we wish to emphasize the exceptional role of regularity first.

J. Picado, A. Pultr, *Separation in Point-Free Topology*,
https://doi.org/10.1007/978-3-030-53479-0_5

because the regularity based proofs are much simpler, transparent and help to see what is actually happening.

Besides regularity, the main topic of this chapter, we discuss here also some of the aspects of the concept of fitness. This important condition, already mentioned in Chap. II, has been (a.o.) originally intended as an improvement of the categorically unsatisfactory subfitness. At the first sight it might not seem to be much stronger: it just extended the subfitness by heredity (and as a pleasant consequence one obtained a reflective subcategory of **Loc**—hence it allowed all the categorical constructions). Since subfitness is classically weaker than T_1 and T_1 is classically hereditary, the intuition suggests that fitness should not be a very strong separation condition. But it is not so (which shows how much richer the system of sublocales is as compared with the system of subspaces): it turns out that this condition is in fact closer to regularity than to the lower separation axioms. Therefore we found it useful to analyse fitness and its relation to regularity here.

1 The Relation Rather Below

1.1 Recall from I.5.5 (and I.3.3) the definition of regularity: We have the relation of a being *rather below* b, written $a \prec b$, if $a^* \vee b = 1$, often used in the form

$$x \prec y \quad \equiv_{\text{df}} \quad \exists z,\ x \wedge z = 0 \text{ and } z \vee y = 1 \qquad (*)$$

and say that L is *regular* if

(reg): *for every* $a \in L,\ a = \bigvee\{x \mid x \prec a\}$.

1.2 From $(*)$ we easily obtain

Proposition *Each sublocale of a regular frame is regular.*

Proof Let $h\colon L \twoheadrightarrow S$ be the frame homomorphism right adjoint to the embedding $j\colon S \subseteq L$ of a sublocale, and let $b \in S$. Then $j(b) = b = \bigvee\{x \in L \mid x \prec b\}$ and we have $b = h(b) = \bigvee\{h(x) \mid x \prec b\}$. But if $x \wedge z = 0$ and $z \vee b = 1$, then $h(x) \wedge h(z) = 0$ and $h(z) \vee h(b) = h(z) \vee b = 1$ so that $h(x) \prec b$ in S. □

1.2.1 Note The point is, of course, in the following general fact:

If $h\colon L \to M$ *is a frame homomorphism and if* $x \prec y$ *in* L*, then* $h(x) \prec h(y)$ *in* M.

1.3 Here are some useful simple rules.

Proposition

(1) *If* $a \prec b$ *then* $a \leqslant b$*; for any* a*,* $0 \prec a \prec 1$.
(2) *If* $x \leqslant a \prec b \leqslant y$*, then* $x \prec y$.
(3) *If* $a \prec b$*, then* $b^* \prec a^*$.

(4) *If* $a \prec b$, *then* $a^{**} \prec b$.
(5) *If* $a \prec b_i$, $i = 1, \dots, n$, *then* $a \prec b_1 \wedge \cdots \wedge b_n$.
(6) *If* $a_i \prec b$, $i = 1, \dots, n$, *then* $a_1 \vee \cdots \vee a_n \prec b$.

Proof

(1) If $a^* \vee b = 1$, then $a = a \wedge (a^* \vee b) = a \wedge b$; moreover, $0^* \vee a = 1$ and $a^* \vee 1 = 1$.
(2) If $a^* \vee b = 1$ then $x^* \vee y \geqslant a^* \vee b = 1$.
(3) If $a^* \vee b = 1$ then $b^{**} \vee a^* \geqslant b \vee a^* = 1$.
(4) is obvious since $a^{***} = a^*$.
(5) If $a^* \vee b_i = 1$ for every i then $a^* \vee (b_1 \wedge \cdots \wedge b_n) = 1$.
(6) If $a_i^* \vee b = 1$ for every i then $(a_1 \vee \cdots \vee a_n)^* \vee b = (a_1^* \wedge \cdots \wedge a_n^*) \vee b = 1$. □

1.3.1 Proposition *In a compact regular frame the relation* $\prec$ *interpolates.*

Proof If $a \prec b$, we have by regularity $a^* \vee \bigvee\{x \mid x \prec b\} = 1$ and hence by compactness and property (6) above there is a $c \prec b$ such that $a^* \vee c = 1$. □

1.4 Aside: "rather below" geometrically The relation $\prec$ is a precise extension of the relation of the open U, V given by $\overline{U} \subseteq V$ in a space X (Chap. I, 3.3). Indeed we have (recall from $(*)$ in II.4.2 that $\mathfrak{c}(a) \subseteq \mathfrak{o}(b)$ iff $a \vee b = 1$ and, dually, that $\mathfrak{o}(a) \subseteq \mathfrak{c}(b)$ iff $a \wedge b = 0$):

1.4.1 Observation *The following statements are equivalent*:

(1) $a \prec b$.
(2) *There is a closed* $\mathfrak{c}(u)$ *such that* $\mathfrak{o}(a) \subseteq \mathfrak{c}(u) \subseteq \mathfrak{o}(b)$.
(3) $\overline{\mathfrak{o}(a)} \subseteq \mathfrak{o}(b)$.

(Recall further from A.6.1.1 the useful explicit formula $\overline{\mathfrak{o}(a)} = \mathfrak{c}(a^*)$, also reminiscent of the geometric situation.)

1.5 Rather below and covers From II.7.2 recall the notions of a cover and refinement $A \leqslant B$, the operation Ax, the relation $\lhd_{\mathcal{A}}$ generated by a system of covers $\mathcal{A}$ and the admissibility of a nearness.

1.5.1 Lemma $Ax \leqslant y$ *implies* $x \prec y$.

Proof We have $1 = Ax \vee \bigvee\{a \in A \mid a \wedge x = 0\} \leqslant y \vee \bigvee\{a \in A \mid a \wedge x = 0\}$ and $x \wedge (\bigvee\{a \in A \mid a \wedge x = 0\}) = 0$. □

1.5.2 Proposition *For any system of covers* $\mathcal{A}$,

$$x \lhd_{\mathcal{A}} y \quad \Rightarrow \quad x \prec y.$$

If $\mathcal{A}$ *contains all two-element covers, then*

$$\lhd_{\mathcal{A}} = \prec .$$

Proof The first statement follows from 1.5.1. Now let $\mathcal{A}$ contain all two-element covers and let $x \prec y$. Then we have for the cover $A = \{x^*, y\}$, $Ax \leqslant y$. □

1.5.3 Corollary *A frame L admits a nearness iff it is regular.*

2 Regularity and Lower Separation

2.1 A frame is obviously regular iff for any $a \nleq b$ in L there is an x such that $x \prec a$ and $x \nleq b$. This is equivalent to

(reg): $\quad \forall a, b \in L, \quad a \nleq b \;\Rightarrow\; \exists c,\ a \vee c = 1$ *and* $c \to 0 \neq b$.

(Indeed, if there is an x as assumed we can set $c = x^*$ to obtain $a \vee c = a \vee x^* = 1$ and $c \to 0 = x^{**} \nleq b$, and if (reg) holds we have for $x = c^*$, $a \vee x^* \geqslant a \vee c = 1$ and $c \to 0 = x \nleq b$.)

Comparing it with the formula for fitness from II.3.1.2,

$$\forall a, b \in L, \quad a \nleq b \;\Rightarrow\; \exists c,\ a \vee c = 1 \text{ and } c \to b \neq b. \tag{fit}$$

we immediately see (since $c \to 0 \leqslant c \to b$) that

$$\text{(reg)} \;\Rightarrow\; \text{(fit)}.$$

We already know that (fit) implies (sfit) and that (sfit) $\not\Rightarrow$ (fit) (from IV.6; note that subfitness follows the same pattern as the formulas above; the formula

$$\forall a, b \in L, \quad a \nleq b \;\Rightarrow\; \exists c,\ a \vee c = 1 \neq c \vee b \tag{sfit}$$

immediately follows from the formula (fit) because if $c \vee b = 1$, then $c \to b = (c \to b) \wedge (b \to b) = (c \vee b) \to b = 1$).

We also have that (fit) $\not\Rightarrow$ (reg), but it is not necessary to prove it now. We will shortly present facts that hold in regular frames but not in general fit ones (see e.g. 2.2.2 below).

2.2 We will use the construction of coproduct as in A.8 (in particular the concept of a cp-ideal of $L \times L$) and the notation introduced there.

2.2.1 Lemma *Let L be a frame, let $U \subseteq L \times L$ be a cp-ideal and let $d_L = \{(x, y) \mid x \wedge y = 0\} \subseteq U$. Let $(a \wedge b, a \wedge b) \in U$. Then for every $x \prec a$ and $y \prec b$, $(x, y) \in U$.*

Proof We have $x^* \vee a = 1 = y^* \vee b$. Hence

$$\begin{aligned}(x, y) \in x \oplus y &= \big(x \wedge (y^* \vee b)\big) \oplus \big(y \wedge (x^* \vee a)\big) = \\ &= \big((x \wedge y^*) \vee (x \wedge b)\big) \oplus \big((y \wedge x^*) \vee (y \wedge a)\big) =\end{aligned}$$

$$= \big((x \wedge y^*) \oplus (y \wedge x^*)\big) \vee \big((x \wedge y^*) \oplus (y \wedge a)\big) \vee$$
$$\vee \big((x \wedge b) \oplus (y \wedge x^*)\big) \vee \big((x \wedge b) \oplus (y \wedge a)\big) \subseteq$$
$$\subseteq (x \oplus x^*) \vee (y^* \oplus y) \vee (x \oplus x^*) \vee ((a \wedge b) \oplus (b \wedge a)) \subseteq U.$$

□

2.2.2 Theorem *Each regular frame is strongly Hausdorff.*

Proof We will use the criterion III.5.3.2. Let $(a \wedge b, a \wedge b) \in U$. For each $y \prec b$ we have, by the lemma,

$$(a, y) = (\bigvee\{x \mid x \prec a\}, y) \in U.$$

Then, again by the lemma, we have $(a, b) = (a, \bigvee\{y \mid y \prec b\}) \in U$. □

2.3 Taking into account that (sH) implies the plain Hausdorff property (H) and that regularity implies fitness we obtain

Corollary *Regularity implies all the properties discussed in previous chapter.*

2.4 (sH) does not imply regularity, a simpler proof We already know from IV.4 that (sH) does not imply subfitness and hence it cannot imply regularity. This is a detour using a rather involved fact. We will present a direct proof (due to Chen [69]) based on a simple criterion of the strong Hausdorff property that is of interest in itself, and can be useful more generally.

2.4.1 Proposition *Let L be strongly Hausdorff and let $h\colon L \to M$ be an epimorphism in* **Frm**. *Then M is also strongly Hausdorff.*

Proof Consider the coproduct injections $\iota_1, \iota_2\colon M \to M \oplus M$. Since h is an epimorphism, the coequalizer of ι_1, ι_2 (that is, the codiagonal map of M) is the same as that of $\iota_1 h, \iota_2 h$. The latter is closed by Theorem IV.2.2, hence the codiagonal map of M is closed and M is strongly Hausdorff. □

Remark IMPORTANT: Note that the Proposition is about *epimorphisms*, not about *onto homomorphisms*. Recall that epimorphisms $h\colon L \to M$ in **Frm** can be very wild and that M can be (even much) bigger than L (such will be, e.g., the situation in 2.4.2 and in the example below). It IS NOT a repetition of the fact that (sH) is inherited by sublocales.

2.4.2 Proposition *Let L be a subframe of M and let M be generated by a subset S such that for each $s \in S$ there is a $t_s \in L$ such that both $s \vee t_s$ and $s \wedge t_s$ are in L. Then the embedding $j\colon L \subseteq M$ is an epimorphism.*

Consequently, if L is strongly Hausdorff, then so is also M.

Proof Let $f, g\colon M \to N$ be frame homomorphisms such that $fj = gj$. Consider an arbitrary $s \in S$ and the t_s from the assumption. Set $a = f(t_s) = g(f_s)$. Then

$f(s) \vee a = g(s) \vee a$ and $f(s) \wedge a = g(s) \wedge a$, and since N is a distributive lattice we have $f(s) = g(s)$. Since S generates M, $f = g$. □

2.4.3 Example Here is the example promised in III.6. Let $(X, \mathcal{T})$ be a topological space. For each closed set F and each open O define a new topology $\mathcal{T}_{OF}$ on X by setting

$$\mathcal{T}_{OF} = \{U \cup ((O \cup F) \cap V) \mid U, V \in \mathcal{T}\}.$$

$\mathcal{T}$ is a subframe of $\mathcal{T}_{OF}$. By Lemma 2.4.2, the embedding $j \colon \mathcal{T} \to \mathcal{T}_{OF}$ is an epimorphism in the category of frames, since $(X \smallsetminus F) \in \mathcal{T}$ satisfies

$$(O \cup F) \cup (X \smallsetminus F) = X \in \mathcal{T} \text{ and } (O \cup F) \cap (X \smallsetminus F) = O \cap (X \smallsetminus F) \in \mathcal{T}.$$

2.4.4 Example Let $\mathbb{I}$ be the closed unit interval with the usual Euclidean topology $\mathcal{T}$. Take

$$O = \bigcup_{n=1}^{\infty} \Big(\frac{1}{n+1}, \frac{1}{n}\Big) \text{ and } F = \{0\}$$

and let $\mathbb{I}_{OF}$ be the closed unit interval equipped with the topology $\mathcal{T}_{OF}$ defined for the choice above of O and F. Since the inclusion $\mathcal{T} \to \mathcal{T}_{OF}$ is an epimorphism and $\mathcal{T}$ is (sH), then $\mathcal{T}_{OF}$ is (sH) by virtue of Lemma 2.4.1.

On the other hand,

$\mathbb{I}_{OF}$ *is not regular.*

Indeed, every $A \in \mathcal{T}_{OF}$ not containing 0 is in $\mathcal{T}$ and $G = \{\frac{1}{n} \mid n = 1, 2, \dots\}$ is closed in $\mathbb{I}_{OF}$. Suppose that the closed sets $\{0\}$ and G are separated in $\mathbb{I}_{OF}$ by disjoint open sets

$$U \cup ((O \cup F) \cap V) \text{ and } W$$

(with $U, V, W \in \mathcal{T}$ such that $0 \in U \cup ((O \cup F) \cap V)$ and $G \subseteq W$). Then $U \cap W = \{0\}$ and $(O \cup F) \cap V \cap W = \{0\}$. The latter means that $V \cap W \subseteq G$ and therefore $V \cap W = \{0\}$. Since 0 is a cluster point of G in $\mathbb{I}$, both U and V cannot contain 0, contradicting $0 \in U \cup ((O \cup F) \cap V)$. Therefore, the space $\mathbb{I}_{OF}$ is not regular.

3 Regularity Replacing Hausdorff I: Density

In this and the following section we will prove some statements we have already proved in more generality.

Since regularity implies the strong Hausdorff property we already know Theorem 3.2 below as a more general Theorem 5.4.1 from Chap. III. Nevertheless, we decided to present it based on the regularity technique to show how it is typically used; besides, as the *Banaschewski's coequalizer theorem* it was among the first results of this nature.

3.1 Recall from III.5.5 that a localic map is *dense* if the image $f[L]$ is dense in M, that is, if $\overline{f[L]} = M$, and that in the frame language, and with $h = f^*$ this property is characterized by the implication

$$h(x) = 0 \Rightarrow x = 0.$$

The role of regularity (or, in fact, any separation) in density appeared first in [25]. For historical reasons we present it in the original formulation with a proof following the original reasoning.

Theorem 3.2 (Banaschewski's coequalizer theorem) *Let L be regular and let $h_1, h_2 \colon L \to M$ be frame homomorphisms. Set*

$$c = \bigvee\{h_1(x) \wedge h_2(y) \mid x \wedge y = 0\}.$$

Then $\breve{c} = (x \mapsto x \vee c) \colon M \to \uparrow c$ is the coequalizer of h_1 and h_2 in **Frm**.

Proof First we will prove that $\breve{c}h_1 = \breve{c}h_2$. Suppose not. Then there is an $a \in M$ such that, say, $h_1(a) \vee c \nleqslant h_2(a) \vee c$, and hence $h_1(a) \nleqslant h_2(a) \vee c$. By regularity there is a $b \prec a$ such that one still has $h_1(b) \nleqslant h_2(a) \vee c$. Choose a u such that $b \wedge u = 0$ and $u \vee a = 1$. Then $h_1(b) \wedge h_2(u) \leqslant c$ and we obtain a contradiction

$$h_1(b) = h_1(b) \wedge (h_2(u) \vee h_2(a)) \leqslant c \vee (h_1(b) \wedge h_2(a)) \leqslant c \vee h_2(a).$$

Now let $k \colon M \to K$ be a frame homomorphism such that $kh_1 = kh_2$. Then

$$k(c) = k(\bigvee\{h_1(x) \wedge h_2(y) \mid x \wedge y = 0\}) = \bigvee\{kh_1(x) \wedge kh_2(y) \mid x \wedge y = 0\} =$$

$$= \bigvee\{kh_1(x) \wedge kh_1(y) \mid x \wedge y = 0\} = \bigvee\{kh_1(x \wedge y) \mid x \wedge y = 0\} = 0$$

and hence for $\overline{k} : \uparrow c \to K$ defined by $\overline{k}(x) = k(x)$ we have $\overline{k}(x \vee c) = k(x)$. □

3.3 Using exactly the same procedure as in III.5.5.1 we now obtain

Proposition *Let L be regular, let S be dense in L and let localic maps $f_i : L \to M$, $i = 1, 2$, coincide on S. Then $f_1 = f_2$.*

3.4 Theorem *The monomorphisms in the category of regular frames are precisely the dense homomorphisms. In the more geometrical localic language, the epimorphisms in the category of regular locales are precisely the localic maps $f : L \to M$ such that $\overline{f[L]} = M$.*

Proof From 3.3 we already know that the dense localic maps (dense homomorphisms) are epimorphisms (monomorphisms).

For the other implication we will adopt the procedure from [220]. Thus let L, M be regular frames and let a homomorphism $h\colon L \to M$ not be dense. Then there is an $a \neq 0$ such that $h(a) = 0$. Set

$$N = \{(x, y) \in L \times L \mid x \vee a = y \vee a\}.$$

This is obviously a subframe of $L \times L$. For each $(u, v) \in L \times L$ define

$$(uv)_1 = u \wedge (v \vee a) \quad \text{and} \quad (uv)_2 = v \wedge (u \vee a).$$

Then $(uv)_1 \vee a = (u \vee a) \wedge (v \vee a) = (uv)_2 \vee a$ so that $((uv)_1, (uv)_2) \in N$. Let $(x, y) \in N$ and $u \prec x$ and $v \prec y$. There are $s, t \in L$ such that

$$s \wedge u = 0,\ s \vee x = 1,\ t \wedge v = 0 \text{ and } t \vee y = 1.$$

Then, for both $i = 1, 2$ we have $(uv)_i \wedge (st)_i = 0$. Moreover

$$(st)_1 \vee x = (s \wedge (t \vee a)) \vee x = (s \vee x) \wedge (t \vee a \vee x) = (s \vee x) \wedge (t \vee y \vee a) = 1$$

and, similarly, $(st)_2 \vee y = 1$, hence $((uv)_1, (uv)_2) \prec (x, y)$ in N. Then

$$\begin{aligned}\bigvee\{((uv)_1, (uv)_2) \mid u \prec x, v \prec y\} &= \Big(\bigvee_{u\prec x, v\prec y} u \wedge (v \vee a), \bigvee_{u\prec x, v\prec y} v \wedge (u \vee a)\Big)\\ &= \Big(\bigvee_{u\prec x} u \wedge (\bigvee_{v\prec y} v \vee a), \bigvee_{v\prec y} v \wedge (\bigvee_{u\prec x} u \vee a)\Big) =\\ &= \big(x \wedge (y \vee a), (y \wedge (x \vee a)\big) = (x, y)\end{aligned}$$

and thus N is regular.

Now define $g_i\colon N \to L$ by setting $g_i(x_1, x_2) = x_i$. Then $g_1(a, 0) = a \neq 0 = g_2(a, 0)$ while, $h(a)$ being 0,

$$hg_1(x_1, x_2) = h(x_1) = h(x_1) \vee h(a) = h(x_1 \vee a) = h(x_2 \vee a) = hg_2(x_1, x_2)$$

and hence h is not a monomorphism. □

4 Regularity Replacing Hausdorff II: Compactness

In this section we will present, again, a proof of a fact that we already know in a more general setting. When restricted to regular frames the facts become much simpler. In particular compare 4.2 (and the even more transparent 4.2.1) with 9.2.1 in Chap. III. While the fact that any compact sublocale of a strongly Hausdorff frame is closed is rather involved, for regular frames it is very easy.

4.1 First, we will prove a (weakened) counterpart to the classical fact that a continuous one-to-one dense mapping $f\colon X \to Y$ with compact X and Hausdorff Y is a homeomorphism.

4.1.1 Recall from A.10.4 that a frame homomorphism $h\colon M \to L$ is said to be *codense* if[2]

$$h(x) = 1 \quad \Rightarrow \quad x = 1.$$

Lemma *Let M be subfit. Then each codense frame homomorphism $h\colon M \to L$ is one-to-one.*

Proof Suppose $a \nleq b$ and $h(a) = h(b)$. Consider a c such that $a \vee c = 1 \neq b \vee c$. Then $h(b \vee c) = h(b) \vee h(c) = h(a) \vee h(c) = h(a \vee c) = 1$, a contradiction. □

Note Here we have needed the subfitness only. Similar features of this property will be discussed in Chap. X.

4.1.2 Theorem *Let L be compact and let M be regular. Then every dense one-to-one localic map $f\colon L \to M$ is an isomorphism.*

Proof Denote by $h\colon M \to L$ the adjoint frame homomorphism. Since f is one-to-one, h is onto and hence, by 4.1.1, it suffices to prove it is codense.

Let $a \in M$ be such that $h(a) = 1$. By regularity, $a = \bigvee\{b \mid b \prec a\}$. Then $1 = h(a) = \bigvee\{h(b) \mid b \prec a\}$ and by compactness there are $b_1, \ldots, b_n \prec a$ such that $\bigvee_{i=1}^{n} h(b_i) = 1$. Thus, setting $b = \bigvee_{i=1}^{n} b_i$ and recalling (6) in 1.3 we see that we have obtained a $b \prec a$ with $h(b) = 1$. We know that h is dense (and want to prove codensity). Hence $h(b^*) = h(b^*) \wedge h(b) = h(b \wedge b^*) = 0$ makes $b^* = 0$, and since $a \vee b^* = 1$ we conclude that $a = 1$. □

4.2 Theorem *Every compact sublocale of a regular frame is closed.*

Proof This is an immediate consequence of 4.1.2: consider the embedding $j\colon S \subseteq L$ with compact S and regular L. Decompose it as

$$S \xrightarrow{\;f\;} \overline{S} \xrightarrow{\;\subseteq\;} L$$

with $f(x) = x$. By 1.2, $\overline{S}$ is regular and hence, by 4.1.2, f is an isomorphism, in particular onto. □

4.2.1 Another proof of 4.2 Let us present another (more direct and perhaps more transparent, by all means more geometric) proof of the previous theorem.

[2] In localic terms, this means that a localic map $f\colon L \to M$ is *codense* if $f[L]$ is a *codense sublocale* of M, that is, $f[L] \subseteq \mathfrak{o}(a) \Rightarrow \mathfrak{o}(a) = L$.

First, let us observe that

(∗) *if S is compact and $S \subseteq \mathfrak{o}(a)$ in a regular L, then there is a $b \prec a$ such that $S \subseteq \mathfrak{o}(b)$.*

(Indeed, $a = \bigvee\{b \mid b \prec a\}$, hence $S \subseteq \mathfrak{o}(a) = \bigvee\{\mathfrak{o}(b) \mid b \prec a\}$ and by compactness and 1.1(5) there is a $b \prec a$ such that $S \subseteq \mathfrak{o}(a)$.)

L is regular and hence fit. Thus, $S = \bigcap\{\mathfrak{o}(a) \mid S \subseteq \mathfrak{o}(a)\}$. Using (∗) choose for each a with $S \subseteq \mathfrak{o}(a)$ some $b_a \prec a$ such that $S \subseteq \mathfrak{o}(b_a) \subseteq \mathfrak{o}(a)$. Then, by 1.4.1,

$$S = \bigcap\{\overline{\mathfrak{o}(b_a)} \mid S \subseteq \mathfrak{o}(a)\}$$

and S is closed. □

5 Prefit (Almost Regular) and Fit

5.1 Subfitness, fitness and regularity are naturally compared in the following table:

$$a \nleq b \;\Rightarrow\; \exists c,\; a \vee c = 1 \neq b \vee c \qquad \text{(sfit)}$$

$$a \nleq b \;\Rightarrow\; \exists c,\; a \vee c = 1 \text{ and } c \to b \nleq b \qquad \text{(fit)}$$

$$a \nleq b \;\Rightarrow\; \exists c,\; a \vee c = 1 \text{ and } c \to 0 \nleq b. \qquad \text{(reg)}$$

(the last implication since $c \to 0 \leqslant c \to b$). We already know from IV.6 that (fit) is strictly stronger than (sfit) and that also (reg) is strictly stronger than (fit).

In this section we will discuss the relation between fitness and regularity. It will be approached via a relation relaxing fitness, different from subfitness (and in fact already quite strong).

5.2 An *almost regular* space, a concept introduced and studied in [261], is a space X such that for every nonempty regular open $U \subseteq X$ there is a nonempty open V such that $\overline{V} \subseteq U$. The expression "almost regular" may sound somewhat strong for this property, seemingly still quite far from regularity. In fact it is close enough, something like "regular up to density"; we will introduce and use a slightly stronger modification for which the regularity-up-to-density will be particularly apparent (see 5.4 below), and then show that fitness is even stronger than that.

5.3 A frame is said to be *prefit* if

(pfit): *for all $a \neq 0$ there exists an $x \neq 0$ with $x \prec a$.*

Note that already this is stronger than the almost regularity above: we do not restrict ourselves to the regular a's (that is, nothing like $a = a^{**}$ is requested[3]).

5.3.1 Prefitness looks deceptively very similar to weak subfitness. It can be rewritten as

$$\forall a,\ 0 < a \Rightarrow \exists x \neq 0 \text{ such that } x^{*} \vee a = 1$$

and this again, setting $c = x^{*}$, as

$$\forall a,\ 0 < a \Rightarrow \exists c = c^{**} \neq 1 \text{ such that } c \vee a = 1.$$

Thus it looks just like weak subfitness with the small difference that the c is assumed regular. This difference is very essential as we will see in the following subsection. But of course,

(pfit) implies (wsfit).

5.4 It turns out that prefitness is already something like "regularity up to density". We have

Proposition *A frame L is prefit iff for each $a \in L$,*

$$a \leqslant \left(\bigvee\{x \mid x \prec a\}\right)^{**}.$$

In other words, this is iff

$$\mathfrak{o}(a) \subseteq \overline{\mathfrak{o}(\bigvee\{x \mid x \prec a\})}.$$

Proof If $a \nleqslant (\bigvee\{x \mid x \prec a\})^{**}$, then $a \wedge (\bigvee\{x \mid x \prec a\})^{*} \neq 0$ and hence there is a $y > 0$ such that $y \prec a \wedge (\bigvee\{x \mid x \prec a\})^{*}$ and hence

$$y^{*} \vee (a \wedge (\bigvee\{x \mid x \prec a\})^{*}) = (y^{*} \vee a) \wedge (y^{*} \vee (\bigvee\{x \mid x \prec a\})^{*}) = 1.$$

In particular, $y \prec a$ and hence $y \leqslant \bigvee\{x \mid x \prec a\}$, and, further, $(\bigvee\{x \mid x \prec a\})^{*} \leqslant y^{*}$. But since also $y^{*} \vee (\bigvee\{x \mid x \prec a\})^{*} = 1$ we see that

$$y^{*} = y^{*} \vee (\bigvee\{x \mid x \prec a\})^{*} = 1,$$

and we have a contradiction $y \leqslant y^{**} = 0$.

On the other hand, if the inequality holds and if $a \neq 0$, then $(\bigvee\{x \mid x \prec a\})^{**} \neq 0$, hence $\bigvee\{x \mid x \prec a\} \neq 0$ and there must be the desired x.

---

[3]We follow the historical terminology in which an open U is called *regular* if $U = \operatorname{int}(\overline{U})$. It has nothing to do with the regularity in separation.

The statement on the closure follows from the fact that $\overline{\mathfrak{o}(u)} = \mathfrak{c}(u^*)$ (A.6.1.1). As a consequence of this equality we have

$$u \leqslant v^{**} \quad \text{iff} \quad \mathfrak{o}(u) \subseteq \overline{\mathfrak{o}(v)}$$

(indeed, $u \leqslant v^{**}$ iff $u \wedge v^* = 0$ iff $\mathfrak{o}(u) \cap \mathfrak{o}(v^*) = \mathsf{O}$ iff $\mathfrak{o}(u) \subseteq \mathfrak{c}(v^*) = \overline{\mathfrak{o}(v)}$).

5.4.1 Note The inequality in the Proposition can be written as

$$a^{**} = (\bigvee\{x \mid x \prec a\})^{**}.$$

This should not be confused with the formula

$$a^{**} = \bigvee\{x \mid x \prec a^{**}\}$$

which holds in the subfit *almost normal* frames (almost normal frames will be discussed (a.o.) in Chap. VII).

5.5 Fit as hereditarily prefit First, we have a simple

5.5.1 Observation *Every fit frame is prefit.*

(Indeed, if $a \nleqslant 0$ we have a c such that $a \vee c = 1$ and $c^* = c \to 0 \neq 0$. Set $x = c^*$.)

5.5.2 In the following it will be of advantage to view the Heyting arrow (similarly like in II.3.3) as the "relative pseudocomplement" (this often used expression is well motivated by the fact that the equivalence $x \wedge y \leqslant b$ iff $y \leqslant x \to b$ can be viewed for $x, y \geqslant b$ as $y \leqslant x \to b$ iff $x \wedge y = 0_{\uparrow b}$; thus we see $x \to b$ as the pseudocomplement x^{*_b} of x in $\uparrow b = \mathfrak{c}(b)$).

5.5.3 Theorem *A frame is fit iff each of its closed sublocales is prefit.*

Proof Let any closed sublocale $\mathfrak{c}(b) = \uparrow b$ of L be prefit. Let $a \nleqslant b$. Then $a \vee b > b = 0_{\mathfrak{c}(b)}$ and hence there is an $x > b$ such that

$$(a \vee b) \vee x^{*_b} = a \vee b \vee (x \to b) = a \vee (x \to b) = 1.$$

Set $c = x \to b = x^{*_b}$. Then $a \vee c = 1$ and $c \to b = c^{*_b} = x^{*_b *_b} \geqslant x > b$.

The converse follows from 5.5.1 since fitness is hereditary. □

5.6 The situation in 5.5.3 fundamentally differs from that in II.5.3 where fit was characterized as hereditarily subfit. There, sublocales like closed or open ones inherited subfitness and the point was in (non)inheriting of subfitness on rather exotic sublocales. Here, the essential sublocales are the closed ones. In spaces, closed sublocales are (correctly represented by) closed subspaces, and hence the theorem is here automatically applicable. We obtain

5.6.1 Theorem *A topological space X is fit (in the sense that the frame $\Omega(X)$ is fit) if and only if for each closed $Y \subseteq X$ and each open U such that $U \cap Y \neq \emptyset$ there is an open V such that $V \cap Y \neq \emptyset$ and $\overline{V \cap Y} \subseteq U \cap Y$.*

5.7 Comparing the properties Now it is perhaps in order to stop for a moment and discuss the relations between the properties we have encountered: weak subfitness, subfitness, prefitness and fitness.

5.7.1 First, it is obvious that

$$\text{(sfit)} \;\not\Rightarrow\; \text{(pfit)} \;:$$

even T_1 does not imply (pfit) in spaces: see the cofinite topology.

5.7.2 The prefitness, however, is a fairly strong property. We will have more trouble showing it does not imply subfitness. It indeed does not, as we will see in the following example.[4]

Example Let $\mathbb{N}$ be the set of natural numbers, $a \notin \mathbb{N} \times \{0, 1\}$. Set

$$X = (\mathbb{N} \times \{0, 1\}) \cup \{a\}$$

and endow it with the following topology:

$$U \subseteq X \text{ is open if } \begin{cases} (\exists n,\ (n, 1) \in U) \;\Rightarrow\; a \in U, \;\text{ and} \\ a \in U \;\Rightarrow\; \exists k\ (n \geqslant k \;\Rightarrow\; (n, 0) \in U). \end{cases}$$

Thus in particular $U_0 = \mathbb{N} \times \{0\}$, $U_1 = (\mathbb{N} \times \{0\}) \cup \{a\}$ and each of $U(n) = (\mathbb{N} \times \{0\}) \cup \{a\} \cup \{(n, 1)\}$ are open and we have $a \in U_1$ and $U_1 \smallsetminus \{a\} = U_0$, $(n, 1) \in U(n)$ and $U(n) \smallsetminus \{(n, 1)\} = U_1$, and finally each $\{(n, 0)\}$ is open and X is T_D. Thus,

$$X \textit{ is not subfit}$$

since otherwise, by II.2.2, it would be T_1, and $\overline{\{a\}}$ contains $\mathbb{N} \times \{1\}$. But

$$X \textit{ is prefit.}$$

Indeed, each $\{(n, 0)\}$ is clopen and each nonempty open set contains some of the $\{(n, 0)\}$.

[4] More examples are given by Dube in [89], within the realm of frames of radical ideals of a ring.

5.7.3 We have already seen that (pfit) (obviously) implies (wsfit). Thus, the situation is as follows:

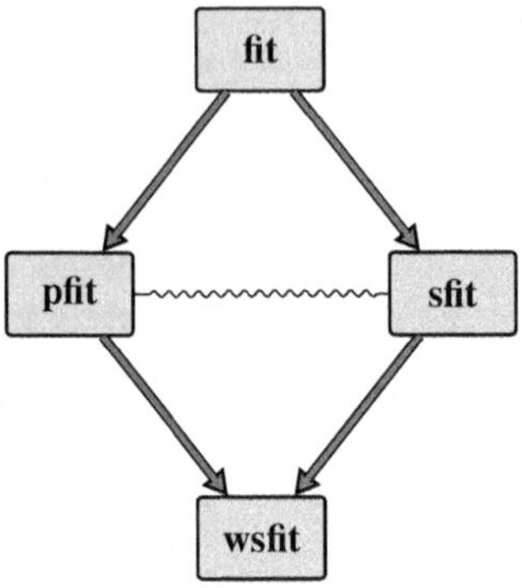

with none of the indicated implications reversible.

5.7.4 Finally let us show that

$$\text{(pfit) \& (sfit)} \quad \not\Rightarrow \quad \text{(fit)}.$$

This will be seen in the following

Example Consider the square $X = \mathbb{I} \times \mathbb{I}$ where $\mathbb{I}$ is the standard closed unit interval and set $Y = \{(x, 1) \mid x \in \mathbb{I}\}$. On X define a topology by declaring U open if

$$\text{either } U \cap Y = \emptyset \text{ and for each } (x, y) \in U \text{ there is a standard } \\ \varepsilon\text{-neighbourhood } V \subseteq U,$$

$$\text{or } U \cap Y \neq \emptyset, \text{ for each } (x, y) \in U \text{ there is a standard } \\ \varepsilon\text{-neighbourhood } V \subseteq U, \text{ and } Y \smallsetminus U \text{ is finite.}$$

Then X is T_1 and hence it is subfit. The space X is also prefit. Indeed, let U be nonempty open. Then $U \cap (X \smallsetminus (\mathbb{I} \times \{0\}))$ is nonempty open and we can choose a nonempty open $U' \subseteq U \cap (X \smallsetminus (\mathbb{I} \times \{0\}))$ such that the standard metric closure of U' does not meet Y (it suffices to take $(x, y) \in U \cap (X \smallsetminus (\mathbb{I} \times \{0\}))$ and an open ε-neighbourhood of (x, y) with ε sufficiently small). Then the closure of any $V \subseteq U'$ in X coincides with the standard metric closure in $\mathbb{I} \times \mathbb{I}$ and the statement follows. On the other hand, X is not fit: Y is closed in X and it is obviously not prefit.

6 Fitness as Relaxed Regularity

In this section we will see that fitness can be defined as an iterated relaxation of regularity. The number of the relaxation steps is unbounded; thus there is a proper class of separation conditions between regularity and fitness (may be not very interesting in themselves, though).

6.1 Let L be an arbitrary frame. For $\alpha \in \{-1\} \cup \mathbf{Ord}$ define maps

$$\rho_\alpha = \rho_\alpha^L : L \to L$$

by setting

$$\begin{aligned}\rho_{-1}(a) &= 0,\\ \rho_{\alpha+1}(a) &= \bigvee\{u \mid \exists x,\ x \vee a = 1,\ u \wedge x \leqslant \rho_\alpha(x)\}, \quad \text{and}\\ \rho_\alpha(a) &= \bigvee_{\beta<\alpha} \rho_\beta(a) \text{ for limit ordinals.}\end{aligned}$$

Observation *For every $a \in L$ and $\alpha \leqslant \beta$,*

$$\rho_\alpha(a) \leqslant \rho_\beta(a) \leqslant a.$$

Proof For the first inequality it suffices to prove that $\rho_{\alpha+1}(a) \geqslant \rho_\alpha(a)$ which follows from $1 \vee a = 1$ and $\rho_\alpha(a) \wedge 1 \leqslant \rho_\alpha(a)$. On the other hand, if $x \vee a = 1$ and $u \wedge x \leqslant \rho_\alpha(x)$, then $u = u \wedge (x \vee a) \leqslant \rho_\alpha(x) \vee a = a$. □

6.2 α-Relaxed regularity A frame is *α-relaxed regular* (briefly, *α-regular*) if

(α-reg): $\rho_\alpha^L = \mathrm{id}$.

Thus, regularity is the same as 0-relaxed regularity, that is, regularity not relaxed at all.

6.3 Proposition *The properties (α-reg) are hereditary, that is, any sublocale of an α-relaxed regular frame is α-relaxed regular.*

Proof Let $h : L \to S$ be the (onto) frame homomorphism adjoint to the embedding $j : S \subseteq L$. We will prove that

$$h(\rho_\alpha^L(a)) \leqslant \rho_\alpha^S(h(a)).$$

Indeed, this is trivially true for ρ_{-1} and if it holds for α, then

$$\begin{aligned}h(\rho_{\alpha+1}^L(a)) &= h(\bigvee\{u \mid \exists x,\ x \vee a = 1,\ u \wedge x \leqslant \rho_\alpha^L(x)\}) =\\ &= \bigvee\{h(u) \mid \exists x,\ x \vee a = 1,\ u \wedge x \leqslant \rho_\alpha^L(x)\} \leqslant\\ &\leqslant \bigvee\{v \mid \exists y,\ y \vee h(a) = 1,\ v \wedge y \leqslant \rho_\alpha^S(x)\} = \rho_{\alpha+1}^S(h(a))\end{aligned}$$

($y = h(x)$ does the job for $v = h(u)$); the limit step is obvious. Now the statement follows since if $\rho^L_\alpha(a) = 1$, then $\rho^S_\alpha(h(a)) = 1$, and h is onto. □

6.4 Proposition *An α-relaxed regular frame is fit.*

Proof By 6.3 and II.5.6 it suffices to prove that such frame is weakly subfit. Suppose $a \neq 0$. Then there is an α such that

$$\rho_\alpha(a) < \rho_{\alpha+1}(a) = \bigvee\{u \mid \exists x,\ x \vee a = 1,\ u \wedge x \leqslant \rho_\alpha(x)\}.$$

Hence there are u and x such that

$$x \vee a = 1,\ u \wedge x \leqslant \rho_\alpha(a),\ \text{and } u \nleqslant \rho_\alpha(a)$$

and hence $x \neq 1$. □

6.5 Theorem *A frame L is fit iff it is α-regular for some α.*

Proof In view of 6.4 it suffices to prove that a fit frame is α-regular for some α, and to prove this it suffices to show that for each individual $a \in L$ there is an α such that $\rho_\alpha(a) = a$.

Suppose for some a there is not. Since the transfinite sequence ρ_α cannot increase unboundedly there is an α such that $b = \rho_\alpha(a) = \rho_{\alpha+1}(a)$. By the assumption on a, $b < a$, and a is a nonzero element in the closed sublocale $\uparrow b$.

By 5.5.1 each closed sublocale of a fit frame is prefit and hence there is an $x > b$ such that $x \prec a$ in $\uparrow b$. That is, we have, further, a y such that $y \vee a = 1$ and $y \wedge x = b = \rho_\alpha(a)$. But then $\rho_{\alpha+1}(a) \geqslant x > \rho_\alpha(a)$, a contradiction. □

6.6 Proposition *If $\alpha < \beta$, then (α-reg) is strictly stronger than (β-reg), even for spaces.*

Proof For an open subset U of a space write $U_\alpha = \rho_\alpha(U)$. We will construct topological spaces $(X(\alpha), \tau_\alpha)$ and open sets U in τ_α such that for any $\beta < \alpha$, $U_\beta \neq U$ and $U_\alpha = U$. Set

$$X(\alpha) = \{(\beta, n) \mid \beta < \alpha, n < \omega\}$$

and endow it with the topology τ_α consisting of the U such that

$$(\beta, n) \in U \quad \Rightarrow \quad \exists m \in \omega,\ \{(\gamma, k) \mid \gamma < \beta,\ k > m\} \subseteq U.$$

Thus, in particular, any neighbourhood of (β, n) meets any set $\{(\gamma, m) \mid m \in \omega\}$ with $\gamma < \beta$ while it does not have to meet $\{(\beta, m) \mid m \in \omega\}$ in anything but the (β, n) itself. Hence if we set $U = \{(\beta, n) \in X(\alpha) \mid n \neq 0\}$, we obtain $U_\gamma = \{(\beta, n) \mid \beta < \gamma,\ n \in \omega\}$ for $\gamma < \alpha$, and $U_\alpha = U$. □

6.7 We know that regularity implies the strong Hausdorff property, while fitness does not imply even the plain Hausdorff one. In fact,

already (2-reg) does not imply (H), even for spaces.

Consider $X = \omega \cup \{a, b\}$ with $a \neq b$, $a, b \notin \omega$ and endow it with the topology consisting of the $U \subseteq X$ such that

$$U \cap \{a, b\} \Rightarrow \exists n \in \omega, \ \{m \mid m \geqslant n\} \subseteq U.$$

Obviously X is not Hausdorff. Observing that the $\{n\}$ with $n \in \omega$ are clopen we easily deduce that

(1) if $U \subseteq \omega$ or $\{a, b\} \subseteq U$, then $U = \rho_1(U)$ and
(2) if $U \cap \{a, b\} = \{x\}$ where $x = a$ or $x = b$, then $\rho_1(U) = U \smallsetminus \{x\}$ and $\rho_2(U) = U$.

Hence X is 2-regular.

7 (Co)reflections

In this section we will prove that the categories of regular frames and of fit frames are coreflective in the category of frames. The former is in the same form as in [220] and we present it here just for convenience. The proof of the latter in [220], however, is incorrect, and presenting it here we pay a debt. It should be noted that the statement appeared first (with another proof) in [152], and that the corrected proof we will present here was published in [222].

7.1 For an arbitrary frame L set

$$L_{\prec} = \{a \in L \mid a = \bigvee\{x \mid x \prec a\}\}.$$

We have

Lemma *$L_{\prec}$ is a subframe of L.*

Proof Obviously $L_{\prec}$ is closed under joins and contains 1. By 1.3 we have for $a, b \in L_{\prec}$ also

$$a \wedge b = \bigvee\{x \wedge y \mid x \prec a, y \prec b\} \leqslant \bigvee\{z \mid z \prec a \wedge b\} \leqslant a \wedge b. \qquad \square$$

7.2 For a frame L and ordinals α define $R_\alpha(L)$ by setting

$$R_0(L) = L,$$

$$R_{\alpha+1}(L) = (R_\alpha(L))_{\prec}, \quad \text{and}$$

$$R_\alpha(L) = \bigcap_{\beta<\alpha} R_\beta(L) \quad \text{for limit ordinals } \alpha.$$

Further set

$$R_\infty(L) = \bigcap_{\beta\in\mathbf{Ord}} R_\beta(L).$$

We have, of course $R_\infty(L) = R_\alpha(L)$ for a sufficiently large α, hence $(R_\infty(L))_< = R_\infty(L)$, and $R_\infty(L)$ is a regular frame.

7.2.1 Theorem *Define $\eta_L : L \to R_\infty(L)$ by setting*

$$\eta_L(a) = \bigvee\{x \in R_\infty(L) \mid x \leqslant a\}.$$

Then the system $(\eta_L)_L$ constitutes an epireflection of the category of locales onto the subcategory of the regular ones.

Proof First, the maps η_L are localic maps: they are adjoints to the frame embeddings $R_\infty(L) \subseteq L$.

Now let M be a regular locale and let $f : L \to M$ be a localic map. By 1.2.1 we have, for each $b \in M$,

$$f^*(b) = \bigvee\{f^*(y) \mid y \prec b\} \leqslant \bigvee\{x \mid x \prec f^*(b)\} \leqslant f^*(b)$$

and hence $f^*(b) \in R_1(L)$, and by III.11.2 we have a localic map

$$f|_{R_1(L)} \colon R_1(L) \to M$$

such that $f|_{R_1(L)} \cdot p_{R_1(L)} = f$. Proceeding by induction we eventually get $f|_{R_\infty(L)} \colon R_\infty(L) \to M$ such that $f|_{R_\infty(L)} \cdot \eta_L = f$. □

7.3 To simplify the notation set

$$\mathfrak{sc}(a) = \bigcap\{\mathfrak{o}(u) \mid \mathfrak{c}(a) \subseteq \mathfrak{o}(u)\}.$$

(The intersection on the right-hand side, the *fitting* of $\mathfrak{c}(a)$, will be studied in more detail, and for general sublocales, in Chap. X. Now it will have only a small technical role. Just note that the definition of fitness as in II.4.3 is claiming that for every a, $\mathfrak{sc}(a) = \mathfrak{c}(a)$.)

7.4 We will use A.6. For a frame L define

$$F_1(L) = \{a \in L \mid \mathfrak{c}(a) = \mathfrak{sc}(a)\} = \{a \in L \mid \mathfrak{c}(a) = \bigcap\{\mathfrak{o}(u) \mid \mathfrak{c}(a) \subseteq \mathfrak{o}(u)\}\}.$$

Explicitly, $a \in F_1(L)$ iff

$$(a \vee u = 1 \;\Rightarrow\; u \to x = x) \quad \Rightarrow \quad x \geqslant a. \qquad (*)$$

7.4.1 Lemma *$F_1(L)$ is a subframe of L, and $F_1(L) = L$ iff L is fit.*

Proof Obviously $0, 1 \in F_1(L)$. Now let $a_i \in F_1(L)$ for $i \in J$. We will show first that $a = \bigvee_{i\in J} a_i$ satisfies the implication (∗). Thus assume that

$$a \vee u = 1 \quad \Rightarrow \quad u \to x = x.$$

If for an individual j, $a_j \vee u = 1$, then $a \vee u = 1$ and hence $u \to x = x$, and $x \geqslant a_j$. Thus $a \geqslant \bigvee_{i\in J} a_i$.

Now consider $a, b \in F_1(L)$. We have by the coframe distribution (A.5.4.1)

$$\begin{aligned}\mathfrak{c}(a \wedge b) &= \mathfrak{c}(a) \vee \mathfrak{c}(b) = \\ &= \bigcap\{\mathfrak{o}(x) \mid \mathfrak{c}(a) \subseteq \mathfrak{o}(x)\} \vee \bigcap\{\mathfrak{o}(y) \mid \mathfrak{c}(b) \subseteq \mathfrak{o}(y)\} = \\ &= \bigcap\{\mathfrak{o}(x \vee y) \mid \mathfrak{c}(a) \subseteq \mathfrak{o}(x),\ \mathfrak{c}(b) \subseteq \mathfrak{o}(y)\} \supseteq \\ &\supseteq \bigcap\{\mathfrak{o}(u) \mid \mathfrak{c}(a) \vee \mathfrak{c}(b) \subseteq \mathfrak{o}(u)\}.\end{aligned}$$

The second statement is immediate from the definition of fitness. □

7.4.2 For a frame L and ordinals α define $F_\alpha(L)$ by setting

$$F_0(L) = L,$$

$$F_{\alpha+1}(L) = F_1(F_\alpha(L)), \quad \text{and}$$

$$F_\alpha(L) = \bigcap_{\beta<\alpha} F_\beta(L) \quad \text{for limit ordinals } \alpha.$$

Since the $F_\alpha(L)$ decrease there is an ordinal $\gamma(L)$ such that $F_1(F_{\gamma(L)}(L)) = F_{\gamma(L)}(L)$ and $F_{\gamma(L)}(L)$ is a fit frame. Set

$$F(L) = F_{\gamma(L)}(L).$$

7.4.3 Theorem *Define $\varphi_L \colon L \to F(L)$ by setting*

$$\varphi_L(a) = \bigvee\{x \in F(L) \mid x \leqslant a\}.$$

Then the system $(\varphi_L)_L$ constitutes an epireflection of the category of locales onto the subcategory of the fit ones.

Proof The maps φ_L are localic maps: they are adjoints to the frame embeddings $F(L) \subseteq L$. Let M be a fit locale and let $f \colon L \to M$ be a localic map. Using III.11.2 and a similar inductive process as in 7.2.1, it suffices to show that $f^*[M] \subseteq F_1(L)$. Let $b \in M$. Since M is fit we have

$$\mathfrak{c}(b) = \bigcap\{\mathfrak{o}(y) \mid \mathfrak{c}(b) \subseteq \mathfrak{o}(y)\}$$

and hence we have for the preimage of f (see A.7.3)

$$\mathfrak{c}(f^*(b)) = f_{-1}[\mathfrak{c}(b)] = f_{-1}[\bigcap\{\mathfrak{o}(y) \mid \mathfrak{c}(b) \subseteq \mathfrak{o}(y)\}] =$$

$$= \bigcap\{f_{-1}[\mathfrak{o}(y)] \mid \mathfrak{c}(b) \subseteq \mathfrak{o}(y)\} =$$

$$= \bigcap\{\mathfrak{o}(f^*(y)) \mid \mathfrak{c}(b) \subseteq \mathfrak{o}(y)\} \supseteq$$

$$\supseteq \bigcap\{\mathfrak{o}(x) \mid \mathfrak{c}(f^*(b)) \subseteq \mathfrak{o}(x)\} \supseteq \mathfrak{c}(f^*(b)).$$

□

Chapter VI
Complete Regularity

In this chapter we will discuss complete regularity. It is very naturally conservatively extended and hence the basic facts will be similar to those in classical spaces. In some respects, however, the point-free approach brings better results. Some of the facts are just more transparent, but there are also results that are entirely out of the classical scope. In particular, in the category of (completely regular) spaces there is no Lindelöf reflection; in the localic extension there is, and quite an interesting one.

We begin with the analysis of the completely below relation, the concept that can be conservatively extended in a quite straightforward fashion. We do not have, of course, a counterpart of a value of a function at a point, but we do have a specific behaviour of localic maps targeting in the frame of reals (or of frame homomorphisms $h\colon \Omega(\mathbb{R}) \to L$) characterizing complete regularity in a satisfactory parallel with the classical real function characterization of completely regular spaces. This is discussed in detail. Then we devote a section to examples of regular but not completely regular spaces, about which the reader will probably naturally ask. Also, we present a useful construction of non-spatial completely regular frames.

Similarly like in spaces, complete regularity is equivalent with admitting a uniformity. The reader will see that this becomes, in the point-free context, a very transparent and natural fact.

Next, we analyse from the point-free perspective the phenomenon of cozero elements. The cozeros are here transparent, and allow developing expedient techniques; in particular we will show how, basing on them, one can produce the above-mentioned Lindelöf reflection.

The relation "completely below" is usually described using a choice-dependent procedure. This procedure results in the largest interpolative subrelation of $\prec$. In the concluding section we show that the definition of $\prec\!\!\prec$ as the largest interpolative subrelation of $\prec$ can be adopted for developing a constructive variant of the theory. As an example we then show that this can be applied for the Banaschewski–Mulvey variant of the Stone–Čech compactification.

J. Picado, A. Pultr, *Separation in Point-Free Topology*,
https://doi.org/10.1007/978-3-030-53479-0_6

1 The Completely Below Relation

1.1 The relation $\lll$ Recall Chap. I, Sect. 4. What has been done there can be extended to general frames. Take the dyadic rationals

$$D = \left\{ \frac{k}{2^n} \mid n \in \mathbb{N},\ k = 0, 1, \dots, 2^n \right\}$$

in the closed unit interval. An element a is said to be *completely below* b *in* L, written

$$a \lll b,$$

if there exist $a_d \in L, d \in D$ such that

$$a_0 = a,\ a_1 = b, \quad \text{and for all} \quad d < e,\ a_d \prec a_e.$$

1.1.1 Assuming the Axiom of Choice—in fact, one needs less, namely the Axiom of Countable Dependent Choice (CDC)—we easily see that

$\lll$ *is the largest interpolative relation contained in* $\prec$.

Indeed: If $a \lll b$, we have

$$a = a_0 \lll a_{\frac{1}{2}} \lll a_1 = b$$

and on the other hand, if $R \subseteq \prec$ interpolates (to simplify the notation write $a(d)$ for a_d), we can construct the a_d inductively, interpolating for already found $a\left(\frac{k}{2^n}\right)$ the elements with next exponent

$$a\left(\frac{k}{2^n}\right) Ra\left(\frac{2k+1}{2^{n+1}}\right) Ra\left(\frac{k+1}{2^n}\right).$$

It is a trivial exercise that we can do the same with any countable $D' \subseteq \mathbb{I}$. The dyadic rationals, however, make the reasoning particularly transparent.

1.1.2 Example For the standard open intervals in the real line we have $(r, s) \lll (p, q)$ whenever $p < r < s < q$. Indeed, $(r, s) \prec (p, q)$ since $(r, s)^* \vee (p, q) = (-\infty, r) \vee (s, \infty) \vee (p, q) = \mathbb{R}$. From this it follows that

$$(r, s) \prec \left(\frac{1}{2}(p + r), \frac{1}{2}(q + s)\right) \prec (p, q)$$

providing an interpolation for the relation $\lll$ that can be continued infinitely, proving that $(r, s) \lll (p, q)$.

Similarly, $(-\infty, s) \lll (-\infty, q)$ and $(r, \infty) \lll (p, \infty)$.

1.2 From V.1.3 we immediately obtain the following basic rules.

Proposition

(1) $a \prec\!\!\prec b \Rightarrow a \leqslant b$, *and for any* a, $0 \prec\!\!\prec a \prec\!\!\prec 1$.
(2) *If* $x \leqslant a \prec\!\!\prec b \leqslant y$, *then* $x \prec\!\!\prec y$.
(3) *If* $a \prec\!\!\prec b$, *then* $b^* \prec\!\!\prec a^*$.
(4) *If* $a \prec\!\!\prec b$, *then* $a^{**} \prec\!\!\prec b$.
(5) *If* $a_i \prec\!\!\prec b_i$, $i = 1, \dots, n$, *then*

$$a_1 \vee \cdots \vee a_n \prec\!\!\prec b_1 \vee \cdots \vee b_n \quad \textit{and} \quad a_1 \wedge \cdots \wedge a_n \prec\!\!\prec b_1 \wedge \cdots \wedge b_n.$$

1.3 Recall V.1.2 and V.1.2.1. From the construction of $\prec\!\!\prec$ from $\prec$ we readily also infer

Proposition *For any frame homomorphism* $h\colon L \to M$, $a \prec\!\!\prec b$ *implies* $h(a) \prec\!\!\prec h(b)$.

1.4 As usual we will denote by $\mathbb{R}$ the real line with the standard topology, and by $\mathbb{I}$ the closed unit interval $[0, 1] \subseteq \mathbb{R}$.

For technical reasons we will work with the subspace

$$\mathbb{D} = \left\{ \frac{k}{2^n} \mid n \in \mathbb{N},\ k \in \mathbb{Z} \right\}$$

of all dyadic rationals, but everything will happen in the $D = \mathbb{I} \cap \mathbb{D}$ used in 1.1.

The frame of reals can be defined in a fully point-free way (see A.10), but we will work here with the $\Omega(\mathbb{R})$ (and $\Omega(\mathbb{I})$, $\Omega(\mathbb{D})$). As a result, some of the statements will be unnecessarily choice dependent. In Sects. 8 and 9 when discussing the choice-free complete regularity we will not speak of the reals at all.

1.4.1 Notation We will write $\mathbb{R}_t$ for $\mathbb{R} \smallsetminus \{t\}$. Hence in particular

$$\mathbb{R}_0 = \mathbb{R} \smallsetminus \{0\} \quad \text{and} \quad \mathbb{R}_1 = \mathbb{R} \smallsetminus \{1\}.$$

1.5 A functional view of $\prec\!\!\prec$ As we observed in I.4.2, $U \prec\!\!\prec V$ in $\Omega(X)$ iff there exists a continuous function $f : X \to \mathbb{I}$ such that

$$f[\overline{U}] = \{0\} \text{ and } f[X \smallsetminus V] = \{1\},$$

that is, a frame homomorphism $h = \Omega(f)\colon \Omega(\mathbb{I}) \to \Omega(X)$ such that

$$h(\mathbb{R}_0) \subseteq U^* \text{ and } h(\mathbb{R}_1) \subseteq V.$$

This can be extended to general frames.

A *scale in* L is a map $\alpha\colon \mathbb{D} \to L$ such that $\alpha(d) \prec \alpha(e)$ whenever $d < e$, $\alpha(d) = 0$ for $d < 0$, and $\alpha(d) = 1$ for $d > 1$.

Thus, scales are the assignments $\alpha(d) = a_d$ from 1.1 trivially extended to the $d < 0$ and $d > 1$.

Note Usually, scales are defined more generally on the rationals, with the extension to $p < 0$ and $p > 1$ not necessarily trivial. The general scales are also referred as *trails* [30].

1.5.1 Lemma *For any scale α in a frame L, the formula*

$$h_\alpha(U) = \bigvee\{\sigma(q) \wedge \sigma(p)^* \mid p < q, [p, q] \subseteq U\}, \quad U \in \Omega(\mathbb{R})$$

defines a frame homomorphism $h_\alpha \colon \Omega(\mathbb{R}) \to L$.

Proof First, $h_\alpha(\emptyset) = 0$ and $h_\alpha(\mathbb{R}) = 1$ are obvious. Let us show that h_α preserves binary meets. For any $U, V \in \Omega(\mathbb{R})$, $h_\alpha(U) \wedge h_\alpha(V)$ is equal to

$$\bigvee\{\alpha(q) \wedge \alpha(p)^* \wedge \alpha(s) \wedge \alpha(r)^* \mid p < q, r < s, [p, q] \subseteq U, [r, s] \subseteq V\}$$

from which it follows immediately that $h_\alpha(U) \wedge h_\alpha(V) = h_\alpha(U \cap V)$, in the case $U \cap V = \emptyset$, and $h_\alpha(U) \wedge h_\alpha(V) \geqslant h_\alpha(U \cap V)$ in general. For the reverse inequality, with $U \cap V \neq \emptyset$, just notice that $[p, q] \subseteq U$ and $[r, s] \subseteq V$ imply $[a, b] = [p \vee r, q \wedge s] \subseteq U \cap V$ and for each of the four possibilities of the pair a, b,

$$\alpha(q) \wedge \alpha(p)^* \wedge \alpha(s) \wedge \alpha(r)^* \leqslant \alpha(b) \wedge \alpha(a)^*.$$

Finally, let $U = \bigcup_{i \in J} U_i$ ($J \neq \emptyset$). The inequality $\bigvee_{i \in J} h_\alpha(U_i) \leqslant h_\alpha(\bigcup_{i \in J} U_i)$ is obvious. For the reverse inequality, after removing redundancies and collecting overlapping intervals, we can assume without loss of generality that each U_i is equal to some open interval (p_i, q_i) and no two of those intervals overlap. Thus, for each pair $p < q$ with $[p, q] \subseteq \bigcup_{i \in J}(p_i, q_i)$, the interval $[p, q]$ is contained in some $(p_j, q_j) = U_j$. Hence $h_\alpha(\bigcup_{i \in J} U_i) = \bigvee_{i \in J} h_\alpha(U_i)$. □

1.5.2 Proposition *The following statements are equivalent in any frame L.*

(1) $a \ll b$.
(2) *There is a frame homomorphism $h \colon \Omega(\mathbb{R}) \to L$ such that $h(\mathbb{R}_0) \leqslant a^*$ and $h(\mathbb{R}_1) \leqslant b$.*

Proof (1)⇒(2): Let $\{c_d \mid d \in D\}$ be a scale from a to b witnessing $a \ll b$ and extend it to a scale α in L. Then the construction in 1.5.1 yields a frame homomorphism $h_\alpha \colon \Omega(\mathbb{R}) \to L$. This is the required homomorphism satisfying (2):

(a) $h_\alpha(\mathbb{R}_0) \leqslant a^*$: Since $[p, q] \subseteq \mathbb{R}_0$ iff $q < 0$ or $p > 0$ and in the former case $c_q \wedge c_p^* = 0$, we have

$$h_\alpha(\mathbb{R}_0) = \bigvee\{c_q \wedge c_p^* \mid 0 < p < q\} \leqslant \bigvee\{c_p^* \mid 0 < p\} \leqslant a^*.$$

(b) $h_\alpha(\mathbb{R}_1) \leqslant b$: Since $[p, q] \subseteq \mathbb{R}_1$ iff $q < 1$ or $p > 1$ and in the latter case $c_q \wedge c_p^* = 0$,

$$h_\alpha(\mathbb{R}_1) = \bigvee\{c_q \wedge c_p^* \mid p < q < 1\} \leqslant \bigvee\{c_q \mid q < 1\} \leqslant b.$$

(2)⇒(1): If h satisfies (2), then we can construct a scale $\{c_d \mid d \in D\}$ from a to b by setting $c_0 = a$, $c_1 = b$, and $c_d = h(-\infty, d)$ for $d \in D$. This is so because $c_p \prec c_q$ for $p < q$ in D since

$$c_p^* \vee c_q = (h(-\infty, p))^* \vee h(-\infty, q) \supseteq h((-\infty, p)^*) \vee h(-\infty, q) =$$
$$= h(p, \infty) \vee h(-\infty, q) = h((-\infty, q) \cup (p, \infty)) = h(\mathbb{R}) = \mathbb{R}. \quad \square$$

1.5.3 Notes

1. The condition $h(\mathbb{R}_0) \leqslant a^*$ is equivalent to

$$a \leqslant h(-\infty, q) \text{ for every } q > 0 \quad \text{and} \quad a \leqslant h(p, \infty) \text{ for every } p < 0.$$

Indeed, if $h(\mathbb{R}_0) \leqslant a^*$, then, for any $q > 0$,

$$a \leqslant a^{**} \leqslant h(\mathbb{R}_0)^* \leqslant h(0, \infty)^* \leqslant h(-\infty, q),$$

where the latter inequality holds since $h(-\infty, q) \vee h(0, \infty) = h(\mathbb{R}) = 1$ and thus $x \wedge h(0, \infty) = 0$ implies $x \wedge h(-\infty, q) = x$. Similarly, $a \leqslant h(p, \infty)$ for every $p < 0$. Conversely,

$$h(\mathbb{R}_0) = h(-\infty, 0) \vee h(0, \infty) = \bigvee_{p<0} h(-\infty, p) \vee \bigvee_{q>0} h(q, \infty) =$$
$$= \bigvee_{p<0} h((p, \infty)^*) \vee \bigvee_{q>0} h((-\infty, q)^*) \leqslant$$
$$\leqslant \bigvee_{p<0} h(p, \infty)^* \vee \bigvee_{q>0} h(-\infty, q)^* \leqslant a^*.$$

2. If $f : X \to \mathbb{R}$ is a continuous map and $U \subseteq V$ in $\Omega(X)$, then we see easily that $\Omega(f)(\mathbb{R}_0) \subseteq U^*$, that is, $f^{-1}[\mathbb{R} \smallsetminus \{0\}] \cap U = \emptyset$ is easily seen to say that $f(x) = 0$ for all $x \in U$, and $\Omega(f)(\mathbb{R}_1) \subseteq V$, that is, $f^{-1}[\mathbb{R} \smallsetminus \{1\}] \subseteq V$ is easily seen to say that $f(x) = 1$ for all $x \notin V$.

2 Complete Regularity

2.1 A frame L is *completely regular* if

(creg): *for every* $a \in L$, $a = \bigvee\{x \mid x \ll a\}$

(recall Chap. I, 5.5 and 4.2.3).

The following result (that appeared originally in the 1953 doctoral dissertation of B. Banaschewski [19]) follows now immediately from I.4.2.3.

2.1.1 Corollary *A space X is completely regular iff the frame $\Omega(X)$ is completely regular.*

2.1.2 Note Proposition 1.5.2 yields an alternative version to I.4.2.2:

For any space X, $U \lll V$ in $\Omega(X)$ if and only if there exists a continuous function $f\colon X \to \mathbb{R}$ such that $f(x) < \frac{1}{2}$ for all $x \in U$ and $f(x) \geqslant 1$ for all $x \notin V$.

Indeed, take $h\colon \Omega(\mathbb{R}) \to \Omega(X)$ for $U \lll V$ as in the proposition and then define $f\colon X \to \mathbb{R}$ by

$$f(x) \in (p, q) \quad \text{iff} \quad x \in h(p, q)$$

whenever $p < q$ in $\mathbb{Q}$. It is easy to check that f is continuous. Moreover:

(a) By Note 1 in 1.5.3, $U \subseteq h(-\infty, \frac{1}{2}) = f^{-1}[(-\infty, \frac{1}{2})]$, showing that $f(x) < \frac{1}{2}$ in U.
(b) $f^{-1}[(-\infty, 1)] = h(-\infty, 1) \subseteq h(\mathbb{R}_1) \subseteq V$, showing that $f(x) \geqslant 1$ outside V. □

2.2 In the following proposition we speak of a frame being generated by a subset. To avoid misunderstanding, the subset in question will be an analogon of a classical subbasis, that is the frame is generated by first forming finite meets and then applying general joins. Just the joins, as it happens in many constructions, would not suffice.

Proposition *A frame L is completely regular iff it is generated by the images of all frame homomorphisms $h\colon \Omega(\mathbb{R}) \to L$.*

Proof Let $x \lll a$. By 1.5.2 and 1.5.3.1, $x \leqslant h(-\infty, \frac{1}{2}) \leqslant h(\mathbb{R}_1) \leqslant a$. Hence, if L is completely regular, then

$$a = \bigvee\{x \mid x \lll a\} \leqslant \bigvee\{h(\mathbb{R}_1) \mid h\colon \Omega(\mathbb{R}) \to L,\ h(\mathbb{R}_1) \leqslant a\} \leqslant a$$

for every $a \in L$.

Conversely, since $\Omega(\mathbb{R})$ is completely regular and frame homomorphisms preserve $\lll$ (recall 1.3), the image of any $h\colon \Omega(\mathbb{R}) \to L$ is a complete regular subframe of L. Finally, any frame L generated by a set of completely regular subframes is itself completely regular. Indeed, whenever $x_i \lll a_i$ $(i = 1, \dots, n)$ in some subframes L_i of L then

$$x_1 \wedge \cdots \wedge x_n \lll a_1 \wedge \cdots \wedge a_n$$

in the subframe generated by $L_1 \cup L_2 \cup \cdots \cup L_n$ and consequently also in L.

(Note that the union of the images is not necessarily closed under finite meets. It is a subbasis, not a basis.)

2.3 Complete regularity of sublocales Let $S \subseteq L$ be a sublocale with embedding $j_S \colon S \to L$ and corresponding frame homomorphism $\nu_S \colon L \to S$. Denote by $\lt\lt_S$ the completely below relation in the locale S. By 1.3,

$$a, b \in S \text{ and } a \lt\lt b \Rightarrow a \lt\lt_S b.$$

Proposition *Each sublocale of a completely regular frame is completely regular.*

Proof Take an $a \in S$. By 1.3, $x \lt\lt a$ implies $\nu_S(x) \lt\lt_S \nu_S(a) = a$ and thus, since $x \leqslant \nu_S(x)$, we have $a = \bigvee\{x \in L \mid x \lt\lt a\} \leqslant \bigvee\{y \mid y \lt\lt_S a\} \leqslant a$. □

3 Complete Regularity is Stronger Than Regularity

3.1 Complete regularity is a stronger property than regularity, but it is not easy to see. In the thirties it was for quite a time an open problem, and the solutions were rather involved. It is a very interesting topic, and new constructions appear.[1] We will present an elegant short one published by Mysior in 1981 ([207]).

3.2 First, however, let us explain why the problem is difficult. The natural intuition suggests, rather, that the two notions should coincide. Take, for instance, the relation $U \prec V$ for open sets (that is, $\overline{U} \subseteq V$): it is hard to visualize a situation depicted in I.3.3 that would prevent interpolation. Admittedly, visual intuition is often deceptive, but here the intuition is in fact really not wide from the target. Typical everyday life spaces, and also specially constructed examples, are countably generated, or if not countably generated, at least Lindelöf, and for such spaces *regularity implies even normality and hence in particular complete regularity* (see VII.1.5.1). Thus, in a dividing example some uncountability has to play an essential role.

3.3 The example The underlying set of the space X is the set

$$\{(x, y) \in \mathbb{R} \times \mathbb{R} \mid y \geqslant 0\} \cup \{\infty\}$$

and its topology is defined as follows:

(R1) All points (x, y) with $y > 0$ are assumed to be isolated, that is, each $\{(x, y)\}$, $y > 0$, is open.

[1] It should be noted that the question can be simpler—although because of the algebraic nature of the approach geometrically less intuitive—in the point-free context, as discussed in [181].

(R2) The basic open neighbourhoods of $(x, 0)$ contain $(x, 0)$ and all but finitely many points from the union of two segments

$$I_x = \{(x, y) \mid 0 \leqslant y \leqslant 2\} \text{ and } J_x = \{(x + y, y) \mid 0 \leqslant y \leqslant 2\}.$$

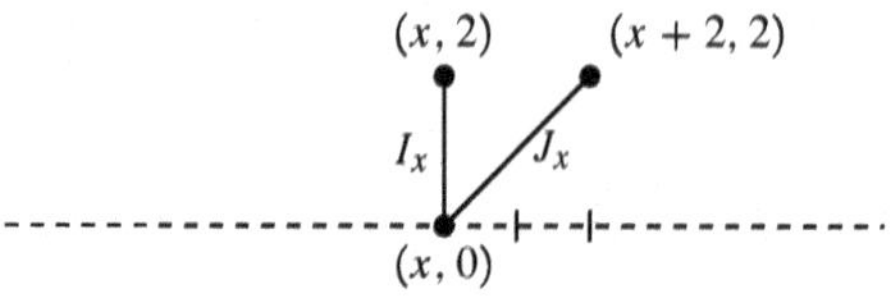

(R3) The basic open neighbourhoods of the point ∞ have the form

$$U_r = \{\infty\} \cup \{(x, y) \mid x > r, y \geqslant 0\} \quad (r \in \mathbb{R}).$$

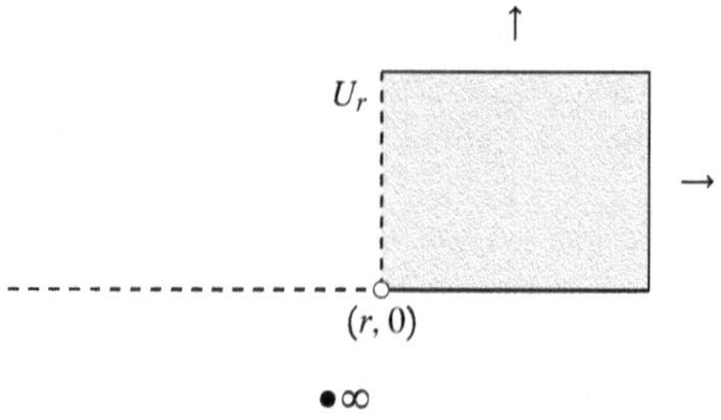

3.3.1 X is regular First, every (x, y) with $y \geqslant 0$ is clopen: it can be divided from each $(u, 0)$ by $I_u \cup J_u \setminus \{(x, y)\}$, and from ∞ by any U_r with $r > x$. Further, each $I_x \cup J_x$ is closed. Indeed,

$$X \setminus I_x = U_x \cup \bigcup\{(a, b) \mid (a,b) \notin I_x, b > 0\} \cup$$
$$\cup \bigcup_{c<x} [(I_c \cup J_c) \setminus (J_c \cap I_x)] \in \Omega(X)$$

while $X \setminus J_x$ is the open

$$U_{x+2} \cup \bigcup\{(a, b) \mid (a, b) \notin J_x, b > 0\} \cup$$
$$\cup \bigcup\{(I_c \cup J_c) \setminus (I_c \cap I_x) \mid c \leqslant x + 2, c \neq x\}.$$

Finally, we provide closed neighbourhoods for ∞ showing that for every $r \in \mathbb{R}$, $\overline{U_{r+2}} \subseteq U_r$:

Indeed,

$$U_{r+2} \subseteq U_{r+2} \cup \{(x,0) \mid r < x \leqslant r+2\} \subseteq U_r$$

and

$$U_{r+2} \cup \{(x,0) \mid r < x \leqslant r+2\} = \Big(X \smallsetminus \bigcup_{r \geqslant a} (I_r \cup J_r)\Big) \cup \Big(X \smallsetminus \bigcup\{(a,b) \mid a \leqslant r+2, b > 0\}\Big)$$

is closed.

3.3.2 *X* is not completely regular Consider $A = \{(x,0) \mid x \leqslant 1\}$. We will prove that for a continuous map $f\colon X \to \mathbb{R}$ such that $f[A] = \{0\}$ also the value $f(\infty)$ will be zero.

Let us introduce the following notation:

$$N = f^{-1}[\{0\}] \quad \text{and, for } n \geqslant 1, \quad L_n = \{(x,0) \mid n-1 \leqslant x \leqslant n\}.$$

Observations

1. *If* $x - 2 \leqslant y \leqslant x$, *then* $J_y \cap I_x \neq \emptyset$.
2. *If* $(y,0) \in N$, *then* $J_y \smallsetminus N$ *is at most countable.*

(For 2:
$N = \bigcap_k f^{-1}[(-\frac{1}{k}, \frac{1}{k})]$, each $J_y \smallsetminus f^{-1}[(-\frac{1}{k}, \frac{1}{k})]$ is finite, and $J_y \smallsetminus N = \bigcup_k J_y \smallsetminus f^{-1}[(-\frac{1}{k}, \frac{1}{k})]$.)

Lemma *For every* n, $N \cap L_n$ *is infinite.*

Proof By induction. $L_1 = \{(x,0) \mid 0 \leqslant x \leqslant 1\}$. Choose a countable $C \subseteq N \cap L_n$. Then, by Observation 2, $\bigcup\{J_c \smallsetminus N \mid (c,0) \in C\}$ is at most countable, and so is

$$P = \{(y,0) \mid (y,t) \in \bigcup\{J_c \smallsetminus N \mid (c,0) \in C\}\}.$$

Consider $L_{n+1} \smallsetminus P$. By Observation 1, for every x, $n \leqslant x \leqslant n+1$ and c with $(c,0) \in C$, $I_x \cap J_c \neq \emptyset$ and hence for $x \in L_{n+1} \smallsetminus P$, $I_x \cap J_c \cap N \neq \emptyset$ so that $I_x \cap N \neq \emptyset$. But N is closed, and every neighbourhood of $(x,0)$ contains I_x and hence $(x,0) \in N$, therefore $L_{n+1} \smallsetminus P \subseteq N$, and $L_{n+1} \cap N$ is infinite. □

Conclusion *Every neighbourhood of* ∞ *meets the closed* N *and hence* $\infty \in N = f^{-1}[\{0\}]$, *that is,* $f(\infty) = 0$. *Since* $f\colon X \to \mathbb{R}$ *was arbitrary such that* $f[A] = \{0\}$, *we cannot separate* A *from* ∞.

3.4 Interpolativity of $\prec$ As we have already mentioned, the relation (as suggestively depicted in the picture in I.3.3) seems to be interpolative. Now we know that this intuition is false. The interpolativity of $\prec$ in a regular frame is in fact even stronger than complete regularity: in a regular frame one can have $\prec \neq \prec\!\!\prec$ and still $a = \bigvee\{x \mid x \prec\!\!\prec a\}$ for all a.

This stronger property is called *almost normality* (see VII.1.3.2).

4 A Class of Non-spatial Completely Regular Frames

4.1 A method for constructing frames from a space We start by describing a method for constructing new frames from spatial frames due to Dube–Iliadis–van Mill–Naidoo [91]. So let κ be an infinite cardinal number and let X be a Tychonoff space of weight at most κ (i.e. $\Omega(X)$ has a basis of cardinality at most κ). Let $\mathcal{I}$ be a κ^+*-complete* ideal of $\mathfrak{P}(X)$. This means that $\mathcal{I}$ is an ideal of $\mathfrak{P}(X)$ closed under joins with indexed set of cardinality $\leqslant \kappa$. Note that the fact that the weight of X is at most κ implies that

$$\forall\, \mathcal{A} \subseteq \Omega(X),\ \exists\, \mathcal{B} \subseteq \mathcal{A}\colon\ |\mathcal{B}| \leqslant \kappa \text{ and } \bigcup \mathcal{A} = \bigcup \mathcal{B}. \qquad (\kappa\text{-weight})$$

Let $\sqsubseteq$ be the relation

$$U \sqsubseteq V \ \equiv_{\mathrm{df}} \ \ U \smallsetminus V \in \mathcal{I}$$

in $\Omega(X)$.

4.1.1 Lemma *For any $U, V, W \in \Omega(X)$ we have:*

(1) $U \sqsubseteq U$.
(2) *If* $U \sqsubseteq V$ *and* $V \sqsubseteq W$, *then* $U \sqsubseteq W$.
(3) *If* $U \sqsubseteq V$ *and* $U \sqsubseteq W$, *then* $U \sqsubseteq V \cap W$.

Proof

(1) $U \smallsetminus U = \emptyset \in \mathcal{I}$.
(2) $U \smallsetminus W \subseteq (V \smallsetminus W) \cup (U \smallsetminus V) \in \mathcal{I}$.
(3) $U \smallsetminus (V \cap W) = (U \smallsetminus V) \cup (U \smallsetminus W) \in \mathcal{I}$. □

Properties (1) and (2) assert that $\sqsubseteq$ is a preorder on $\Omega(X)$. We follow the standard procedure to make it a partial order. Define the equivalence relation $\sim$ on $\Omega(X)$ by

$$U \sim V \ \equiv_{\mathrm{df}} \ \ U \sqsubseteq V \text{ and } V \sqsubseteq U,$$

that is,

$$U \sim V \ \text{ iff } \ U \Delta V = (U \cup V) \smallsetminus (U \cap V) \in \mathcal{I}.$$

Denote by $[U]$ the equivalence class of any $U \in \Omega(X)$ and consider the set $\Omega(X)/\sim$ partially ordered by

$$[U] \leqslant [V] \equiv U \sqsubseteq V.$$

4.1.2 Lemma *For any $U, V, U_i \in \Omega(X)$ $(i \in J)$ we have the following.*

(1) $\bigvee_{i \in J}[U_i]$ *exists and is equal to* $[\bigcup_{i \in J} U_i]$.
(2) $[U] \wedge [V]$ *exists and is equal to* $[U \cap V]$.

Proof

(1) Clearly $[U_j] \leqslant [\bigcup_{i \in I} U_i]$ for every $j \in I$. Now assume that $[W]$ satisfies $[U_i] \leqslant [W]$ for every $i \in I$. By $(*)$, there is some $J \subseteq I$ of cardinality at most κ such that $\bigcup_{j \in J} U_j = \bigcup_{i \in I} U_i$. Since $U_j \smallsetminus W \in \mathcal{I}$ for every $j \in J$ and $\mathcal{I}$ is κ^+-complete, we have

$$\Big(\bigcup_{i \in I} U_i\Big) \smallsetminus W = \Big(\bigcup_{j \in J} U_i\Big) \smallsetminus W = \bigcup_{j \in J}(U_j \smallsetminus W) \in \mathcal{I}.$$

Hence $[\bigcup_{i \in I} U_i] \leqslant [W]$.
(2) $[U \cap V] \leqslant [U], [V]$ is obvious and $[W] \leqslant [U \cap V]$ for any $[W] \leqslant [U], [V]$ by (3) in 4.1.1. □

This shows that $\Omega(X)/\sim$ is a complete lattice, actually a frame. Denote it by

$$L_{X,\mathcal{I}}$$

and consider the surjective frame homomorphism

$$h = (U \mapsto [U]) \colon \Omega(X) \to L_{X,\mathcal{I}}.$$

Since $\Omega(X)$ is completely regular we conclude, by 2.3, that

4.1.3 Proposition *$L_{X,\mathcal{I}}$ is a completely regular frame.* □

4.1.4 Remark It is shown in [91] that $L_{X,\mathcal{I}}$ is also *countably compact* (that is, every countable cover has a finite subcover).

4.2 Non-spatial examples Let X be a compact Hausdorff space with weight $\mathfrak{c}$ and such that

if $F \subseteq X$ is closed, then either F is finite or $|F| = 2^{\mathfrak{c}}$.

Further, set $\mathcal{I} = \{S \subseteq X : |S| \leqslant \mathfrak{c}\}$. It is clearly a $\mathfrak{c}^+$-complete ideal. Then

4.2.1 Theorem *The frame $L_{X,\mathcal{I}}$ is not spatial.*

Proof We will show this by checking that $L_{X,\mathcal{I}}$ has no prime elements.

Assume that $[U] \neq [X]$, that is, $X \smallsetminus U$ has cardinality at least $\mathfrak{c}^+$. Let $\mathcal{U}$ be the collection of all relatively open subsets of $X \smallsetminus U$ of cardinality at most $\mathfrak{c}$. Then $|\bigcup \mathcal{U}| \leqslant \mathfrak{c}$. Indeed, the weight of $X \smallsetminus U$ is at most $\mathfrak{c}$ and hence, by (κ-weight) in 4.1, there is a subcollection $\mathcal{V}$ of $\mathcal{U}$ such that $|\mathcal{V}| \leqslant \mathfrak{c}$ and $\bigcup \mathcal{V} = \bigcup \mathcal{U}$.

Now, since $|X \smallsetminus U| \geqslant \mathfrak{c}^+$, there are distinct $p, q \in (X \smallsetminus U) \smallsetminus \bigcup \mathcal{U}$. Let A and B be disjoint open neighbourhoods in X of p, respectively, q. By 2 in 4.1.2,

$$[A \cup U] \wedge [B \cup U] = [(A \cup U) \cap (B \cup U)] = [U]. \qquad (*)$$

But $|A \cap (X \smallsetminus U)| \geqslant \mathfrak{c}^+$ (otherwise $A \cap (X \smallsetminus U) \in \mathcal{U}$ would contradict $p \notin \bigcup \mathcal{U}$). Hence $[A \cup U] \nleqslant [U]$. Likewise $[B \cup U] \nleqslant [U]$. It then follows from $(*)$ that $[U]$ is not a prime element of $L_{X,\mathcal{I}}$.

Examples There are many spaces X with the properties from 4.2. One such example is $\beta\omega \smallsetminus \omega$, the Stone–Čech remainder of the countable discrete space ω [112, 9.12]. Note that $\beta\omega \smallsetminus \omega$ has no isolated points.

5 Complete Regularity and Uniformity

5.1 Recall the notions and notation from II.7.2 and V.1.5 (covers, refinements, etc.). Further, for covers A, B set

$$AB = \{Ab \mid b \in B\}.$$

In this and the next chapter we will need a few rules, easy to check:

(1) $x \leqslant Ax$.
(2) $A \leqslant B$ and $x \leqslant y$ imply $Ax \leqslant By$.
(3) $A(Bx) \leqslant (AB)x = A(B(Ax))$.
(4) $(A_1 \wedge \cdots \wedge A_n)(B_1 \wedge \cdots \wedge B_n) \leqslant (A_1B_1) \wedge \cdots \wedge (A_nB_n)$.
(5) $A(\bigvee_{i \in J} x_i) = \bigvee_{i \in J} Ax_i$.

5.1.1 The mapping $(x \mapsto Ax)\colon L \to L$ obviously preserves meets and hence has a right adjoint, denoted y/A (that is, we have $Ax \leqslant y$ iff $x \leqslant y/A$). Thus

$$y/A = \bigvee \{x \mid Ax \leqslant y\}.$$

Recall the concept of nearness discussed in II.7.2. Since $A(x/A) \leqslant x$, the admissibility condition of II.7.2 can be rewritten as

$$\forall a,\ a = \bigvee_{A \in \mathcal{A}} a/A \quad (\text{with } a/A \lhd_{\mathcal{A}} a).$$

5.2 Uniformity A *uniformity* on a frame L is an admissible nearness $\mathcal{A}$ such that

(U) for any $A \in \mathcal{A}$ there is some $B \in \mathcal{A}$ such that $BB \leqslant A$

(one speaks of such B as of *star refinements* of A).

A *subbasis* of a uniformity is a system of covers $\mathcal{S}$ satisfying (U) such that

$$\mathcal{A} = \{A \mid \exists A_1, \dots, A_n \in \mathcal{S} \text{ such that } A_1 \wedge \cdots \wedge A_n \leqslant A\}$$

is admissible. A *basis* of a uniformity is a system of covers $\mathcal{B}$ satisfying (U) and

(U') for any $A, B \in \mathcal{B}$ there is some $C \in \mathcal{B}$ such that $C \leqslant A \wedge B$

such that

$$\mathcal{A} = \{A \mid \exists B \in \mathcal{B} \text{ such that } B \leqslant A\}$$

is admissible. Then, in each case, $\mathcal{A}$ is the smallest uniformity containing $\mathcal{S}$ resp. $\mathcal{B}$.

5.3 Lemma *If $\mathcal{A}$ is a (basis of) uniformity, then $\lhd_{\mathcal{A}}$ interpolates. Consequently,*[2]

$$a \lhd_{\mathcal{A}} b \;\Rightarrow\; a \lll b.$$

Proof Let $a \lhd_{\mathcal{A}} b$. Then $Aa \leqslant b$ for some $A \in \mathcal{A}$ and if we choose a $B \in \mathcal{A}$ such that $BB \leqslant A$ we have $a \lhd_{\mathcal{A}} Ba$. Then, by the rules in 5.1, $B(Ba) \leqslant (BB)a \leqslant Aa \leqslant b$ and hence $Ba \lhd_{\mathcal{A}} b$. □

5.4 Theorem *A frame L admits a uniformity iff it is completely regular.*

Proof If L admits a uniformity $\mathcal{A}$, then it is completely regular by the lemma above. Conversely, let L be completely regular. For any sequence

$$a_0 = 0 \lll a_1 \lll a_2 \lll \cdots \lll a_n \lll a_{n+1} = 1 \quad (n > 1)$$

let

$$A(a_1, a_2, \dots, a_n) = \{a_2, a_3 \wedge a_1^*, a_4 \wedge a_2^*, \dots, a_n \wedge a_{n-2}^*, a_{n-1}^*\}.$$

This is a cover: one shows inductively that

$$\bigvee\{a_2, a_3 \wedge a_1^*, a_4 \wedge a_2^*, \dots, a_k \wedge a_{k-2}^*\} = a_k$$

using that $a_i \prec a_{i+1}$. On the other hand,

$$(a_i \wedge a_{i-2}^*) \wedge (a_k \wedge a_{k-2}^*) = 0$$

[2]Here we use the Axiom of Countably Dependent Choice (CDC)—recall 1.1.1.

whenever $i \leqslant k-2$ or $i \geqslant k+2$. Therefore, for any $k \geqslant 3$ with $k = 3i$, $k = 3i+1$ or $k = 3i+2$, we have

$$\begin{aligned} A(a_1, \dots, a_n)(a_k \wedge a_{k-2}^*) &= (a_{k-1} \wedge a_{k-3}^*) \vee (a_k \wedge a_{k-2}^*) \vee (a_{k+1} \wedge a_{k-1}^*) \\ &= a_{k+1} \wedge a_{k-3}^* \leqslant a_{3(i+1)} \wedge a_{3(i-1)}^* \end{aligned}$$

while $A(a_1, \dots, a_n)a_2 = a_3 \leqslant a_6 = a_6 \wedge a_0^*$. Hence, if we interpolate

$$a_1 \lll u_1 \lll v_1 \lll a_2 \lll u_2 \cdots \lll v_{n-1} \lll a_n$$

and set $B = A(a_1, u_1, v_1, a_2, u_2, \dots, v_{n-1}, a_n)$ we see that

$$BB \leqslant A(a_1, a_2, \dots, a_n).$$

Thus, the system $\mathcal{S}$ of all the $A(a_1, a_2, \dots, a_n)$ is a subbasis of a uniformity $\mathcal{A}$. Now $\mathcal{S}$ is admissible because $x \lll a$ implies $A(x, a)x = \{a, x^*\}x \leqslant a$; since $\mathcal{S} \subseteq \mathcal{A}$, $\mathcal{A}$ is admissible as well. □

6 Cozero Elements

6.1 Cozeros Recall that an open set U in a space X is said to be a cozero set if there is a continuous map $f\colon X \to \mathbb{R}$ such that $f^{-1}[\mathbb{R}_0] = U$. More generally, a *cozero element* of a frame L (briefly, a *cozero* of L) is an element of a form $h(\mathbb{R}_0)$ where $h\colon \Omega(\mathbb{R}) \to L$ is a frame homomorphism. The cozero element associated with a homomorphism h, that is, $h(\mathbb{R}_0)$, will be denoted by $\operatorname{coz} h$.

The set of all cozero elements of L—the *cozero part of* L— will be denoted by

$$\operatorname{Coz}(L).$$

For any topological space X, the cozero part of the frame $\Omega(X)$ consists exactly of the usual cozero sets of X.

6.2 Internal characterizations Cozero elements can be characterized in L internally in terms of the relation $\lll$, without mentioning (the homomorphisms representing) real functions. For that, we first need a lemma whose proof is very similar to the proof of 1.5.1.

6.2.1 Lemma *For each frame homomorphism $h\colon \Omega(\mathbb{R}) \to L$ and each $t \in \mathbb{Q}$, the formula*

$$(h-t)(U) = \bigvee\{h(p,q) \mid \exists \varepsilon > 0\colon (p-t-\varepsilon, q-t+\varepsilon) \subseteq U\}, \quad U \in \Omega(\mathbb{R})$$

defines a frame homomorphism

$$h - t\colon \Omega(\mathbb{R}) \to L$$

such that $(h-t)(\mathbb{R}_0) = h(\mathbb{R}_t)$.

Proof Clearly, $(h-t)(\emptyset) = h(0) = 0$ and

$$(h-t)(\mathbb{R}) = \bigvee\{h(p,q) \mid p,q \in \mathbb{Q}\} = h(\bigvee\{(p,q) \mid p,q \in \mathbb{Q}\}) = h(\mathbb{R}) = 1.$$

Further, for any $U, V \in \Omega(\mathbb{R})$, $(h-t)(U) \wedge (h-t)(V)$ is equal to

$$\bigvee\Big\{h(p,q) \wedge h(r,s) \mid \exists \varepsilon, \delta > 0\colon (p-t-\varepsilon, q-t+\varepsilon) \subseteq U, (r-t-\delta, s-t+\delta) \subseteq V\Big\}.$$

Obviously, $(h-t)(U) \wedge (h-t)(V) \geqslant (h-t)(U \cap V)$. For the reverse inequality, notice first that

$$h(p,q) \wedge h(r,s) = h((p,q) \cap (r,s)) = h(p \vee r, q \wedge s);$$

moreover

$$(p-t-\varepsilon, q-t+\varepsilon) \subseteq U \;\text{ and }\; (r-t-\delta, s-t+\delta) \subseteq V$$

imply

$$\Big((p-t-\varepsilon) \vee (r-t-\delta), (q-t+\varepsilon) \wedge (s-t+\delta)\Big) \subseteq U \cap V.$$

Thus, for $\gamma = \varepsilon \wedge \delta$,

$$((p \vee r) - t - \gamma, (q \wedge s) - t + \gamma) \subseteq U \cap V$$

and $(h-t)(U) \wedge (h-t)(V) = (h-t)(U \cap V)$.

Let $U = \bigcup_{i \in J} U_i$ $(I \neq \emptyset)$. The inequality

$$\bigvee_{i \in J} (h-t)(U_i) \leqslant (h-t)(\bigcup_{i \in J} U_i)$$

is obvious. For the reverse inequality, after removing redundancies and collecting overlapping intervals, we may assume without loss of generality that each U_i is equal to some open interval (p_i, q_i) and no two of those intervals overlap. Let $p < q$ and $\varepsilon > 0$ be such that

$$(p-t-\varepsilon, q-t+\varepsilon) \subseteq \bigcup_{i \in J} (p_i, q_i).$$

Then $(p - t - \varepsilon, q - t + \varepsilon) \subseteq (p_j, q_j) = U_j$ for some j and thus

$$(h - t)(\bigcup_{i \in J} U_i) = \bigvee_{i \in J} (h - t)(U_i).$$

Finally,

$$(h - t)(\mathbb{R}_0) = \bigvee\{h(p, q) \mid p < q, \exists\varepsilon > 0 \colon (p - t - \varepsilon, q - t + \varepsilon) \subseteq \mathbb{R}_0\},$$

and since

$$(p - t - \varepsilon, q - t + \varepsilon) \subseteq \mathbb{R}_0 \text{ iff } p - t - \varepsilon \geqslant 0 \text{ or } q - t + \varepsilon \leqslant 0,$$

we have $(h - t)(\mathbb{R}_0) = \bigvee\{h(p, q) \mid p < q, (p, q) \subseteq \mathbb{R}_t\} = h(\mathbb{R}_t)$. □

6.2.2 Proposition *The following statements on an element a of a frame L are equivalent.*

(1) $a \in \mathrm{Coz}\,(L)$.
(2) $a = \bigvee_{q<1} a_q$ *for some scale* $\{a_q \mid q \in \mathbb{D}\}$.
(3) $a = \bigvee_{n=1}^{\infty} x_n$ *with* $x_n \lll a$ *for all* $n = 1, 2, \ldots$.
(4) $a = \bigvee_{n=1}^{\infty} y_n$ *with* $y_n \lll y_{n+1}$ *for all* $n = 1, 2, \ldots$.

Proof (1)⇒(4): Let $a = h(\mathbb{R}_0)$ for some frame homomorphism $h \colon \Omega(\mathbb{R}) \to L$. Then

$$a = \bigvee_{n=1}^{\infty} h\Big((-\infty, -\frac{1}{n}) \cup (\frac{1}{n}, \infty)\Big)$$

and by 1.1.2 and the fact that any frame homomorphism preserves $\lll$, $y_n = h((-\infty, -\frac{1}{n}) \cup (\frac{1}{n}, \infty))$ is completely below y_{n+1} for every n.

(4)⇔(3): (4)⇒(3) is obvious. Conversely, let $a = \bigvee_{n=1}^{\infty} x_n$ with $x_n \lll a$ and define $y_1 = x_1 \lll a$. Then take a c_2 with $x_1 \lll c_2 \lll a$ and set $y_2 = x_2 \vee c_2$ so that $y_1 \lll y_2 \lll a$ (since $\lll$ is stable under binary joins) and $y_2 \geqslant x_2$. If $y_1 \lll \cdots \lll y_n \lll a$ with $y_k \geqslant x_k$ are already chosen, take a c_{n+1} with $y_n \lll c_{n+1}$ and set $y_{n+1} = x_{n+1} \vee c_{n+1}$.

(4)⇒(2): From $y_1 \lll y_2 \lll \cdots \lll y_n \cdots \leqslant a$ and the fact that $\lll$ interpolates it easily follows that there is a scale $\{a_q \mid q \in \mathbb{D}\}$ such that $\bigvee_{q<1} a_q = \bigvee_{n=1}^{\infty} y_n$.

(2)⇒(1): If $a = \bigvee_{q<1} a_q$ for some scale $\{a_q \mid q \in \mathbb{D}\}$, then, by changing a_1 to a if necessary, we may assume that the scale is from a_0 to a and 1.5.2 yields a frame homomorphism $h \colon \Omega(\mathbb{R}) \to L$ such that $h(\mathbb{R}_1) \leqslant a$. The proof that (1) implies (2) in 1.5.2 shows that $h(\mathbb{R}_1) = \bigvee_{q<1} a_q = a$. Finally, take the $h - 1$ from Lemma 6.2.1. □

6.2.3 Hence, in any frame L, we have

Corollary

1. *If* $a \prec\!\!\prec b$, *then there exists* $c \in \mathrm{Coz}(L)$ *such that* $a \prec\!\!\prec c \prec\!\!\prec b$.
2. *Each* a *in* $\mathrm{Coz}(L)$ *can be written as* $a = \bigvee_{n=1}^{\infty} y_n$ *with* $y_n \in \mathrm{Coz}(L)$ *and* $y_n \prec\!\!\prec y_{n+1}$ *for all* $n \in \mathbb{N}$.

Proof

1. If $\{c_q \mid q \in \mathbb{D}\}$ is a scale between a and b, then $c = \bigvee_{q<\frac{1}{2}} c_q$ is a cozero element by the proposition, and clearly $a \prec\!\!\prec c \prec\!\!\prec b$.
2. is obvious after 1. □

6.2.4 Lindelöf frames Recall that an $a \in L$ is called *Lindelöf* if $a = \bigvee S$ for $S \subseteq L$ implies $a = \bigvee T$ for some countable $T \subseteq S$. The L itself is *Lindelöf* whenever its top element 1 is Lindelöf.

Proposition *Let L be a completely regular Lindelöf frame and* $a \in L$. *Then* a *is a cozero element iff it is Lindelöf.*

Proof Let $a \in \mathrm{Coz}(L)$. Using the characterization (3) in 6.2.2 we have $a = \bigvee_{n=1}^{\infty} x_n$ with $x_n \prec\!\!\prec a$ for all $n \in \mathbb{N}$. Consider an $S \subseteq L$ such that $a \leqslant \bigvee S$. For each n, $S \cup \{x_n^*\}$ is a cover of L. Since L is Lindelöf, we may choose for each n, using the Axiom of Countable Choice (CC), a countable $T_n \subseteq S$ such that $T_n \cup \{x_n^*\}$ is still a cover. It follows that $x_n \leqslant \bigvee T_n$ and therefore $a \leqslant \bigvee T$ for $T = \bigcup_{n=1}^{\infty} T_n$ and T is countable by CC.[3]

Conversely, a is equal to $\bigvee\{x \in L \mid x \prec\!\!\prec a\}$ by complete regularity, and if a is Lindelöf, then it is a join of countably many $x \prec\!\!\prec a$. This makes it a cozero element by 6.2.2. □

6.3 σ-Frames In this and the next chapters it will be expedient to use the following modification of the concept of frame.

A *σ-frame* is a lattice with *countable* joins with the distribution law

$$a \wedge \bigvee_{i \in J} b_i = \bigvee_{i \in J} (a \wedge b_i)$$

for countable J (while a distribution over a possible bigger join is not assumed), and a *σ-frame homomorphism* between σ-frames preserves countable joins and finite meets. The resulting category is denoted by

$$\boldsymbol{\sigma\mathrm{Frm}}.$$

Some of the notions concerning frames can be automatically adopted for σ-frames, like, for instance, the subfitness. Since we do not necessarily have here

[3]Recall that CC is weaker than Countably Dependent Choice.

pseudocomplements, the relation "rather below" has to be understood as defined by

$$a \prec b \quad \equiv_{\mathrm{df}} \quad \exists x,\ a \wedge x = 0 \text{ and } x \vee b = 1.$$

We then say that x *witnesses* $a \prec b$.

6.3.1 σ-Regular σ-frames A σ-frame is *σ-regular* if, for each $a \in L$, there is a sequence $(a_n)_{n\in\mathbb{N}}$ in L with $a_n \prec a$ and $a = \bigvee_{n=1}^{\infty} a_n$. We shall denote by **RegσFrm** the category of σ-regular σ-frames.

6.3.2 Notes

1. Often the σ-frame we will deal with will be actually a frame. A σ-regular frame is obviously regular as a frame, but regularity does not imply σ-regularity. In the literature, when speaking on σ-frames, one sometimes uses the expression "regular" meaning "σ-regular". Usually it does not create confusion, but one has to be careful.
2. It is an easy exercise to show that σ-regularity implies subfitness even if the σ-frame in question is not a frame.

6.3.3 Lemma *Let L be a σ-regular σ-frame. For every $a, b \in L$ there are $u, v \in L$ such that $a \vee u = a \vee b = v \vee b$ and $u \wedge v = 0$.*

Proof By σ-regularity, $a = \bigvee_{n=1}^{\infty} a_n$ and $b = \bigvee_{n=1}^{\infty} b_n$ with $a_n \prec a$ and $b_n \prec b$ for all n (and we may assume that $a_n \leqslant a_{n+1}$ and $b_n \leqslant b_{n+1}$). Hence for each n we have u_n and v_n such that $a_n \wedge u_n = 0$, $u_n \vee a = 1$, $b_n \wedge v_n = 0$ and $v_n \vee b = 1$. Let

$$u = \bigvee_{n=1}^{\infty} (b_n \wedge u_n) \quad \text{and} \quad v = \bigvee_{n=1}^{\infty} (a_n \wedge v_n).$$

Then

$$a \vee u = a \vee \bigvee_{n=1}^{\infty} (b_n \wedge u_n) = \bigvee_{n=1}^{\infty} (a \vee (b_n \wedge u_n)) = \bigvee_{n=1}^{\infty} (a \vee b_n) = a \vee b.$$

Likewise $v \vee b = a \vee b$, and finally

$$u \wedge v = \bigvee_{n=1}^{\infty} \bigvee_{k=1}^{\infty} (u_n \wedge b_n \wedge a_k \wedge v_k) = 0$$

since $u_n \wedge a_k = 0$ if $n \geqslant k$ and $b_n \wedge v_k = 0$ if $k \geqslant n$. □

(It follows in particular that any σ-regular σ-frame is normal.[4])

6.3.4 Proposition *In any σ-regular σ-frame, $\prec$ is interpolative.*

[4]The reader has certainly observed a similarity with the well-known classical fact on Lindelöf regular spaces being normal—see VII.1.5.1 below.

Proof Let $a \prec b$ in a σ-regular σ-frame L. Then $a \wedge t = 0$ and $t \vee b = 1$ for some t. The latter condition implies, using the lemma, that there are u, v satisfying $t \vee u = 1 = b \vee v$ and $u \wedge v = 0$. Then $a \prec u$ (since $t \vee u = 1$ and $a \wedge t = 0$) and $u \prec b$ (since $b \vee v = 1$ and $u \wedge v = 0$). □

6.4 The σ-frame Coz (L) It follows immediately from the characterization (3) in 6.2.2 that

- Coz (L) is a σ-frame,
- it is a σ-regular sub-σ-frame of L and
- generates L iff L is completely regular.

6.4.1 Observation Let $x_0 \prec x_1 \prec x_2$ be witnessed by y_1 and y_2 (that is, $x_0 \wedge y_1 = 0$, $x_1 \vee y_1 = 1$, $x_1 \wedge y_2 = 0$ and $x_2 \vee y_2 = 1$). Then, immediately, $y_2 \prec y_1$ (witnessed by x_1).

As a consequence, if $x_0 \prec x_1 \ll x_2$ is witnessed by y_1, y_2, then $y_2 \ll y_1$.

6.4.2 Proposition *Let $a \ll b$ in a frame L. Then $a \prec b$ is witnessed by some $u \in$ Coz (L). In particular, if $a, b \in$ Coz (L), $a \ll b$ in L implies $a \prec b$ in* Coz (L).

Proof Consider a' such that $a \prec a' \ll b$ and x_1, x_2 witnessing it. Then $x_2 \ll x_1$ by 6.4.1, and by 6.2.3 there is a $c \in$ Coz (L) such that $x_2 \ll c \ll x_1$. Then $a \wedge u \leqslant a' \wedge u \leqslant a' \wedge x_1 = 0$ and $b \vee u \geqslant b \vee x_2 = 1$. □

6.4.3 Corollary *The σ-frame* Coz (L) *is σ-regular.*

Proof For any $a \in$ Coz (L), if $a = \bigvee_{n=1}^{\infty} x_n$ with $x_n \ll a$ for $n = 1, 2, \ldots$, take $c_n \in$ Coz (L) such that $x_n \ll c_n \ll a$. Then $a = \bigvee_{n=1}^{\infty} a_n$ and $a_n \prec a$ in Coz (L) by the preceding proposition. □

6.5 The converse to 1 in 6.2.3 does not hold as easy examples show. Next result provides the characterization of $\ll$ in terms of cozero elements.

6.5.1 Corollary *The following are equivalent for $a, b \in L$.*

(1) $a \ll b$.
(2) *There exist $c, d \in$* Coz (L) *such that $a \leqslant c$, $d \leqslant b$ and $c \prec d$ in* Coz (L).

Proof (1)⇒(2): Apply 6.2.3 twice to get $c, d \in$ Coz (L) such that

$$a \ll c \ll d \ll b.$$

The conclusion follows by the particular case in 6.4.2.

(2)⇒(1): The relation $\prec$ in Coz (L) is interpolative (by 6.3.4). Hence $c \prec d$ in Coz (L) implies $c \ll d$ in L. □

6.6 Cozeros in metrizable frames In [152], Isbell defined a frame to be *metrizable* if it admits a uniformity with a countable basis, that is, if there is an admissible uniformity $\mathcal{A}$ and A_n, $n = 1, 2, \ldots$, in $\mathcal{A}$ such that for every $A \in \mathcal{A}$ there is an $A_n \leqslant A$. Later on it has turned out that this is equivalent with a fairly geometric metrizability condition, namely with the existence of a *diameter* with natural properties following the properties of diameters in metric spaces (see, e.g., Chapter XI in [220]).

6.6.1 Theorem *In a metrizable frame every element is cozero.*

Proof Recall the formula $a = \bigvee_{A \in \mathcal{A}} a/A$ from 5.1.1 and consider a countable system A_n, $n = 1, 2, \ldots$, that generates $\mathcal{A}$. If $A \geqslant A_n$, then $a/A_n \geqslant a/A$ so that $a = \bigvee_{n=1}^{\infty} a/A_n$. Since $a/A_n \lhd_{\mathcal{A}} a$ we have by 5.3, $a/A_n \ll a$. Use (3) in 6.2.2. □

6.6.2 Note The property that all open sets are cozeros is in fact not much more general than metrizability. In spaces (and now also in frames) it is called *perfect normality*. By Vedenisoff's Theorem (see [93]) a space X is perfectly normal iff

each open set is a union of countably many closed ones,

or, iff

any two disjoint closed sets A, B can be precisely divided by a continuous real function (that is, a continuous $f \colon X \to \mathbb{R}$ such that $f^{-1}[\{0\}] = A$ and $f^{-1}[\{1\}] = B$).

There are examples of non-metrizable perfectly normal spaces, even of compact ones. Such is the Alexandroff's *double-arrow space*, also called *split interval* (see [93]).

Perfect normality in the point-free context will be discussed in Sect. 1 of Chap. VIII below.

7 The Regular Lindelöf Reflection

By a 1977 result of Dowker and Strauss [84], the Lindelöf property is preserved under coproducts of regular frames. This is somewhat surprising since (even binary) products of (regular) Lindelöf spaces need not be Lindelöf. As we will see in this section, one even has a Lindelöf coreflection for frames, constructed by Schlitt in [246].

7.1 The functor $\mathcal{H}$ Let A be a σ-frame. A *σ-ideal* of A is an ideal closed under countable joins. Denote by

$$\mathcal{H}(A)$$

the set of all σ-ideals in A ordered by inclusion.

Proposition *$\mathcal{H}(A)$ is a Lindelöf frame. It is completely regular whenever A is (completely) σ-regular.*

Proof Any intersection of σ-ideals is clearly a σ-ideal and thus $\mathcal{H}(A)$ is a complete lattice (with meets given by intersection). That $\mathcal{H}(A)$ is in fact a frame follows from the following description of joins in $\mathcal{H}(A)$: $a \in \bigvee_{i\in J} I_i$ iff $a = \bigvee_{n=1}^{\infty} a_n$, where $a_n \in I_{i_n}$ for suitable i_n. Actually, this implies that the elements of a meet $I \cap \bigvee_{i\in J} I_i$ are of the form

$$b \wedge \bigvee_{n=1}^{\infty} a_n = \bigvee_{n=1}^{\infty} (b \wedge a_n),$$

where $b \in I$ and $a_n \in I_{i_n}$ for some i_n, and the latter obviously belongs to $\bigvee_{i\in J}(I \cap I_i)$.

For Lindelöfness, if $\bigvee \mathcal{S} = \downarrow 1$ in $\mathcal{H}(A)$, then $1 = \bigvee S$ for some countable subset S of the ideal generated by $\mathcal{S}$. Then S is already contained in the ideal generated by some countable subset $\mathcal{T}$ of $\mathcal{S}$ so that $\bigvee \mathcal{T} = \downarrow 1$.

Finally, for complete regularity, observe first that $a \prec b$ in A trivially implies $\downarrow a \prec \downarrow b$ in $\mathcal{H}(A)$, and the same then holds for $a \prec\prec b$. Now, by the (complete) regularity of A, $a = \bigvee_{n=1}^{\infty} a_n$, where $a_n \prec\prec a$ for $n \in \mathbb{N}$, for any $a \in A$. Hence also $\downarrow a = \bigvee_{n=1}^{\infty} \downarrow a_n$ in $\mathcal{H}(A)$, where $\downarrow a_n \prec\prec \downarrow a$, and since the $\downarrow a$'s generate $\mathcal{H}(A)$ the proof is finished. □

The correspondence $A \mapsto \mathcal{H}(A)$ is functorial. One can define

$$\mathcal{H}(h)\colon \mathcal{H}(A) \to \mathcal{H}(B)$$

for an $h\colon A \to B$ in $\boldsymbol{\sigma}$-**Frm** as the frame homomorphism that takes a σ-ideal I in A to the σ-ideal generated by $h[I]$. By the proposition, it restricts to a functor from **Reg$\boldsymbol{\sigma}$Frm** to **CRegFrm**.

7.2 The Lindelöfication of a completely regular frame The correspondence $L \mapsto \mathrm{Coz}\,(L)$ is clearly functorial.

We know already from 6.4.3 that $\mathrm{Coz}\,(L)$ is a σ-regular σ-frame. Hence we have a functor Coz : **CRegFrm** → **Reg$\boldsymbol{\sigma}$Frm**.

7.2.1 Theorem *The functor* $\mathcal{H}$: **Reg$\boldsymbol{\sigma}$Frm** → **CRegFrm** *is left adjoint to* Coz : **CRegFrm** → **Reg$\boldsymbol{\sigma}$Frm**, *with adjunction maps*

$$\eta_A\colon A \to \mathrm{Coz}\,\mathcal{H}(A),\ a \mapsto \downarrow a$$

and

$$\varepsilon_L\colon \mathcal{H}(\mathrm{Coz}\,(L)) \to L,\ I \mapsto \bigvee I \quad (\textit{with join taken in } L).$$

Moreover, η_A *is an isomorphism for all* A, *and* ε_L *is an isomorphism iff* L *is Lindelöf.*

Proof $\mathcal{H}(A)$ is a completely regular Lindelöf frame. Hence η_A really maps A into $\mathrm{Coz}\,(\mathcal{H}(A))$ since $\mathrm{Coz}\,(\mathcal{H}(A))$ consists of the Lindelöf elements of $\mathcal{H}(A)$ (by 6.2.4), and these are obviously the principal ideals $\downarrow a$.

It is clear that η_A is a σ-frame homomorphism and ε_L is a frame homomorphism, and that they are natural in A and L, respectively, so it remains to check the adjunction identities:

$$\varepsilon_{\mathcal{H}(A)}\mathcal{H}(\eta_A)(\downarrow a) = \varepsilon_{\mathcal{H}(A)}\big(\{I \in \mathrm{Coz}\,(\mathcal{H}(A)) \mid I \subseteq \downarrow a\}\big) = \downarrow a$$

for any $a \in A$, and

$$\mathrm{Coz}\,(\varepsilon_L)\eta_{\mathrm{Coz}\,(L)}(c) = \mathrm{Coz}\,(\varepsilon_L)(\downarrow c) = \varepsilon_L(\downarrow c) = \bigvee(\downarrow c) = c$$

for any $c \in \mathrm{Coz}\,(L)$.

Concerning the final part of the proposition, it is clear that η_A is an isomorphism and L is Lindelöf whenever ε_L is an isomorphism (because any $\mathcal{H}(A)$ is Lindelöf). Regarding the converse of the latter, it suffices to show that ε_L is codense whenever L is Lindelöf (by V.4.1.1). For any σ-ideal I of $\mathrm{Coz}\,(L)$, if $\bigvee I = 1$, then $\bigvee S = 1$ for some countable $S \subseteq I$ but then $1 = \bigvee S \in I$ and thus $\bigvee I = \downarrow 1$. □

Corollary 7.2.2 (The regular Lindelöf (co)reflection) *The category of completely regular Lindelöf frames is coreflective in* **CRegFrm** *with coreflection functor* $\mathcal{H}\mathrm{Coz}$ *and coreflection map* $\varepsilon_L \colon \mathcal{H}(\mathrm{Coz}\,(L)) \to L$ *given by join.*

Proof For any $h \colon L \to M$ in **Frm**, with L completely regular Lindelöf, ε_L is an isomorphism and therefore $\varepsilon_M \cdot \mathcal{H}\mathrm{Coz}\,(h) \cdot \varepsilon_L^{-1} = h$.

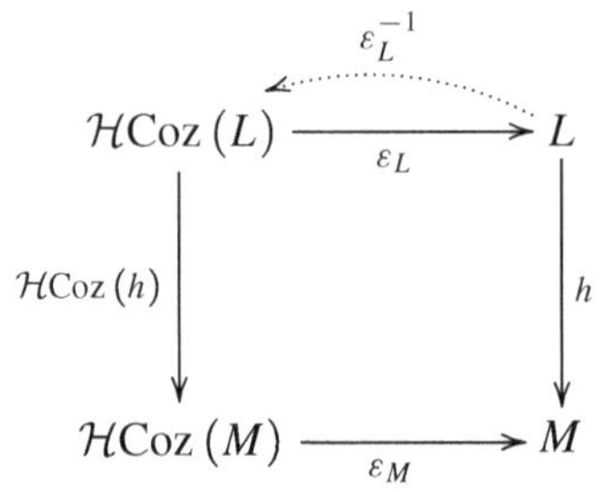

Hence h factors through ε_M.

Moreover this factorization is unique. Indeed, take any $g \colon L \to \mathcal{H}\mathrm{Coz}\,(M)$ such that $\varepsilon_M g = h$. For any $a \in \mathrm{Coz}\,(L)$, $g(a)$ is a cozero element of $\mathcal{H}\mathrm{Coz}\,(M)$, hence a principal ideal, say $g(a) = \downarrow b$ for some $b \in \mathrm{Coz}\,(M)$. But then $h(a) = \varepsilon_M g(a) = \varepsilon_M(\downarrow b) = b$, hence

$$g(a) = \downarrow h(a) = \mathcal{H}\mathrm{Coz}\,(h) \cdot \varepsilon_L^{-1}(a).$$

Since $\mathrm{Coz}\,(L)$ generates L (recall 6.4), $g = \mathcal{H}\mathrm{Coz}\,(h) \cdot \varepsilon_L^{-1}$. □

7.2.3 Notes

1. This corollary is a surprising result that transcends classical topology: there is no Lindelöf reflection in the category of Tychonoff spaces. This follows from the fact that any uncountable product of copies of $\mathbb{R}$ is not normal and hence not Lindelöf because regular Lindelöf spaces are paracompact and thus normal.
2. It also follows from 7.2.1 that

   Coz *and* $\mathcal{H}$ *induce an equivalence between the category of completely regular Lindelöf frames and* ***Regσ-Frm***

   and

   for completely regular frames L*, the* σ*-frames* $\mathrm{Coz}\,(L)$ *are precisely the* σ*-regular* σ*-frames.*

3. The fact that the category of completely regular Lindelöf frames is coreflective in **CRegFrm** implies, in particular, that

Corollary *The coproduct of a family of completely regular Lindelöf frames is Lindelöf.*

This is another surprising result that does not have counterpart in classical topology: Lindelöf property is not preserved under (even binary) products of spaces [265]. But also surprising is the fact that, unlike the case with the standard Tychonoff's Theorem for frames (see 9.2 below), one cannot get rid of some choice principle in the (co)product theorem for Lindelöf frames (it lies close to the axiom of countable choice, as proved by Schlitt in [246]).

8 Choice-Free Complete Regularity

In this section, following [53], we will show that in the point-free approach to complete regularity all choice principles can be avoided. This can be done by working with the largest interpolative relation $\prec_\circ$ contained in $\prec$ instead of $\prec\!\!\prec$ (under CDC $\prec_\circ$ and $\prec\!\!\prec$ coincide, in general not).

8.1 The interpolative part of a relation The union of any set of interpolative relations is clearly interpolative, and hence any binary relation R on a frame L contains a largest interpolative relation, the *interpolative part* $R_\circ$ of R.

Let R_∞ be the relation on L defined by

$$aR_\infty b \equiv_{\mathrm{df}} \quad \exists (c_{n,k})_{n=0,1,\dots;\ k=0,1,\dots,2^n} \quad \text{such that}$$

$$c_{0,0} = a, \quad c_{0,1} = b, \quad c_{n,k} = c_{n+1,2k} \quad \text{and} \quad c_{n,k} R c_{n,k+1}$$

for all $n = 0, 1, \dots$ and $k = 0, 1, \dots, 2^n$.

Clearly $R_\infty \subseteq R$ (case $n = 0$) and R_∞ is interpolative (since $aR_\infty b$ implies $aR_\infty c_{11} R_\infty b$). Hence

$$R_\infty \subseteq R_\circ.$$

On the other hand, if CDC is assumed, successive choices of interpolating elements show that $S_\infty = S$ for any interpolative relation S. Hence, for any relation R,

$$R_\circ = (R_\circ)_\infty \subseteq R_\infty$$

so that $R_\circ = R_\infty$.

This suggests to view $R_\circ$ as the choice-free version of R_∞.

8.1.1 Proposition *Let $h: L \to M$ be a frame homomorphism. If R is a relation on L and S a relation on M such that $h[R] \subseteq S$, then $h[R_\circ] \subseteq S_\circ$.*

Proof It suffices to check that the relation

$$h[R_\circ] = \{(h(a), h(b)) \mid aR_\circ b\} \subseteq S$$

is interpolative. □

In particular, for the rather below relation (recall V.1.2.1) we have

$$a \prec b \Rightarrow h(a) \prec h(b)$$

and hence we obtain

8.1.2 Corollary *For any frame homomorphism $h: L \to M$,*

$$a \prec_\circ b \Rightarrow h(a) \prec_\circ h(b).$$

8.2 Entailments A binary relation on a frame L is an *entailment* if

1. $aRb \Rightarrow a \leqslant b$,
2. $a' \leqslant aRb \leqslant b' \Rightarrow a'Rb'$,
3. R is a (bounded) sublattice of $L \times L$ and
4. $aRb \Rightarrow b^* Ra^*$.

8.2.1 Proposition *The interpolative part of any entailment is an entailment.*

Proof Let R be an entailment. For any interpolative $S \subseteq R$, the relation

$$T = \{(a, b) \mid a \leqslant a' \text{ and } b \geqslant b' \text{ for some } a'Sb'\} \subseteq R$$

is also interpolative. Indeed, if aTb with $a \leqslant a'$, $a'Sb'$ and $b' \leqslant b$, then $a'ScSb'$ for some c and hence $aTcTb$. Moreover,

$$U = \{(a, b) \mid a \leqslant a_1 \wedge a_2,\ a_i S b_i,\ b_1 \wedge b_2 \leqslant b,\ \text{for some } a_1, a_2, b_1, b_2\} \subseteq R$$

is also interpolative: given aUb with $a \leqslant a_1 \wedge a_2$, $a_i S b_i$ and $b_1 \wedge b_2 \leqslant b$, there are c_1 and c_2 such that $a_i S c_i S b_i$ and hence $aU(c_1 \wedge c_2)Ub$. Likewise,

$$V = \{(a, b) \mid a \leqslant a_1 \vee a_2,\ a_i S b_i,\ b_1 \vee b_2 \leqslant b,\ \text{for some } a_1, a_2, b_1, b_2\} \subseteq R$$

is also interpolative. Finally,

$$W = \{(b^*, a^*) \mid aSb\} \subseteq R$$

is another interpolative relation (the fact that aSb implies $aScSb$ for some c says that $b^*Sc^*Sa^*$ whenever b^*Wa^*) and $\{(0,0), (1,1)\} \subseteq R$ is trivially interpolative. Hence, all the new relations above are contained in $R_\circ$. The case $S = R_\circ$ means that the properties of entailments above hold for $R_\circ$. Property 1 for $R_\circ$ follows trivially from the corresponding property for R. □

8.3 Constructive complete regularity The rather below relation $\prec$ on any frame is an entailment. By 8.2, $\prec_\circ$ is also an entailment. Given 8.1, $\prec_\circ$ is considered as the constructive form of $\prec\!\!\prec$ and a frame L is called *strongly regular* if

(sreg): $\forall a \in L,\ a = \bigvee\{x \in L \mid x \prec_\circ a\}$.

Of course, any regular frame L with interpolating $\prec$ is strongly regular. This is the case for compact regular L (recall V.1.3.1; in next chapter we will see—in VII.1.5—that this is true, more generally, for Lindelöf regular frames[5]). Hence, constructively,

any compact regular frame is strongly regular.

But, as it is well known, in the constructive setting, a compact regular frame need not be completely regular. Hence, constructively,

strong regularity is strictly weaker than complete regularity.

8.3.1 Notes

1. Recall Theorem 5.4. Avoiding the use of CDC there (which is used in the second part of the lemma), we may only conclude that, for any uniformity $\mathcal{A}$, $\lhd_{\mathcal{A}} \subseteq \prec_\circ$. Hence, the constructive content of 5.4 is that

   a frame L admits a uniformity iff it is strongly regular. $(*)$

[5] And even more generally for normal regular frames.

2. Since the existence of uniformities on frames is preserved by coproducts and homomorphic images, it follows from $(*)$ that

   the strongly regular frames form a coreflective subcategory of ***Frm***

   (the clue is to see that any frame has a largest subframe which has a uniformity, and this provides the coreflection, see [53, Cor. 2] for more information).

9 Choice-Free Compactification

9.1 Let L be a frame. Denote by

$$\mathfrak{J}(L)$$

the set of all ideals in L ordered by inclusion.

9.1.1 Proposition *$\mathfrak{J}(L)$ is a compact frame.*

Proof Clearly, any intersection of ideals is an ideal. Moreover, any system $I_i, i \in J$, of ideals has a supremum in $\mathfrak{J}(L)$, namely

$$\bigvee_{i\in J} I_i = \{\bigvee F \mid F \text{ finite}, \; F \subseteq \bigcup_{i\in J} I_i\}.$$

Indeed, this set is an ideal (it is closed under binary joins and if $x \leqslant \bigvee F$, then by distributivity $x = \bigvee\{a \wedge x \mid a \in F\}$ with $\{a \wedge x \mid a \in F\} \subseteq \bigcup_{i\in J} I_i$) containing all the I_i, and that if K is an ideal containing all the I_i, then necessarily $\bigvee F \in K$ for any finite $F \subseteq \bigcup_{i\in J} I_i$.

Now if K, I_i are ideals, then trivially $K \cap (\bigvee_{i\in J} I_i) \supseteq \bigvee_{i\in J}(K \cap I_i)$, and if $F \subseteq \bigcup_{i\in J} I_i$ and $x = \bigvee F$ is in K, then $F \subseteq \bigcup_{i\in J}(I_i \cap K)$ and $x \in \bigvee_{i\in J}(K \cap I_i)$.

Finally, $\mathfrak{J}(L)$ is compact. Indeed, if $\bigvee I_i = L = 1_{\mathfrak{J}(L)}$, then, in particular, $1 \in \bigvee I_i$ and thus $1 = \bigvee F$ for some finite $F \subseteq \bigcup I_i$. But then $F \subseteq \bigcup_{i\in J_0} I_i$ for some finite $J_0 \subseteq J$, hence $1 \in \bigvee_{i\in J_0} I_i$ and $L = \bigvee_{i\in J_0} I_i$. □

9.1.2 Regular ideals An ideal I is said to be *regular* if

$$\text{for every } a \in I, \text{ there is a } b \in I \text{ such that } a \prec_\circ b. \tag{R}$$

Proposition *The set*

$$\mathfrak{R}(L)$$

of regular ideals in L is a subframe of $\mathfrak{J}(L)$. Hence it is a compact frame.

Proof Let I_1, I_2 be regular ideals and let $a \in I_1 \cap I_2$. There are $b_i \in I_i$ such that $a \prec_\circ b_i$. Then, by 8.2, $a \prec_\circ b_1 \wedge b_2 \in I_1 \cap I_2$.

Regarding joins, let I_i, $i \in J$, be regular, $F \subseteq \bigcup_{i \in J} I_i$ finite. For $x \in F$ choose $i(x)$ such that $x \in I_{i(x)}$ and an $x' \in I_{i(x)}$ such that $x \prec_\circ x'$. Set $F' = \{x' \mid x \in F\}$. Then $F' \subseteq \bigcup_{i \in J} I_i$ and by 8.2, $\bigvee F \prec_\circ \bigvee F'$. □

9.1.3 The regular ideals $\sigma(a)$ For $a \in L$ set

$$\sigma(a) = \{x \mid x \prec_\circ a\}.$$

Obviously

each $\sigma(a)$ is a regular ideal.

Lemma *If $a \prec_\circ b$, then $\sigma(a) \prec \sigma(b)$.*

Proof Interpolate $a \prec_\circ x \prec_\circ y \prec_\circ b$. Then $y \in \sigma(b)$ and, by 8.2, $x^* \in \sigma(a^*)$. Thus, $1 = x^* \vee y \in \sigma(a^*) \vee \sigma(b)$ and $\sigma(a^*) \vee \sigma(b) = L = 1_{\mathfrak{J}(L)}$. If $x \in \sigma(a^*) \cap \sigma(a)$, then $x \leqslant a^* \wedge a = 0$, hence $x = 0$, and $\sigma(a^*) \cap \sigma(a) = \{0\} = 0_{\mathfrak{J}(L)}$. Use V.1.1 to obtain the statement. □

9.1.4 Proposition *If L is strongly regular, then $\mathfrak{R}(L)$ is strongly regular.*

Proof As $\mathfrak{R}(L)$ is compact it suffices to prove it is regular (by 8.3). For an ideal I we have, by the interpolation property of $\prec_\circ$,

$$I = \bigcup\{\sigma(a) \mid a \in I\} = \bigvee\{\sigma(a) \mid a \in I\}$$

and since L is strongly regular, each a is the join $\bigvee\{b \mid b \prec_\circ a\}$ and hence

$$\sigma(a) = \bigcup\{\sigma(b) \mid b \prec_\circ a\} = \bigvee\{\sigma(b) \mid b \prec_\circ a\}.$$

By the lemma, $I = \bigcup\{\sigma(b) \mid b \prec_\circ a,\ a \in I\} \subseteq \bigvee\{K \mid K \prec I\}$. □

9.1.5 The dense embedding $\sigma : L \to \mathfrak{R}(L)$ Let L be a strongly regular frame and consider the mappings

$$\upsilon = (I \mapsto \bigvee I) \colon \mathfrak{R}(L) \to L \quad \text{and} \quad \sigma = (a \mapsto \sigma(a)) \colon L \to \mathfrak{R}(L).$$

We obviously have

$$\upsilon\sigma(a) = a \quad \text{and} \quad I \subseteq \sigma\upsilon(I) \tag{$*$}$$

and hence these maps are adjoint, υ to the left and σ to the right; hence

υ preserves joins.

It preserves finite meets as well:

$$\begin{aligned}\upsilon(I_1) \wedge \upsilon(I_2) = \bigvee I_1 \wedge \bigvee I_2 = \bigvee\{a_1 \wedge a_2 \mid a_i \in I_i\} \leqslant \\ \leqslant \bigvee\{a \mid a \in I_1 \cap I_2\} = \upsilon(I_1 \cap I_2) \leqslant \upsilon(I_1) \wedge \upsilon(I_2).\end{aligned}$$

Hence, υ is a frame homomorphism, and σ is a localic map.

Proposition *σ is a dense localic embedding.*

Proof It is obviously one-to-one and $\sigma(0) = \downarrow 0 = \{0\}$. □

9.1.6 The functor $\mathfrak{R}$ We are now ready to define a functor

$$\mathfrak{R}\colon \mathbf{SRegFrm} \to \mathbf{KRegFrm}$$

from the category of strongly regular frames to the category of compact regular frames by mapping an $L \in \mathbf{SRegFrm}$ to $\mathfrak{R}(L)$ and a frame homomorphism $h\colon L \to M$ to

$$\mathfrak{R}(h)\colon \mathfrak{R}(L) \to \mathfrak{R}(M)\colon I \mapsto \downarrow h[I].$$

Indeed, the set $h[I]$ is obviously closed under finite meets, hence $\downarrow h[I]$ is an ideal, and it follows from 8.1.2 that $\downarrow h[I]$ is a regular ideal. Moreover, $\mathfrak{R}(h)$ is a frame homomorphism:

(a) Clearly,

$$\mathfrak{R}(h)(I_1 \cap I_2) = \downarrow h[I_1 \cap I_2] \subseteq \downarrow h[I_1] \cap \downarrow h[I_2] = \mathfrak{R}(h)(I_1) \cap \mathfrak{R}(h)(I_2).$$

Conversely, for any $y \in \downarrow h[I_1] \cap \downarrow h[I_2]$, there are $x_i \in I_i$ $(i = 1, 2)$ such that $h(x_1) = h(x_2) = y$ so that $h(x_1 \wedge x_2) = y$; since $x_1 \wedge x_2 \in I_1 \cap I_2$ then $y \in \downarrow h[I_1 \cap I_2]$.

(b) Finally,

$$\begin{aligned}\mathfrak{R}(h)(\bigvee_{i \in J} I_i) &= \downarrow h[\{\bigvee F \mid F \text{ finite}, F \subseteq \bigcup_{i \in J} I_i\}] \\ &= \downarrow\{\bigvee F \mid F \text{ finite}, F \subseteq \bigcup_{i \in J} h[I_i]\} \\ &= \{\bigvee F \mid F \text{ finite}, F \subseteq \bigcup_{i \in J} \downarrow h[I_i]\} \\ &= \bigvee_{i \in J} \mathfrak{R}(h)(I_i).\end{aligned}$$

If we specify the L in the definition above writing $\upsilon_L : \mathfrak{R}(L) \to L$, we obtain a natural transformation $\upsilon : \mathfrak{R} \stackrel{\cdot}{\to} \mathrm{Id}$. Indeed, the diagrams

$$\begin{array}{ccc} \mathfrak{R}(L) & \xrightarrow{\upsilon_L} & L \\ {\scriptstyle \mathfrak{R}(h)}\downarrow & & \downarrow{\scriptstyle h} \\ \mathfrak{R}(M) & \xrightarrow{\upsilon_M} & M \end{array}$$

commute: $\upsilon_M \mathfrak{R}(h)(I) = \bigvee {\downarrow} h[I] = \bigvee h[I] = h(\bigvee I) = h\upsilon_L(I)$.

9.1.7 $\mathfrak{R}(L)$ as a compactification

Proposition *Let L be compact. Then υ and σ are mutually inverse isomorphisms and $L \cong \mathfrak{R}L$.*

Proof Let $x \in \sigma\upsilon(I)$. Thus, in particular, $x \prec \bigvee I$, and $x^* \vee \bigvee I = 1$. By compactness there are $y_1, \dots, y_n \in I$ such that $x^* \vee y_1 \vee \cdots \vee y_n = 1$. But $y = y_1 \vee \cdots \vee y_n \in I$ and we have $x \prec y$, hence $x \in I$. Thus, $\sigma\upsilon(I) \subseteq I$ and the statement follows from $(*)$ in 9.1.5. □

Summing up, we have:

Theorem (Stone–Čech compactification) *There is a functor*

$$\mathfrak{R} \colon \mathbf{SRegLoc} \to \mathbf{SRegLoc}$$

and a natural transformation $\sigma : \mathrm{Id} \stackrel{\cdot}{\longrightarrow} \mathfrak{R}$ such that

1. *each $\mathfrak{R}(L)$ is (regular and) compact,*
2. *each σ_L is a dense localic embedding and*
3. *σ_L is an isomorphism iff L is compact.* □

Corollary *The category* **KRegLoc** *of compact regular locales is a reflective subcategory of the category* **SRegLoc** *of strongly regular locales, with $\mathfrak{R}$ the reflection functor and $\upsilon_L : \mathfrak{R}(L) \to L$ the reflective arrow at L.*

9.1.8 Note The construction above follows step by step the compactification procedure presented by Banaschewski and Mulvey in [43, 44], with the only difference that we work with strongly regular locales instead of the completely regular ones. On close scrutiny, the reader can observe that in the constructions and proofs there was never used any choice principle, and neither was used the excluded middle. Thus the facts are fully constructive. In the original proof in [43] the use of any choice principle and the excluded middle was avoided as well with the small proviso that the delimitation of the objects for which one could apply the result contains a concept flawed by the Countably Dependent Choice; in the new variant one does not need even that.

9.2 Choice-free product of compact frames Checking the proof of the standard fact that a reflective subcategory is closed under all limits we observe that also this is fully constructive (no choice principle, no excluded middle) and we can infer from 9.1.7 that

> *products of compact regular locales are compact, and this is a fully constructive fact.*

This is in fact true generally, that is, for arbitrary locales, but, as can be expected, the proof is much more difficult—see, e.g., [160, 180] or [24].

It is well known that Tychonoff's Theorem, the claim that all products of compact topological spaces are compact, is equivalent with the Axiom of Choice. The restriction to compact Hausdorff spaces is equivalent with an only slightly weaker choice principle, namely with the Boolean Ultrafilter Theorem. This may lead, at the first sight, to a suspicion that the explanation might be in the fact that the product of compact spaces as locales does not correspond to the product in **Top** (which does happen with more general spaces). But this is not the case: at least for the regular compact case the correspondence is precise: regular compact locales are always spatial (recall the Hofmann–Lawson duality, [150], or e.g. [220]), hence everything happens as if in spaces). The explanation is in the fact that the proof of this spatiality needs the axiom of choice. Thus, what happens without AC is that the product is compact anyway, but may not have enough points.

Chapter VII
Normality

Of the classical separation axioms, normality is the easiest to extend. There is, basically, nothing "pointy" about it. This however does not mean that there is not much interest about it in the extended context. On the contrary. Besides the new view one gains of the plain normality itself and of its relations to the other axioms one has natural strengthenings (and, in a smaller extent also weakenings) that are not so obviously point-free and the behaviour of which is of an independent interest.

In this chapter we start with plain normality and its basic properties. Among the axioms of the T_k-sequence it stands aside not only because of the worse behaviour with respect to constructions (in particular, normality is not hereditary, which will be discussed in a special section later). A less conspicuous distinction is that it does not follow the implication pattern (recall that in this book T_0 is automatically assumed and hence we have $T_1 \Leftarrow T_2 \Leftarrow T_3 \Leftarrow T_{3\frac{1}{2}}$); normality does not imply the lower ones, even with the help of T_0. One needs the (ubiquitous) subfitness, which is one of the first (very simple) facts introduced, together with the influence of normality on the relation $\prec$.

Next, we discuss the behaviour of finite covers (the extremely interesting case of general covers is mentioned only briefly: the associated *full normality*, or *paracompactness*, is given a special chapter in our preceding monograph [220]).

We have already mentioned (at least twice) the pioneering Wallman's article [282] in connection with subfitness and the origins of point-free thinking. In this chapter we will discuss another aspect of this paper, namely the fact that under normality the introduced compactification, even extended to the general point-free spaces, coincides with the (point-free) Stone–Čech compactification (VI.9).

In the next chapter we will introduce, besides some more involved facts, also some conditions stronger than normality. One of them, however, the *complete normality* (the hereditary variant of normality), is still included in the current one.

J. Picado, A. Pultr, *Separation in Point-Free Topology*,
https://doi.org/10.1007/978-3-030-53479-0_7

1 Normal Frames

1.1 Recall from I.5.5 (and 4.3.1) the definition: a frame L is *normal* if

(norm): *for every* $a, b \in L$ *such that* $a \vee b = 1$
there are $u, v \in L$ *such that* $a \vee u = b \vee v = 1$ *and* $u \wedge v = 0$.

(Equivalently, $a \vee b = 1$ implies the existence of $u \in L$ such that $a \vee u = 1 = b \vee u^*$.)

We immediately obtain

Proposition *A space X is normal iff the frame $\Omega(X)$ is normal.*

1.2 Proposition *In a normal frame the relation $\prec$ interpolates.*

Proof If $a \prec b$, we have $a^* \vee b = 1$. Hence there is a u such that $a^* \vee u = 1$ and $u^* \vee b = 1$, that is, $a \prec u \prec b$.

1.2.1 Note In fact, the interpolativeness of $\prec$ can be reformulated as a weaker form of normality. See 1.3.2 below.

1.3 Lemma *A normal subfit frame is regular.*

Proof Suppose not. Then there is an a such that $a \nleqslant b = \bigvee\{x \mid x \prec a\}$ and by subfitness, $a \vee c = 1 \neq b \vee c$ for some c. Now by normality there is a u such that $a \vee u^* = 1 = u \vee c$. Then, however, $u \leqslant b$, and $1 = u \vee c \leqslant b \vee c$, a contradiction. □

1.3.1 Corollary *A normal subfit frame is completely regular.*

(It is regular, and since $\prec$ interpolates, $\prec = \prec\!\prec$.)

1.3.2 Note A closer look to the proof of 1.2 above reveals that $\prec$ already interpolates in the broader class of *almost normal* frames.[1] In fact, the converse is also true and the almost normality condition is a characterization of the interpolativeness of $\prec$: if $\prec$ interpolates and $a^* \vee b = 1$, then $a \prec b$ and $a \prec u \prec b$ for some u; then $a^* \vee u = 1$ and $u^* \vee b = 1$.

Then, similar to the implication (norm) & (sfit) ⇒ (reg) we have that

in any almost normal and subfit frame,

$$a^* = \bigvee\{x \mid x \prec a^*\} \quad \textit{for every } a \in L.$$

[1]A frame is *almost normal* whenever the (norm) condition holds for elements a, b at least one of which is regular (recall that an element x of a frame is *regular* if $x^{**} = x$).

1.4 Here are two useful characterizations of normal frames.

Proposition *The following statements about a frame L are equivalent.*

(1) *L is normal.*
(2) *For any $(a_i)_{i=1}^{\infty}$ and $(b_i)_{i=1}^{\infty}$ in L such that*

$$a_i \vee (\bigwedge_{j=1}^{\infty} b_j) = 1 \quad \textit{and} \quad b_i \vee (\bigwedge_{j=1}^{\infty} a_j) = 1 \quad (i = 1, 2, \ldots)$$

there is a $u \in L$ such that $a_i \vee u = 1$ and $b_i \vee u^* = 1$ $(i = 1, 2, \ldots)$.
(3) *For any $a, b \in L$ such that $a \vee b = 1$, there is a countable $(u_i)_{i=1}^{\infty} \subseteq L$ such that*

$$\bigvee_{i=1}^{\infty} (a \vee u_i) = 1 \quad \textit{and} \quad \bigwedge_{i=1}^{\infty} (b \vee u_i^*) = 1.$$

Proof (1)⇒(2): Let L be a normal frame. Then, for each i, $a_i \vee (\bigwedge_{j=1}^{\infty} b_j) = 1$ implies by normality the existence of $u_i \in L$ satisfying

$$a_i \vee u_i = 1 \quad \text{and} \quad (\bigwedge_{j=1}^{\infty} b_j) \vee u_i^* = 1.$$

Similarly, from $b_i \vee (\bigwedge_{j=1}^{\infty} a_j) = 1$ it follows that there is some $v_i \in L$ such that $b_i \vee v_i = 1$ and $(\bigwedge_{j=1}^{\infty} a_j) \vee v_i^* = 1$. Then we have, for each i,

$$a_i \vee (\bigwedge_{k=1}^{i} v_k^*) \geqslant (\bigwedge_{j=1}^{\infty} a_j) \vee (\bigwedge_{k=1}^{i} v_k^*) = \bigwedge_{k=1}^{i} ((\bigwedge_{j=1}^{\infty} a_j) \vee v_k^*) = 1.$$

Similarly, $b_i \vee (\bigwedge_{k=1}^{i} u_k^*) - 1$. Now set

$$u_i' = u_i \wedge \bigwedge_{k=1}^{i} v_k^*, \quad v_i' = v_i \wedge \bigwedge_{k=1}^{i} u_k^*, \quad u = \bigvee_{i=1}^{\infty} u_i' \text{ and } v = \bigvee_{i=1}^{\infty} v_i'.$$

Clearly,

$$a_i \vee u \geqslant a_i \vee u_i' = (a_i \vee u_i) \wedge \big(a_i \vee (\bigwedge_{k=1}^{i} v_k^*)\big) = 1$$

and, similarly, $b_i \vee v = 1$. Moreover

$$u \wedge v = \bigvee_{i=1}^{\infty} \bigvee_{j=1}^{\infty} (u_i' \wedge v_j') = \bigvee_{i=1}^{\infty} \bigvee_{j=1}^{\infty} \Big(u_i \wedge v_j \wedge \bigwedge_{k=1}^{i} v_k^* \wedge \bigwedge_{l=1}^{j} u_l^*\Big) = 0.$$

Hence $v \leqslant u^*$ and thus $b_i \vee u^* \geqslant b_i \vee v = 1$.

(2)⇒(3): This is obvious. In fact, if $a \vee b = 1$, then there is a $u \in L$ such that $a \vee u = b \vee u^* = 1$; set $u_i = u$ for all $i \in \mathbb{N}$.

(3)⇒(1): Let $a \vee b = 1$. There are $(u_i)_{i=1}^{\infty}$ and $(v_i)_{i=1}^{\infty}$ such that

$$\bigvee_{i=1}^{\infty}(a \vee u_i) = 1 = \bigwedge_{i=1}^{\infty}(b \vee u_i^*) \quad \text{and} \quad \bigvee_{i=1}^{\infty}(b \vee v_i) = 1 = \bigwedge_{i=1}^{\infty}(a \vee v_i^*).$$

Set

$$u = \bigvee_{i=1}^{\infty}\Big(u_i \wedge \bigwedge_{j\leqslant i} v_j^*\Big) \quad \text{and} \quad v = \bigvee_{i=1}^{\infty}\Big(v_i \wedge \bigwedge_{j\leqslant i} u_j^*\Big).$$

Then

$$a \vee u = \bigvee_{i=1}^{\infty}\Big((a \vee u_i) \wedge \Big(a \vee \bigwedge_{j\leqslant i} v_j^*\Big)\Big) = \bigvee_{i=1}^{\infty}(a \vee u_i) = 1.$$

Similarly, $b \vee v = 1$. Finally,

$$u \wedge v = \bigvee_{i=1}^{\infty}\bigvee_{j=1}^{\infty}\Big(u_i \wedge \bigwedge_{k\leqslant i} v_k^* \wedge v_j \wedge \bigwedge_{l\leqslant j} u_l^*\Big) = 0.$$ □

1.5 Normality and compactness Recall from VI.6.2.4 that a frame L is *Lindelöf* if every cover of L contains a countable subcover. Similarly like in spaces we have

1.5.1 Proposition *Any regular Lindelöf frame (in particular, every compact regular frame) is normal, hence (being subfit) completely regular.*

Proof Let $a \vee b = 1$. By regularity, $a = \bigvee\{x \mid x \prec a\}$ and $b = \bigvee\{y \mid y \prec b\}$. Using the Lindelöf property we can see that $\bigvee_{n=1}^{\infty}(x_n \vee y_n) = 1$ for some $x_n \prec a$ and $y_n \prec b$. By (6) in V.1.3 we may assume that $x_n \leqslant x_{n+1}$ and $y_n \leqslant y_{n+1}$ for all n. Set

$$u = \bigvee_{n=1}^{\infty}(x_n^* \wedge y_n) \quad \text{and} \quad v = \bigvee_{n=1}^{\infty}(x_n \wedge y_n^*).$$

Then we have

$$a \vee u = \bigvee_{n=1}^{\infty} a \vee (x_n^* \wedge y_n) = \bigvee_{n=1}^{\infty}(a \vee y_n) \geqslant \bigvee_{n=1}^{\infty}(x_n \vee y_n) = 1.$$

Similarly, $b \vee v = 1$. Moreover

$$u \wedge v = \bigvee_{n=1}^{\infty}\bigvee_{m=1}^{\infty}(x_n^* \wedge y_n \wedge x_m \wedge y_m^*) = 0$$

since $x_n^* \wedge x_m = 0$ if $n \geqslant m$ while $y_n \wedge y_m^* = 0$ if $n \leqslant m$. □

1.5.2 Recall from III.9.1.1 that any compact strongly Hausdorff frame is regular.

Corollary *Any compact strongly Hausdorff frame is normal, and hence completely regular.*

1.6 Mapping invariance theorem A localic map $f: L \to M$ is *closed*[2] if the image of any closed sublocale of L is a closed sublocale of M. Since f is monotone, $f(a)$ is obviously smallest in $f[\mathfrak{c}(a)]$. Thus, if $f[\mathfrak{c}(a)]$ should be closed, it must be the closed sublocale $\mathfrak{c}(f(a))$; that is, f is closed iff

$$\forall a \in L,\ f[\mathfrak{c}(a)] = \mathfrak{c}(f(a)).$$

This is equivalent to

$$\forall a \in L,\ \forall b \in M,\ f(a \vee f^*(b)) = f(a) \vee b. \tag{$*$}$$

(We always have $f(a \vee f^*(b)) \geqslant f(a) \vee ff^*(b) \geqslant f(a) \vee b$. If f is closed, then $f(a) \vee b = f(x)$ for some $x \geqslant a$ and, as $f(x) \geqslant b$, we have $x \geqslant f^*(b) \vee a$, and $f(a \vee f^*(b)) \leqslant f(x) = f(a) \vee b$. Conversely, if $x \geqslant f(a)$, then $x = f(a) \vee x = f(a \vee f^*(x))$.)

Note As an easy application it follows that the embedding $j: S \subseteq L$ is a closed map iff S is a closed sublocale.

Proposition *Let $f: L \to M$ be a surjective*[3] *localic map. If f is closed and L is normal, then M is also normal.*

Proof Let $a \vee b = 1$ in M. Then $f^*(a) \vee f^*(b) = 1$ and by normality there are $u, v \in L$ such that $u \wedge v = 0$ and $f^*(a) \vee u = 1 = f^*(b) \vee v$. Then, by $(*)$, $a \vee f(u) = f(f^*(a) \vee u) = f(1) = 1$ and similarly $b \vee f(v) = 1$. Moreover, $f(u) \wedge f(v) = f(0) = 0$ as f is dense. □

As a consequence,

the image of any normal locale under a closed localic map is normal.

This is the point-free version of the Hausdorff mapping invariance theorem [137, 283].

2 Normality and the System of Finite Covers

2.1 Let L be a normal frame. If $a_1 \vee a_2 = 1$, then there is, first, a u_1 such that $a_2 \vee u_1 = 1 = a_1 \vee u_1^*$. From the first identity we have, further, a u_2 such that

[2] Closed localic maps are the localic maps corresponding to closed continuous maps—see, e.g., [220].

[3] The reader will certainly notice that we use the density of f only; but closed dense maps are onto anyway.

$u_1 \vee u_2 = 1 = a_2 \vee u_2^*$. Thus we get, from a cover $\{a_1, a_2\}$, a cover $\{u_1, u_2\}$ with $u_i \prec a_i$. This holds for general finite covers.

2.1.1 Proposition *A frame L is normal iff for each finite system $\{a_i\}_{i=1}^n \subseteq L$ such that $\bigvee_{i=1}^n a_i = 1$ there are $u_i \prec a_i$ such that $\bigvee_{i=1}^n u_i = 1$.*

Proof Let $\bigvee_{i=1}^n a_i = 1$. By normality, there is a $u_1 \in L$ such that

$$\bigvee_{i=2}^{n} a_i \vee u_1 = 1 = a_1 \vee u_1^*.$$

The latter identity means that $u_1 \prec a_1$, whereas applying normality to the former yields a $u_2 \in L$ such that $a_2 \vee u_2^* = 1$, that is, $u_2 \prec a_2$, and

$$\bigvee_{i=3}^{n} a_i \vee u_1 \vee u_2 = 1.$$

Proceeding this way we get $u_1, u_2, \dots, u_{n-1} \in L$ such that

$$u_i \prec a_i \quad \text{and} \quad a_n \vee \bigvee_{i=1}^{n-1} u_i = 1.$$

Applying normality again we ultimately obtain a $u_n \in L$ such that $u_n \prec a_n$ and $\bigvee_{i=1}^n u_i = 1$. The converse implication is obvious. □

2.2 Recall again the notions and notations from II.7.2 and V.1.5 (covers, refinements, etc.). A cover A is said to be *normal* if there are covers A_n, $n = 1, 2, \dots$, such that

$$A = A_1 \quad \text{and} \quad A_{n+1}A_{n+1} \leqslant A_n \quad \text{for all } n.$$

2.2.1 Proposition *In a normal frame, each finite cover has a finite star refinement. Consequently, each finite cover is normal.*

Proof Let $A = \{a_1, a_2, \dots, a_n\}$ be a finite cover. By 2.1.1 pick a cover $B = \{b_1, b_2, \dots, b_n\}$ with $b_i \prec a_i$. Now $C_i = \{b_i^*, a_i\}$ $(i = 1, 2, \dots, n)$ are covers and we have a finite cover

$$C = B \wedge C_1 \wedge \cdots \wedge C_n.$$

We claim that $CC \leqslant A$. Indeed, pick $c, x \in C$ such that $x \wedge c \neq 0$. Since

$$c = b_i \wedge \bigwedge_{j=1}^{n} c_j \;\; (c_j \text{ either } b_j^* \text{ or } a_j) \quad \text{and} \quad x = b_k \wedge \bigwedge_{j=1}^{n} x_j \;\; (x_j \text{ either } b_j^* \text{ or } a_j)$$

for some $i, k \in \{1, 2, \dots, n\}$, then $x_i \neq b_i^*$, hence $x_i = a_i$ and $x \leqslant a_i$. □

2.2.2 Proposition *Normal frames are precisely the frames in which each finite cover is normal.*

Proof Let each finite cover have a star-refinement. In particular, if $a \vee b = 1$, there is a cover C with $CC \leqslant \{a, b\}$. Set

$$x = \bigvee\{c \in C \mid Cc \leqslant a\}, \quad y = \bigvee\{c \in C \mid Cc \leqslant b\} \quad \text{and} \quad u = y^*, \ v = u^*.$$

Then $Cx \leqslant a$, $Cy \leqslant b$ and hence, by V.1.5.1, $x^* \vee a = 1$ and $u \vee b = y^* \vee b = 1$. Moreover, $x^* = x^* \wedge (x \vee y) \leqslant y^{**} = v$ (since $x \vee y = 1$). Hence $a \vee v = 1$ and the frame is normal. □

2.3 Fully normal frames If not only the finite covers but all covers of a regular frame are normal one speaks of a

fully normal frame.

(Note that we have formally assumed just regularity; however, such a frame is normal by 2.2.2 and hence completely regular by 1.3.1.)

Notes

1. Full normality is equivalent to the well-known *paracompactness* (see [220, Thm. IX.2.3.4]). Let us recall the most standard definition of this property:
   A subset S of a frame L is *locally finite* if there is a cover A of L such that for each $a \in A$ there are only finitely many $s \in S$ such that $a \wedge s \neq 0$. A frame is *paracompact* if it is regular and each of its covers has a locally finite refinement.
   Paracompactness has several equivalent characterizations, full normality being one of them (for a point-free treatment of the variety see, e.g., [82, 174, 241, 270]). It is one of the most important stronger variants of normality (by the way, better behaving in the point-free context than in the classical setting). In [220] it was given a chapter, therefore we do not discuss it here in more detail.
2. All metrizable frames are known to be fully normal. In [174] it was shown in detail and as directly as possible how the existence of a countable admissible system of covers (the metrizability) provided an arbitrary cover with a star-refinement (the full normality). The point is in the existence of a (downwards) well-ordered admissible system of covers, countability being a special case.
3. Recall VI.5.2. Full normality is claiming precisely that the system of all covers is a uniformity.

3 The Wallman Compactification

3.1 Let L be an arbitrary frame. For $x, a \in L$, x is said to be *a-small* if for any $y \in L$,

$$x \vee y = 1 \Rightarrow a \vee y = 1.$$

3.1.1 For $a \in L$ set

$$\sigma(a) = \{x \mid x \text{ is } a\text{-small}\}.$$

Obviously

each $\sigma(a)$ is an ideal (containing a).

3.1.2 Here are some useful simple facts.

Proposition

(1) *If $a \leqslant b$, then $\sigma(a) \subseteq \sigma(b)$.*
(2) *$x \in \sigma(a)$ iff $\sigma(x) \subseteq \sigma(a)$.*
(3) *If $y \in \sigma(x \to a)$, then $x \wedge y \in \sigma(a)$.*

Proof (1) and (2) are obvious.
(3) Let $(x \wedge y) \vee z = 1$, that is, $x \vee z = 1 = y \vee z$. Then $(x \to a) \vee z = 1$ and we have $1 = (x \wedge (x \to a)) \vee z = (x \wedge a) \vee z$. Hence $a \vee z = 1$. □

3.2 A curious sublocale For each $a \in L$ let $s(a) = \bigvee \sigma(a)$ (the *saturation* of a). Trivially, $s(a) \geqslant a$.

Proposition *The set $L_s = \{a \in L \mid s(a) = a\}$ of saturated elements of L is a sublocale of L.*

Proof Let $a_i \in L_s$ $(i \in J)$. By (1) in 3.1.2, $s(\bigwedge_{j\in J} a_j) \leqslant s(a_i) = a_i$ for every i. Hence $s(\bigwedge_{i\in J} a_i) = \bigwedge_{i\in J} a_i$ and L_s is closed under arbitrary meets.
Furthermore, let $a \in L_s$ and $x \in L$. To show that $x \to a \in L_s$ we need to check that $s(x \to a) \leqslant x \to a$, that is, $x \wedge s(x \to a) \leqslant a$. The latter follows immediately from (3) in 3.1.2. □

The joins in L_s, that we denote by $\bigsqcup a_i$, are given by $s(\bigvee a_i)$.

3.3 Another characterization of subfitness Note that a frame L is subfit iff $a \nleqslant b$ implies that $a \notin \sigma(b)$, that is,

$$a \nleqslant b \;\Rightarrow\; \sigma(a) \nsubseteq \sigma(b).$$

3.3.1 Proposition *The following statements about a frame L are equivalent.*

(1) *L is subfit.*
(2) *$\sigma(a) = {\downarrow}a$ for each $a \in L$.*
(3) *$L = L_s$.*

Proof (1) $\Rightarrow$ (2): Since $\sigma(a)$ is an ideal containing a, trivially ${\downarrow}a \subseteq \sigma(a)$. The converse inclusion follows from subfitness by 3.3.
(2) $\Rightarrow$ (3) is obvious.
(3) $\Rightarrow$ (1): Assuming (3), if $a \in \sigma(b)$, then $a \leqslant s(b) = b$ and the subfitness of L follows by 3.3. □

3.4 Proposition *Let L be a compact frame. Then:*

(1) *for any $a \in L$, $s(a) \in \sigma(a) \cap L_s$,*
(2) $s(a) = 1 \;\Rightarrow\; a = 1$,
(3) *L_s is compact and*
(4) *L_s is subfit.*

Proof

(1) If $s(a) \vee y = 1$, then, by compactness, $x \vee y = 1$ for some $x \in \sigma(a)$ and hence $a \vee y = 1$. This shows that $s(a)$ is a-small. The fact that $s(a) \in L_s$ is a consequence of (2) in 3.1.2.
(2) Since L is compact and $\sigma(a)$ is an ideal, $s(a) = 1$ implies that $1 \in \sigma(a)$. Then, in particular, $1 \vee 0 = 1$ implies $a \vee 0 = 1$.
(3) Let $\{a_i\}_{i\in J} \subseteq L_s$ such that $\bigsqcup_{i\in J} a_i = s(\bigvee_{i\in J} a_i) = 1$. By (2), $\bigvee_{i\in J} a_i = 1$. Since L is compact, $\bigvee_{i\in F} a_i = 1$ for some finite $F \subseteq J$. Hence

$$\bigsqcup_{i\in F} a_i = s(\bigvee_{i\in F} a_i) = s(1) = 1.$$

(4) Let $a \nleqslant b$ in L_s. Then $\sigma(a) \nsubseteq \sigma(b)$, that is, there is an $x \in L$ which is a-small but not b-small. Hence there exists a $y \in L$ such that $x \vee y = 1$, $a \vee y = 1$ and $b \vee y \neq 1$. Hence $s(a \vee s(y)) = 1$. On the other hand, by (2), $1 \neq s(b \vee y) \geqslant b \vee s(y)$ and thus $s(b \vee s(y)) \neq 1$ (again by (2)). Set $c = s(y) \in L_s$. We have $a \sqcup c = s(a \vee c) = 1$ and $b \sqcup c = s(b \vee c) \neq 1$. □

By statement (1) we have a mapping

$$s = (a \mapsto s(a))\colon L \to L_s$$

for any compact L. By (2)–(4) this is a codense[4] frame homomorphism onto a compact subfit frame.

Note that for any $a \in L$ and $x \in L_s$ with $a \leqslant x$, $s(a) \leqslant s(x) = x$. Therefore

$$s(a) = \bigwedge\{x \in L_s \mid x \geqslant a\}$$

and s is the left adjoint of the localic embedding $L_s \subseteq L$ hence a frame homomorphism.

3.5 Wallman frames Following [42] we will refer to compact subfit frames as *Wallman frames*.

[4]Recall V.4.1.1. The fact that s is a codense homomorphism means that the sublocale L_s is a *codense sublocale* of L, that is, $(L_s)^\circ = L$ for the *fitting* operator

$$S^\circ = \bigcap\{\mathfrak{o}(a) \mid S \subseteq \mathfrak{o}(a)\}$$

on sublocales [75, 2.5]. See also Chap. X.

Proposition *For any compact frame L, $s\colon L \to L_s$ is universal among the codense frame homomorphisms onto Wallman frames.*

Proof Let $h\colon L \to M$ be any codense frame homomorphism onto a Wallman frame. We first show for any $a, x \in L$ that $h(x) \in \sigma(h(a))$ whenever $x \in \sigma(a)$. Let $h(x) \vee z = 1$ for $z \in M$. Then $z = h(y)$ for some $y \in L$, and hence $h(x \vee y) = h(x) \vee z = 1$. Since h is codense this implies $x \vee y = 1$, hence $a \vee y = 1$ and finally $h(a) \vee z = 1$. This shows in particular that $h(a) = h(s(a))$ for any $a \in L$ because M is Wallman. Set

$$\widetilde{h} = h_{|L_s} \colon L_s \to M.$$

The diagram

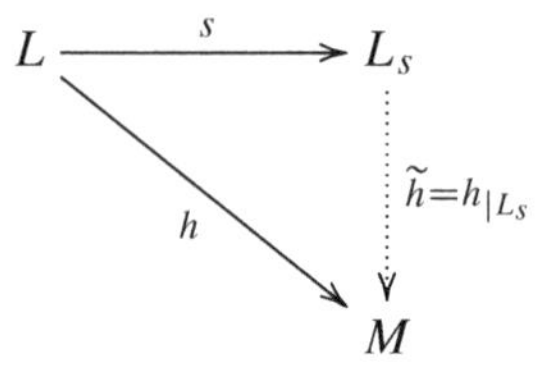

clearly commutes and $\widetilde{h}$ is indeed a frame homomorphism:

$$\widetilde{h}(\bigsqcup_{i\in J} a_i) = \widetilde{h}(s(\bigvee_{i\in J} a_i)) = h(s(\bigvee_{i\in J} a_i)) =$$

$$= h(\bigvee_{i\in J} a_i) = \bigvee_{i\in J} h(a_i) = \bigvee_{i\in J} \widetilde{h}(s(a_i)) = \bigvee_{i\in J} \widetilde{h}(a_i).$$

Finally, $\widetilde{h}$ is necessarily unique because s maps L onto L_s. □

3.6 Saturated ideals Recall the frame $\mathfrak{J}(L)$ of ideals of L from VI.9. It is compact hence the results from 3.4 apply to it and we have a compact subfit sublocale $\mathfrak{J}(L)_s$ consisting of all saturated ideals of L. Let us compute the saturation of an arbitrary principal ideal.

3.6.1 Lemma *We have for any $a \in L$ that*

(1) *$\sigma(a)$ is $\downarrow a$-small and*
(2) *$s(\downarrow a) = \sigma(a)$.*

Proof

(1) Let $\sigma(a) \vee J = 1$. Then there are $b \in \sigma(a)$ and $y \in J$ such that $b \vee y = 1$. Since b is a-small it follows that $a \vee y = 1$. Hence $\downarrow a \vee J = 1$.
(2) By definition, $s(\downarrow a) = \bigvee\{I \in \mathfrak{J}(L) \mid I \text{ is } \downarrow a\text{-small}\}$. Hence $s(\downarrow a) \geqslant \sigma(a)$ by (1). To show that $\sigma(a)$ is an upper bound, we take an ideal I of L which is $\downarrow a$-small and show that $I \subseteq \sigma(a)$. So let $x \in I$ and suppose that $x \vee y = 1$ for

some $y \in L$. Then $I \vee \downarrow y = 1$ which implies $\downarrow a \vee \downarrow y = 1$. This means that $a \vee y = 1$ and x is a-small as required. □

Combining this with the equivalence (1)⇔(2) of 3.3.1 we obtain

3.6.2 Corollary *A frame L is subfit iff every principal ideal of L is a saturated element of $\mathfrak{I}(L)$.*

3.7 The subfit compactification Let L be a subfit frame. Consider the mappings

$$\upsilon = (I \mapsto \bigvee I)\colon \mathfrak{I}(L)_s \to L \quad \text{and} \quad \sigma = (a \mapsto \sigma(a) = \downarrow a)\colon L \to \mathfrak{I}(L)_s.$$

We obviously have

$$\upsilon\sigma(a) = a \quad \text{and} \quad I \subseteq \sigma\upsilon(I) \tag{$*$}$$

and hence these maps are adjoint, υ to the left and σ to the right; hence υ preserves joins. It also preserves finite meets:

$$\upsilon(I) \wedge \upsilon(J) = \bigvee I \wedge \bigvee J = \bigvee\{a \wedge b \mid a \in I, b \in J\} \leqslant$$
$$\leqslant \bigvee\{x \mid x \in I \cap J\} = \upsilon(I \cap J) \leqslant \upsilon(I) \wedge \upsilon(J).$$

Thus, υ is a frame homomorphism, and σ is a localic map.

3.7.1 Theorem *Let L be a subfit frame. The map $\sigma : L \to \mathfrak{I}(L)_s$ is a compactification of L (called the Wallman compactification).*

Proof σ is a dense localic embedding (it is obviously one-to-one and $\sigma(0) = \downarrow 0 = \{0\}$). Moreover $\mathfrak{I}(L)_s$ is a compact subfit frame (by 3.4). □

This is the point-free analogue of the classical compactification of an arbitrary T_1-space constructed by Wallman in 1938 ([282]; generalized by Shanin in [251]). It was first constructed by Johnstone in [163] (see also [42]).

3.7.2 Lemma *If L is compact, then $\downarrow(\bigvee I)$ is I-small for any $I \in \mathfrak{I}(L)$.*

Proof If $\downarrow(\bigvee I) \vee J = 1$, then $\bigvee I \vee b = 1$ for some $b \in J$ and by compactness $a \vee b = 1$ for some $a \in I$. Hence $I \vee J = 1$. □

3.7.3 Proposition *Let L be compact. Then υ and σ are mutually inverse isomorphisms and $L \cong \mathfrak{I}(L)_s$.*

Proof Let $I \in \mathfrak{I}(L)_s$. By the lemma, $\downarrow(\bigvee I) \in \sigma(I)$. Hence $\bigvee I \in I$ (since $I = s(I) \supseteq \downarrow(\bigvee I)$). This implies that $\sigma\upsilon(I) = \downarrow(\bigvee I) \subseteq I$ and the statement follows from $(*)$ in 3.7. □

3.8 The case of a normal L Wallman also showed that his compactification is Hausdorff if (and only if) the original space is normal. In the following we analyse the compactification σ for the case of a normal L. Recall that a sublattice B of a frame L is a *basis* for L if every $a \in L$ is a join of elements of B.

3.8.1 Lemma *A compact frame L with a basis B is normal iff the lattice B is normal.*

Proof Suppose L is normal. Let $a, b \in B$ with $a \vee b = 1$. Since L is normal, there are $u, v \in L$ such that $u \wedge v = 0$ and $a \vee u = b \vee v = 1$. Further, we may write $u = \bigvee u_i$ and $v = \bigvee v_j$ with $u_i, v_j \in B$. Then, by compactness, it is obvious that we may assume $u, v \in B$. Hence B is normal.

Conversely, consider $a, b \in L$ such that $a \vee b = 1$. Since B generates L, we may rewrite this as $\bigvee a_i \vee \bigvee b_j = 1$ with $a_i, b_j \in B$ for all i, j. Then by compactness we get $x, y \in B$ such that $x \vee y = 1$ with $x \leqslant a$ and $y \leqslant b$. Since B is normal, there exist $u, v \in B$ such that $u \wedge v = 0$ and $x \vee u = y \vee v = 1$. Hence $a \vee u = b \vee v = 1$. □

Using 3.7.1, 3.7.3 and 3.8.1 we obtain

3.8.2 Theorem *Let L be a normal subfit frame. The map $\sigma : L \to \mathfrak{J}(L)_s$ is a completely regular compactification of L. It is an isomorphism iff L is compact.*

Proof $B = \{\downarrow a \mid a \in L\}$ is a basis of $\mathfrak{J}(L)_s$. It is normal since L is normal. Hence $\mathfrak{J}(L)_s$ is also normal by the lemma and then it is completely regular by 1.3.1. □

We conclude with the proof that the Wallman compactification of a normal subfit frame coincides with its Stone–Čech compactification (recall functor $\mathfrak{R}$ from VI.9.1.6). For that we first need a lemma.

3.8.3 Lemma *For any $I \in \mathfrak{J}(L)$,*

$$s(I) = \{a \in L \mid \forall b \in L\ (a \vee b = 1 \Rightarrow \exists c \in I : c \vee b = 1)\}. \qquad (*)$$

Proof $\subseteq$: If $a \in s(I)$ and $a \vee b = 1$ for any $b \in L$, then $s(I) \vee \downarrow b = 1$ in $\mathfrak{J}(L)$. By (1) in 3.4, $s(I)$ is an I-small element of $\mathfrak{J}(L)$. Hence $I \vee \downarrow b = 1$ and, therefore, $c \vee b = 1$ for some $c \in I$.

$\supseteq$: It suffices to check that the ideal in the right side of $(*)$ which we denote by J is I-small. To see this, let $J \vee K = 1$ for any ideal K. Then $a \vee b = 1$ for some $a \in J$ and $b \in K$, whence $c \vee b = 1$ for some $c \in I$ (by the definition of J). Hence $I \vee K = 1$. □

3.8.4 Theorem *If L is a normal subfit frame, then $\mathfrak{J}(L)_s \cong \mathfrak{R}(L)$.*

Proof For $I \in \mathfrak{J}(L)$ set

$$\rho(I) = \{a \in L \mid \exists b \in I : a \prec b\} \subseteq I.$$

Recall the properties of $\prec$ from V.1.3. Obviously $\rho(I)$ is an ideal and $\rho(-)$ is a monotone operator in $\mathfrak{J}(L)$. Moreover, since L is normal and subfit, $\prec$ interpolates and $\rho(I)$ is a regular ideal.

Let $I \in \mathfrak{R}(L)$. We show that $I = \rho(s(I))$. The inclusion $I \subseteq \rho(s(I))$ is clear (from $I \subseteq s(I)$ it follows that $\rho(I) \subseteq \rho(s(I))$ and $I = \rho(I)$ since I is regular). For any $a \in \rho(s(I))$ let $b \in s(I)$ such that $a \prec b$. Then $a^* \vee b = 1$ and by normality there are $u, v \in L$ such that $u \wedge v = 0$ and $a^* \vee u = 1 = b \vee v$ (in particular, $a \prec u$). Since $b \vee v = 1$ and $b \in s(I)$, it follows from the lemma that there is a $w \in I$ such that $w \vee v = 1$. But $v \leqslant u^*$ and therefore $w \vee u^* = 1$, that is, $u \prec w$. Hence $a \prec u \prec w \in I$ and $a \in I$.

Finally we show that $s(\rho(I)) = I$ for any $I \in \mathfrak{I}(L)_s$. The inclusion $s(\rho(I)) \subseteq I$ is obvious (since $\rho(I) \subseteq I$, by the definition of ρ, and I is saturated). For the converse inclusion we use the lemma to show that any $a \in I$ is in $s(\rho(I))$. If $a \vee b = 1$ for $b \in L$, there are $u, v \in L$ such that $u \wedge v = 0$ and $a \vee u = 1 = b \vee v$. In particular, $v \prec a$. Hence $v \in \rho(I)$, and $a \in s(\rho(I))$. □

4 Complete Normality: Heredity

4.1 John Isbell [155] It is easy to see that in the language of sublocales, normality can be equivalently expressed by the requirement that (quite like in spaces)

for any disjoint closed sublocales A, B there are disjoint open sublocales U, V such that $A \subseteq U$ and $B \subseteq V$.

In 1985, J. Isbell introduced *complete normality* by requiring that (in analogy with the classical definition)

for any separated sublocales A, B, that is, sublocales such that

$$\overline{A} \cap B = \mathsf{O} = A \cap \overline{B},$$

there are disjoint open sublocales U, V such that $A \subseteq U$ and $B \subseteq V$.

(In other words, "*separated sublocales can be separated by open sublocales*".)

4.2 Harold Simmons [255] Previously, already in 1978, H. Simmons had proved that the classical complete normality in spaces X was equivalent with the following (point-free) formula about $L = \Omega(X)$.

(cnorm): *For every $a, b \in L$ there exist $x, y \in L$ such that*
$x \wedge y = 0,\ x \leqslant b \leqslant a \vee x$ and $y \leqslant a \leqslant b \vee y$.

4.2.1 Notes

1. The conditions $x \leqslant b$ and $y \leqslant a$ in (cnorm) are redundant since

$$\forall\, a, b \in L\ \ \exists\, x, y \in L\colon\ x \wedge y = 0,\ b \leqslant a \vee x \text{ and } a \leqslant b \vee y \qquad (*)$$

already implies (cnorm). Indeed, given $(*)$, the elements $\tilde{x} = x \wedge b$ and $\tilde{y} = y \wedge a$ clearly satisfy $\tilde{x} \wedge \tilde{y} = 0$, $\tilde{x} \leqslant b \leqslant a \vee \tilde{x}$ and $\tilde{y} \leqslant a \leqslant b \vee \tilde{y}$.

2. Also note that for an arbitrary frame L, (cnorm) is equivalent to

(cnorm'): *For every $a, b \in L$ there exists $x \in L$ such that* $x \leqslant b \leqslant a \vee x$ *and* $x^* \wedge (a \vee b) \leqslant a \leqslant b \vee x^*$.

Proof (cnorm) $\Rightarrow$ (cnorm'): For each $a, b \in L$ let $x \in L$ given by (cnorm). Then $x^* \wedge (a \vee b) = (x^* \wedge a) \vee (x^* \wedge b) \leqslant a$ since $x^* \wedge b \leqslant x^* \wedge (a \vee x) = x^* \wedge a$. Moreover $a \leqslant b \vee y \leqslant b \vee x^*$.

(cnorm') $\Rightarrow$ (cnorm): For each $a, b \in L$ let $x \in L$ given by (cnorm') and set $y = x^* \wedge (a \vee b)$. Then $x \wedge y = 0$, $y \leqslant a$ and $b \vee y = b \vee (x^* \wedge (a \vee b)) = (b \vee x^*) \wedge (b \vee a) \geqslant a$. □

The next result shows that Simmons' condition (cnorm) characterizes general completely normal frames (hence Isbell's definition is a conservative extension of the classical definition).

4.2.2 Proposition *The following statements about a frame L are equivalent.*

(1) *L is completely normal (in the sense of Isbell).*
(2) *For every $a, b \in L$ there are $x, y \in L$ such that $x \wedge y = 0$, $b \leqslant a \vee x$ and $a \leqslant b \vee y$.*

Proof (1) $\Rightarrow$ (2): Given $a, b \in L$ set

$$A = \mathfrak{o}(a) \cap \mathfrak{c}(b) \quad \text{and} \quad B = \mathfrak{c}(a) \cap \mathfrak{o}(b).$$

The sublocales A and B are separated: $A \cap \overline{B} \subseteq A \cap \mathfrak{c}(a) = \mathsf{O}$ and $\overline{A} \cap B \subseteq \mathfrak{c}(b) \cap B = \mathsf{O}$. Thus, by complete normality there exist $x, y \in L$ such that

$$\mathfrak{o}(x) \cap \mathfrak{o}(y) = \mathsf{O}, \quad A \subseteq \mathfrak{o}(y) \quad \text{and} \quad B \subseteq \mathfrak{o}(x).$$

These are the elements x and y we are looking for. Indeed, $\mathfrak{o}(x) \cap \mathfrak{o}(y) = \mathsf{O}$ means that $x \wedge y = 0$; furthermore, $A \subseteq \mathfrak{o}(y)$ means that $\mathfrak{o}(a) \cap \mathfrak{c}(b) \subseteq \mathfrak{o}(y)$, that is, $\mathfrak{c}(y) \cap \mathfrak{o}(a) \cap \mathfrak{c}(b) = \mathsf{O}$. Equivalently, $\mathfrak{o}(a) \cap \mathfrak{c}(y \vee b) = L$, that is, $\mathfrak{c}(y \vee b) \subseteq \mathfrak{c}(a)$. Hence $a \leqslant y \vee b$. By a similar reasoning, $B \subseteq \mathfrak{o}(x)$ implies $b \leqslant x \vee a$.

(2) $\Rightarrow$ (1): Let A, B be a pair of separated sublocales of L with $\mathfrak{c}(a) = \overline{A}$ and $\mathfrak{c}(b) = \overline{B}$. Then there are $x, y \in L$ such that $x \wedge y = 0$, $b \leqslant a \vee x$ and $a \leqslant b \vee y$. It follows immediately that

$$\mathfrak{o}(x) \cap \mathfrak{o}(y) = \mathsf{O} \quad \text{and} \quad A \cap \mathfrak{c}(b) \supseteq A \cap \mathfrak{c}(a) \cap \mathfrak{c}(x) = A \cap \mathfrak{c}(x).$$

Likewise $B \cap \mathfrak{c}(a) \supseteq B \cap \mathfrak{c}(y)$. Hence $A \cap \mathfrak{c}(x) = \mathsf{O} = B \cap \mathfrak{c}(y)$, that is, $A \subseteq \mathfrak{o}(x)$ and $B \subseteq \mathfrak{c}(y)$. □

4.2.3 Corollary *A space X is completely normal iff the frame $\Omega(X)$ is completely normal.*

4.3 Completeness and heredity Normality behaves badly under constructions: in particular, it is not hereditary. The *hereditarily normal* frames, that is, the frames in which every sublocale is normal are precisely the completely normal ones [107]. To prove this we need to recall 1.4 and observe that by a similar procedure one may conclude that

4.3.1 Proposition *A frame L is completely normal iff for any countable sets $\{a_i\}_{i=1}^{\infty}$ and $\{b_i\}_{i=1}^{\infty}$ in L there is a $u \in L$ such that*

$$\bigwedge_{i=1}^{\infty} b_i \leqslant a_k \vee u \text{ and } \bigwedge_{i=1}^{\infty} a_i \leqslant b_k \vee u^* \text{ for every } k \in \mathbb{N}.$$

4.3.2 Theorem *The following statements about a frame L are equivalent.*

(1) *L is hereditarily normal.*
(2) *Every open sublocale of L is normal.*
(3) *L is completely normal.*

Proof (2) $\Rightarrow$ (1): Let L be a frame whose open sublocales are normal and consider an arbitrary sublocale S of L. Let j_S denote the corresponding localic embedding with left adjoint $\nu_S \colon L \twoheadrightarrow S$. In order to prove that S is normal, consider $a, b \in S$ such that $a \overset{S}{\vee} b = 1$. Let $t = a \vee b \in L$. By the assumption, $\mathfrak{o}(t)$ is normal and we have

$$(t \to a) \overset{\mathfrak{o}(t)}{\vee} (t \to b) = t \to (a \vee b) = 1$$

(recall that $u \overset{S}{\vee} v$ is $\nu_S(u \vee v)$; here we have used $\nu_{\mathfrak{o}(t)} = (x \mapsto (t \to x))$ so that $(t \to a) \overset{\mathfrak{o}(t)}{\vee} (t \to b) = (a \vee b) \to ((t \to a) \vee (t \to b)) = (a \vee b) \to ((b \to a) \vee (a \to b)) = 1)$.

Hence there are $c, d \in L$ such that

$$(t \to a) \overset{\mathfrak{o}(t)}{\vee} (t \to c) = 1 = (t \to b) \overset{\mathfrak{o}(t)}{\vee} (t \to d)$$

and

$$(t \to c) \wedge (t \to d) = 0_{\mathfrak{o}}(t),$$

that is,

$$t \leqslant (a \vee c) \wedge (b \vee d), \tag{$*$}$$

and

$$0 = t \wedge c \wedge d = (a \wedge c \wedge d) \vee (b \wedge c \wedge d). \tag{$**$}$$

Set $u = \nu_S(c)$ and $v = \nu_S(d)$ in S. Then:

I. $u \wedge v = \nu_S(c \wedge d) = 0_S$. In fact, it follows from $(**)$ that

$$a \wedge \nu_S(c \wedge d) = \nu_S(a \wedge c \wedge d) = 0_S$$

and, similarly, $b \wedge \nu_S(c \wedge t) = 0_S$. This combined with $a \overset{S}{\vee} b = 1$ yields $\nu_S(c \wedge d) = 0_S$.

II. $a \overset{S}{\vee} u = 1$ (and similarly $b \overset{S}{\vee} v = 1$). Indeed, if $s \geqslant a \vee u$ (with $s \in S$), then $s \geqslant a \vee c \geqslant t$ by $(*)$. Hence $s = 1$ (since $\nu_S(t) = \nu_S(a \vee b) = 1$).

I and II show that S is normal.

(3) $\Rightarrow$ (1): Let S be a sublocale of L and let $a \overset{S}{\vee} b = 1$. Then (recall A.6.5) $\mathfrak{c}_S(a) = S \cap \mathfrak{c}(a)$, $\mathfrak{c}_S(b) = S \cap \mathfrak{c}(b)$, and

$$\overline{\mathfrak{c}_S(a)} \cap \mathfrak{c}_S(b) = \overline{S \cap \mathfrak{c}(a)} \cap S \cap \mathfrak{c}(b) \subseteq \mathfrak{c}(a) \cap S \cap \mathfrak{c}(b) =$$

$$= \mathfrak{c}_S(a) \cap \mathfrak{c}_S(b) = \mathfrak{c}_S(a \overset{S}{\vee} b) = \mathsf{O}.$$

Similarly, $\mathfrak{c}_S(a) \cap \overline{\mathfrak{c}_S(b)} = \mathsf{O}$. Hence there are $u, v \in L$ such that

$$u \wedge v = 0, \quad \mathfrak{c}_S(a) \subseteq \mathfrak{o}(u) \quad \text{and} \quad \mathfrak{c}_S(b) \subseteq \mathfrak{o}(v).$$

Consider the open sublocales

$$\mathfrak{o}_S(\nu_S(u)) \quad \text{and} \quad \mathfrak{o}_S(\nu_S(v))$$

of S. Clearly, $\mathfrak{c}_S(a) \subseteq \mathfrak{o}_S(\nu_S(u))$ and $\mathfrak{c}_S(b) \subseteq \mathfrak{o}_S(\nu_S(v))$, that is, $a \overset{S}{\vee} \nu_S(u) = 1$ and $b \overset{S}{\vee} \nu_S(v) = 1$. Moreover, $\nu_S(u) \wedge \nu_S(v) = \nu_S(u \wedge v) = \nu_S(0) = 0_S$. This shows that S is normal.

(1) $\Rightarrow$ (3): If $A \cap \overline{B} = \mathsf{O} = \overline{A} \cap B$ with $\overline{B} = \mathfrak{c}(b)$ and $\overline{A} = \mathfrak{c}(a)$, then $A \subseteq \mathfrak{o}(b)$ and $B \subseteq \mathfrak{o}(a)$. Set

$$U = \mathfrak{o}(a) \vee \mathfrak{o}(b) = \mathfrak{o}(a \vee b).$$

U is normal by a assumption. Further, $\overline{A} \cap U = \mathfrak{c}(a) \cap \mathfrak{o}(b)$ and $\overline{B} \cap U = \mathfrak{c}(b) \cap \mathfrak{o}(a)$. By A.6.5,

$$\overline{A} \cap U = \mathfrak{c}_U(\nu_A(a)) \quad \text{and} \quad \overline{B} \cap U = \mathfrak{c}_U(\nu_B(b)).$$

These are disjoint closed sublocales of U. Hence

$$\mathfrak{c}_U(\nu_A(a) \overset{U}{\vee} \nu_B(b)) = (\overline{A} \cap U) \cap (\overline{B} \cap U) = \mathsf{O}$$

that is $\nu_A(a) \overset{U}{\vee} \nu_B(b) = 1$. Then, by the normality of U, there exist $u, v \in U$ satisfying $u \wedge v = 0_U$ and $\nu_A(a) \overset{U}{\vee} u = 1 = \nu_B(b) \overset{U}{\vee} v$. In particular,

$$u \wedge v = 0_U \Leftrightarrow u \wedge v = (t \vee s) \to 0 \Leftrightarrow u \wedge v \wedge (t \vee s) = 0. \qquad (*)$$

On the other hand, by A.6.3.1,

$$\nu_A(a) \overset{U}{\vee} u = 1 \Leftrightarrow \mathfrak{c}_U(\nu_A(a)) \subseteq \mathfrak{o}_U(u) = U \cap \mathfrak{o}(u) = \mathfrak{o}(u \wedge (b \vee a))$$

and similarly

$$\nu_B(b) \overset{U}{\vee} v = 1 \Leftrightarrow \mathfrak{c}_U(\nu_B(b)) \subseteq \mathfrak{o}(v \wedge (b \vee a)).$$

Set $c = u \wedge (b \vee a)$ and $d = v \wedge (b \vee a)$. By $(*)$, $c \wedge d = 0$, that is, $\mathfrak{o}(c) \wedge \mathfrak{o}(d) = \mathsf{O}$. Finally,

$$A \subseteq \overline{A} \cap \mathfrak{o}(b) = \overline{A} \cap U = \mathfrak{c}_U(\nu_A(a)) \subseteq \mathfrak{o}(c)$$

and, similarly, $B \subseteq \mathfrak{o}(d)$. □

Chapter VIII
More on Normality and Related Properties

Here we start with two more variants of normality. There is the *perfect normality*, which turns out to be a conjunction of the classical *perfectness* (which is slightly different in the point-free context due to the different behaviour of sublocales and subspaces) and normality; in a way it can be viewed as a weaker form of metrizability. Next we deal with the technically important *collectionwise normality*. Then, in the penultimate section we prove and discuss the Katětov-Tong insertion theorem [123, 129, 217], using (to advantage) the techniques of the point-free real line. We finish with a certain duality between normality and *extremal disconnectedness* that allows to translate several results concerning normality to facts about extremal disconnected frames [131].

1 Perfect Normality

1.1 Perfectly normal frames In the point-free context, perfect normality was introduced by Charalambous in 1974 as a property of σ-frames.[1] In particular, a frame L is *perfectly normal* if

(pnorm): *it is normal and for each $a \in L$ there is a sequence $(a_n)_{n=1}^{\infty} \subseteq L$ such that*

$$\forall b, c \in L\ (b \wedge a = c \wedge a \quad \text{iff} \quad b \vee a_n = c \vee a_n \textit{ for all } n).$$

1.1.1 Lemma *A frame L is perfectly normal iff*

(pnorm'): $\forall a \in L,\ \exists (a_n)_{n=1}^{\infty} \subseteq L\colon\ a = \bigvee_{n=1}^{\infty} a_n$ and $a_n \prec a$ for all n.

Proof $\Rightarrow$: Let $a \in L$. The property (pnorm) for $b = 1$ and $c = a$ yields a sequence $(a_n)_{n=1}^{\infty}$ such that $a \vee a_n = 1$ for all n. Then by normality there are b_n and x_n

[1] In [68]; see also Gilmour [114].

J. Picado, A. Pultr, *Separation in Point-Free Topology*,
https://doi.org/10.1007/978-3-030-53479-0_8

such that $b_n \vee a_n = 1$, $x_n \vee a = 1$ and $x_n \wedge b_n = 0$ (so that $b_n \prec a$). Since $(\bigvee_{k=1}^{\infty} b_k) \geqslant b_n$ for all n we have

$$(\bigvee_{k=1}^{\infty} b_k) \vee a_n = 1 = a \vee a_n,$$

and, by (pnorm) again, we get $(\bigvee_{k=1}^{\infty} b_k) \wedge a = a \wedge a = a$. Hence $a \leqslant \bigvee_{k=1}^{\infty} b_k \leqslant a$.

$\Leftarrow$: First the normality. Let $a \vee b = 1$. In the equations $a = \bigvee_{n=1}^{\infty} x_n$ and $b = \bigvee_{n=1}^{\infty} y_n$ with $x_n \prec a$ and $y_n \prec b$ we may assume the sequences (x_n) and (y_n) to be increasing so that for the elements u_n and v_n satisfying

$$x_n \wedge u_n = 0,\ u_n \vee a = 1,\ y_n \wedge v_n = 0 \text{ and } v_n \vee b = 1$$

we have

$$x_n \wedge u_m = 0 \text{ and } y_n \wedge v_m \text{ for } n \leqslant m. \qquad (*)$$

Set $u = \bigvee_{n=1}^{\infty}(u_n \wedge y_n)$ and $v = \bigvee_{n=1}^{\infty}(v_n \wedge x_n)$. By $(*)$

$$u \wedge v = \bigvee_{n=1}^{\infty} \bigvee_{m=1}^{\infty} (u_n \wedge y_n \wedge v_m \wedge x_m) = 0,$$

and

$$a \vee u = \bigvee_{n=1}^{\infty} (a \vee u_n) \wedge (a \vee y_n) = \bigvee_{n=1}^{\infty} (a \vee y_n) = a \vee b = 1.$$

Likewise $v \vee b = 1$.

Finally, let $a = \bigvee_{n=1}^{\infty} d_n$ with $d_n \prec a$ witnessed by a_n, that is,

$$d_n \wedge a_n = 0 \text{ and } a_n \vee a = 1.$$

If $b \wedge a = c \wedge a$, then $(b \wedge a) \vee a_n = (c \wedge a) \vee a_n$, hence $b \vee a_n = c \vee a_n$ for all n. On the other hand, if $b \vee a_n = c \vee a_n$ for all n, then

$$b \wedge a = \bigvee_{n=1}^{\infty} (b \wedge d_n) = \bigvee_{n=1}^{\infty} (b \wedge d_n) \vee (a_n \wedge d_n) =$$

$$= \bigvee_{n=1}^{\infty} (b \vee a_n) \wedge d_n = \bigvee_{n=1}^{\infty} (c \vee a_n) \wedge d_n = c \wedge a. \qquad \square$$

1.2 By VI.6.2.2 we obtain

1.2.1 Corollary *A frame is perfectly normal iff every element is a cozero.*

Hence, by VI.6.6.1,

1.2.2 Corollary *Every metrizable frame is perfectly normal.*

1.3 Perfect frames *Perfect spaces* were introduced by Heath and Michael in [143]. They are the spaces in which each open set is a union of countably many closed sets (that is, all the open sets are F_σ). By De Morgan formulas this is equivalent with the assumption that all the closed sets are G_δ.

These two equivalent formulations of perfectness can be immediately extended to the point-free setting as assumptions on open resp. closed sublocales [125, 127]. In the coframe of sublocales of a locale we do not have, however, one of the De Morgan formulas (recall A.2.1.1) that would make the two resulting concepts generally equivalent. We will speak of the former as of *F-perfectness* and of the latter, generally stronger, as of the *G-perfectness*.

Under normality, however, they coincide.

1.3.1 Proposition *A normal frame is F-perfect iff it is G-perfect.*

Proof Suppose L is a normal F-perfect frame and let $a \in L$. There is a countable subset $\{a_n\}_{n=1}^{\infty}$ of L such that $\mathfrak{o}(a) = \bigvee_{n=1}^{\infty} \mathfrak{c}(a_n)$. Then $a \vee a_n = 1$ for all n and hence there are u_n and v_n in L such that

$$a \vee u_n = 1 = a_n \vee v_n \text{ and } u_n \wedge v_n = 0 \quad \text{for all } n.$$

In particular, $\mathfrak{c}(a_n) \subseteq \mathfrak{o}(v_n)$ (recall A.6.3.1). Moreover, $v_n \prec a$. Thus

$$\mathfrak{o}(a) = \bigvee_{n=1}^{\infty} \mathfrak{c}(a_n) \subseteq \bigvee_{n=1}^{\infty} \mathfrak{o}(v_n) = \mathfrak{o}(\bigvee_{n=1}^{\infty} v_n) \subseteq \mathfrak{o}(a).$$

Hence $a = \bigvee_{n=1}^{\infty} v_n$ with $v_n \prec a$ for all n and

$$\mathfrak{c}(a) \subseteq \bigcap_{n=1}^{\infty} \mathfrak{o}(u_n) \subseteq \bigcap_{n=1}^{\infty} \mathfrak{c}(v_n) = \mathfrak{c}(\bigvee_{n=1}^{\infty} v_n) = \mathfrak{c}(a). \qquad \square$$

1.3.2 Note By II.4.4.1, G-perfect frames are fit while F-perfect frames are generally only subfit. The cofinite topology $\Omega(\mathbb{N})$ on $\mathbb{N}$ is an example of an F-perfect frame that is not fit—in fact, the only closed sublocales of $\Omega(\mathbb{N})$ which are meets of open sublocales are $\mathfrak{c}(\mathbb{N})$ and $\mathfrak{c}(\emptyset)$—hence it is not G-perfect.

1.4 Now we will prove that perfect normality is the combination of the two properties, normality and perfectness.

1.4.1 Proposition *The following statements about a frame L are equivalent.*

(1) *L is perfectly normal.*
(2) *L is normal and F-perfect.*
(3) *L is normal and G-perfect.*

Proof We already know that (2) and (3) are equivalent. Assume L satisfies (pnorm'). In particular, it is normal. Now by perfectness, for each $a \in L$, $a = \bigvee_{n=1}^{\infty} b_n$ with $b_n \prec a$. The latter implies $\mathfrak{c}(b_n^*) \cap \mathfrak{c}(a) = 0$, that is, $\overline{\mathfrak{o}(b_n)} \cap \mathfrak{c}(a) = 0$. Then $\overline{\mathfrak{o}(b_n)} \subseteq \mathfrak{o}(a)$ hence

$$\mathfrak{o}(a) = \bigvee_{n=1}^{\infty} \mathfrak{o}(b_n) \subseteq \bigvee_{n=1}^{\infty} \overline{\mathfrak{o}(b_n)} \subseteq \mathfrak{o}(a).$$

Conversely, let $\mathfrak{o}(a) = \bigvee_{d \in D} \mathfrak{c}(d)$ with a countable $D \subseteq L$. For each $d \in D$, the inclusion $\mathfrak{c}(d) \subseteq \mathfrak{o}(a)$ means that $a \vee d = 1$. By normality there is a $B = \{b_d \mid d \in D\}$ such that $d \vee b_d = 1$ (that is, $\mathfrak{c}(d) \subseteq \mathfrak{o}(b_d)$) and $b_d \prec a$ (that is, $\bigvee B \leqslant a$). Then

$$\mathfrak{o}(a) = \bigvee_{d \in D} \mathfrak{c}(d) \subseteq \bigvee_{d \in D} \mathfrak{o}(b_d) = \mathfrak{o}(\bigvee B),$$

that is, $a \leqslant \bigvee B$, and hence $a = \bigvee B$ with $B \prec a$. □

1.5 The spatial case The space $(\mathbb{N}, \Omega(\mathbb{N}))$ from 1.3.2, being countable and T_1, is a perfect space. Thus G-perfectness is not a conservative extension of topological perfectness. Regarding F-perfectness we have, however

1.5.1 Proposition *If a space X is perfect, then $\Omega(X)$ is F-perfect.*

Proof For each open $A \subseteq X$ there is a sequence $(A_n)_{n=1}^{\infty}$ of opens such that $A = \bigcup_{n=1}^{\infty}(X \smallsetminus A_n)$. Then $A_n \cup A = X$ for all n, and hence $\bigvee_{n=1}^{\infty} \mathfrak{c}(A_n) \subseteq \mathfrak{o}(A)$. On the other hand, let $B \in \mathfrak{o}(A)$ and $B_n = A_n \cup B \in \mathfrak{c}(A_n)$ for all n. Then

$$B = A \to B = \mathrm{Int}((X \smallsetminus A) \cup B) = \mathrm{Int}\big(\big(\bigcap_{n=1}^{\infty} A_n\big) \cup B\big) =$$

$$= \mathrm{Int}\big(\bigcap_{n=1}^{\infty} (A_n \cup B)\big) = \mathrm{Int}\big(\bigcap_{n=1}^{\infty} B_n\big) = \bigwedge_{n=1}^{\infty} B_n \in \bigvee_{n=1}^{\infty} \mathfrak{c}(A_n).$$

Hence $\mathfrak{o}(A) \subseteq \bigvee_{n=1}^{\infty} \mathfrak{c}(A_n)$ and we see that $\Omega(X)$ is F-perfect. □

The reverse implication, however, does not hold in general, as the following example taken from [125] shows.

Let X be a T_1-space such that $\emptyset$ is a meet-irreducible element of $\Omega(X)$ (that is, $A \cap B = \emptyset$ in $\Omega(X)$ only if $A = \emptyset$ or $B = \emptyset$). Consider $Y = X \cup \{\infty\}$ with $\infty \notin X$, endowed with the topology

$$\Omega(Y) = \{\emptyset\} \cup \{A \cup \{\infty\} \mid \emptyset \neq A \in \Omega(X)\}.$$

The space Y is perfect iff the topology in X is the trivial one, that is, $\{\emptyset, X\}$. On the other hand, the frames $\Omega(Y)$ and $\Omega(X)$ are isomorphic and hence $\Omega(Y)$ is F-perfect iff so is $\Omega(X)$. Therefore, if the given X is a perfect space with a nontrivial topology where $\emptyset$ is meet-irreducible (such as, e.g. the space $\mathbb{N}$ endowed with the cofinite topology), it follows from 1.5.1 that $\Omega(X)$ and $\Omega(Y)$ are F-perfect. However, Y is not perfect.

Note that Y is not T_D. In fact, for T_D-spaces we have

1.5.2 Proposition *Let X be a T_D-space. Then X is perfect iff $\Omega(X)$ is F-perfect.*

Proof Let X be a T_D-space. Pick an arbitrary $A \in \Omega(X)$. If X is perfect, then there is a sequence $(A_n)_{n=1}^{\infty} \subseteq \Omega(X)$ such that $\mathfrak{o}(A) = \bigvee_{n=1}^{\infty} \mathfrak{c}(A_n)$. Then $\mathfrak{c}(A_n) \subseteq \mathfrak{o}(A)$, that is, $A \cup A_n = X$ for all n. Hence $\bigcup_{n=1}^{\infty}(X \smallsetminus A_n) \subseteq A$.

Conversely, let $x \in A$. By axiom T_D there is an open $B \ni x$ such that $C = B \smallsetminus \{x\}$ is also open. Since

$$A \to C \in \mathfrak{o}(A) = \bigvee_{n=1}^{\infty} \mathfrak{c}(A_n),$$

there are $B_n \in \Omega(X)$, $n = 1, 2, \ldots$, such that $A_n \subseteq B_n$ for all n and

$$A \to C = \bigwedge_{n=1}^{\infty} B_n = \operatorname{Int}\Big(\bigcap_{n=1}^{\infty} B_n\Big).$$

Since $x \in A \cap B$ we have $A \cap B \nsubseteq C$. Then $B \nsubseteq A \to C$, from which it follows that $x \notin A \to C$. Hence

$$x \in X \smallsetminus \operatorname{Int}\Big(\bigcap_{n=1}^{\infty} B_n\Big) = \overline{\bigcup_{n=1}^{\infty} (X \smallsetminus B_n)}.$$

Since B is an open neighbourhood of x it follows that $B \cap \big(\bigcup_{n=1}^{\infty}(X \smallsetminus B_n)\big) \neq \emptyset$. But $A \to C \subseteq \bigcap_{n=1}^{\infty} B_n$. Hence

$$x \in \bigcup_{n=1}^{\infty} (X \smallsetminus B_n) \subset \bigcup_{n=1}^{\infty} (X \smallsetminus A_n). \qquad \square$$

Adding normality we have

1.5.3 Proposition

1. *If a space X is perfectly normal, then $\Omega(X)$ is perfectly normal.*
2. *If X is T_0, then X is perfectly normal iff $\Omega(X)$ is perfectly normal.*

Proof

1. follows from 1.5.1.
2. It remains to show that a T_0-space X is perfect whenever $\Omega(X)$ is perfectly normal. First, $\Omega(X)$ is subfit (recall 1.3.2). Hence X is subfit and, by II.2.4, it is regular. Being T_0, this makes X a T_2-space hence a perfect space (by 1.5.2). $\square$

1.6 Mapping invariance theorem Recall VII.1.6. Now we will show that closed localic maps preserve perfect normality.

First, the image of any F-perfect locale under a closed localic map is F-perfect.

1.6.1 Proposition *Let $f: L \to M$ be a surjective closed localic map. If L is F-perfect, then M is also F-perfect.*

Proof Consider the left adjoint f^* of f and let $b \in M$. Since L is perfect we have

$$\mathfrak{o}(f^*(b)) = \bigvee_{n=1}^{\infty} \mathfrak{c}(a_n) \quad \text{for some} \quad \{a_n\}_{n=1}^{\infty} \subseteq L.$$

We refer to Sect. 7 in the Appendix for background facts about images and preimages that are relevant here. First, as f is onto, we have

$$\mathfrak{o}(b) = ff_{-1}[\mathfrak{o}(b)]$$

(the inclusion "$\supseteq$" follows from the adjunction $f[-] \dashv f_{-1}[-]$; conversely, for each $b \to y \in \mathfrak{o}(b)$, we have $b \to y = b \to f(a) = f(f^*(b) \to a)$ for some $a \in L$, where $f^*(b) \to a \in \mathfrak{o}(f^*(b)) = f_{-1}[\mathfrak{o}(b)]$).

Then, finally,

$$\mathfrak{o}(b) = ff_{-1}[\mathfrak{o}(b)] = f[\mathfrak{o}(f^*(b))] =$$

$$= f[\bigvee_{n=1}^{\infty} \mathfrak{c}(a_n)] = \bigvee_{n=1}^{\infty} f[\mathfrak{c}(a_n)] = \bigvee_{n=1}^{\infty} \mathfrak{c}(f(a_n)). \qquad \square$$

Generally $f[-]$ does not preserve countable meets. Thus the argument of this proof does not work for G-perfectness. Nevertheless, assuming normality we have

1.6.2 Corollary *Let $f: L \to M$ be a surjective closed localic map. If L is perfectly normal, then M is also perfectly normal.*

Proof Combine 1.6.1 with the fact that normality is invariant under closed localic maps (VII.1.6). $\square$

1.7 Mild normality and Oz frames Recall from A.2.1 the basic properties of pseudocomplements, in particular the identity

$$(a \wedge b)^{**} = a^{**} \wedge b^{**}.$$

1.7.1 In the literature we meet, besides stronger variants of normality (such as the ones we discuss now, and have discussed in Chap. VII) also some weaker ones. We have already mentioned the almost normal frames (VII.1.3.2). Relaxing (norm) a bit further, requiring it for the regular elements only, defines the class of the so-called *mildly normal frames*.[2] A frame L is *mildly normal* if

[2]These are the point-free extensions of the κ-*normal spaces* introduced by Schepin in 1972 in [244] (and introduced as *mildly normal spaces* in [263]). Clearly, a space X is κ-normal if and only if $\Omega(X)$ is mildly normal.

(mnorm): *for any* **regular** $a, b \in L$ *such that* $a \vee b = 1$
there are $u, v \in L$ *such that* $u \wedge v = 0$ *and* $a \vee u = 1 = b \vee v$.

(Elements u and v can be also assumed to be regular since $u \wedge v = 0$ iff $u^{**} \wedge v^{**} = 0$.)

Combining it with perfectness we then have the *perfectly mildly normal frames* (briefly, *pm-normal* frames) where (pnorm') is assumed for the regular elements only. The proof of 1.4.1 can then be adapted for pm-normal frames and we get

1.7.2 Proposition *The following statements about a frame L are equivalent.*

(1) *L is pm-normal.*
(2) *Every regular $a \in L$ is a cozero element.*
(3) *L is mildly normal and for each regular element a in L there is a countable family $(a_n)_{n=1}^{\infty}$ of regular elements in L such that $\mathfrak{o}(a) = \bigvee_{n=1}^{\infty} \mathfrak{c}(a_n)$.*
(4) *L is mildly normal and for each regular element a in L there is a countable family $(a_n)_{n=1}^{\infty}$ of regular elements in L such that $\mathfrak{c}(a) = \bigcap_{n=1}^{\infty} \mathfrak{o}(a_n)$.*

1.7.3 Notes

1. The equation (2) appears in the literature as a characterization of the class of *Oz frames* (see [34, Proposition 2.2]). Thus, the Oz property is weaker than perfect normality and is intrinsically characterized by pm-normality.
2. As an aside, we note that Oz frames L are also characterized by the σ-regularity of the sub-σ-frame ϱL of L generated by the set of regular elements.[3] Indeed, we have

Proposition *A frame L is Oz iff the σ-frame ϱL is σ-regular.*

Proof If L is Oz and if $a \in \varrho L$, then a is a cozero element and $a = \bigvee_{n=1}^{\infty} a_n$ for some sequence $(a_n)_{n=1}^{\infty}$ in L with $a_n \lll a$. Then $a_n \leqslant a_n^{**} \prec a$ so that $a = \bigvee_{n=1}^{\infty} a_n^{**}$ with $a_n^{**} \prec a$. Conversely, if ϱL is σ-regular, then the relation $\prec$ interpolates (recall VI.6.3.4) so that $\varrho L \subseteq \mathrm{Coz}\,(L)$ (assuming Countable Dependent Choice). □

For more about Oz frames see [34, 38, 90].

2 Collectionwise Normality

Collectionwise normality [235] is another important stronger variant of normality, weaker than paracompactness. To introduce it we need to recall a few notions.

[3]The σ-frame ϱL consists precisely of all countable joins of regular elements of L since any finite meet of regular elements is regular by (4) in A.2.1.1.

2.1 A set $\{x_i\}_{i\in J}$ of elements of L is *disjoint* if $x_i \wedge x_j = 0$ for every $i \neq j$. It is *co-discrete* if there is a cover A of L such that for any $a \in A$, $a \leqslant x_i$ for all i but at most one $j \in J$. Note in particular that a pair $\{x, y\}$ is co-discrete iff $x \vee y = 1$. On the other hand, a finite $\{x_1, x_2, \ldots, x_n\}$ is co-discrete only if $x_1 \vee x_2 \vee \cdots \vee x_n = 1$ but the converse does not hold for $n > 2$.

2.2 Collectionwise normal frames A frame L is *collectionwise normal* if

(clnorm): *for any co-discrete system* $\{x_i\}_{i\in J}$ *there is a disjoint* $\{u_i\}_{i\in J}$ *such that* $x_i \vee u_i = 1$ *for every* $i \in J$.

2.2.1 Note Let $a \vee b = 1$ in a collectionwise normal frame L. Then $\{a, b\}$ is clearly co-discrete and so there is a disjoint $\{u_1, u_2\}$ such that $u_1 \vee a = 1 = u_2 \vee b$. Hence L is normal.

2.3 More generally, for a cardinal $\kappa \geqslant 2$, a frame L is *κ-collectionwise normal* if it satisfies the definition of collectionwise normality for index sets J with cardinality $|J| \leqslant \kappa$. Hence collectionwise normality is κ-collectionwise normality for any cardinality κ and κ-collectionwise normality coincides with normality for every $\kappa \leqslant \aleph_0$.

For any two cardinalities $\kappa \leqslant \lambda$, λ-collectionwise normality implies κ-collectionwise normality. Hence, κ-collectionwise normality implies normality for every κ.

2.3.1 Examples

1. Examples of κ-collectionwise normal frames are the κ-hedgehog frames of [128]. The notion of κ-collectionwise normality is the necessary and sufficient condition under which Urysohn's separation and Tietze's extension type results hold for hedgehog-valued continuous functions [128].
2. A space is *zero-dimensional* (or, *totally disconnected*) if each neighbourhood of each point contains a clopen one. In other words, X is zero-dimensional if $\Omega(X)$ has a basis consisting of clopen sets, and this definition is naturally extended to all frames. In particular (see A.6.4), the frame $\mathsf{S}(L)^{\mathrm{op}}$ (isomorphic to the frame of congruences $\mathfrak{C}(L)$, or to the frame of nuclei $\mathcal{N}(L)$) is zero-dimensional. This property, a sort of "high dispersedness", can be useful in modelling discrete modification of a frame (the more disconnected a frame resp. space is, the closer it is to being Boolean resp. discrete). In [232] Plewe proved that the frames of sublocales $\mathsf{S}(L)^{\mathrm{op}}$ are collectionwise normal, and also that they are *ultranormal*, another concept of normality type. A frame L is *ultranormal* if

   (unorm): *for any* $a, b \in L$ *such that* $a \vee b = 1$ *there is a complemented element* $c \in L$ *such that* $c \leqslant a$ *and* $c^* \leqslant b$.

   and one has that

   subfit ultranormal frames are zero-dimensional,

   and very strongly so.

(Let BL denote the *Boolean part* of L, that is, the set of complemented elements of L. If $a \nleqslant \bigvee\{c \in BL \mid c \leqslant a\}$, then by subfitness there would exist an $x \in L$ such that $a \vee x = 1 \neq x \vee \bigvee\{c \in BL \mid c \leqslant a\}$, and then, by ultranormality, there would exist a complemented d such that $d \leqslant a$ and $d^* \leqslant x$, contradicting the fact that $x \vee \bigvee\{c \in BL \mid c \leqslant a\} < 1$.)

This shows that sublocale frames are highly disconnected point-free spaces close to Boolean. It still remains an open problem how close to Boolean they are, and, more specifically, which frames are isomorphic to $\mathsf{S}(L)^{\mathrm{op}}$. See also 3.7.1 below.

2.4 In the definition of collectionwise normality in [235] disjoint families are replaced by the so-called *discrete* families. The latter are the $\{x_i\}_{i\in J} \subseteq L$ for which there is a cover A of L such that for any $a \in A$, $a \wedge x_i = 0$ for all i with at most one exception.

2.4.1 Remark Any discrete system is clearly disjoint: if there is a cover A such that for every $a \in A$, $a \wedge y_i = 0$ for all i with at most one exception, then $a \wedge y_i \wedge y_j = 0$ for all $a \in A$ and $i \neq j$, that is, $y_i \wedge y_j = 0$. However, the converse is not true.

Thus the definition in 2.2 looks formally weaker than the one in [128, 235]. Nonetheless they coincide. To show this we need the following:

2.4.2 Lemma *For any co-discrete $\{x_i\}_{i\in J}$ in L and $y \in L$,*

$$y \vee \bigwedge_{i\in J} x_i = \bigwedge_{i\in J} (y \vee x_i).$$

Proof It suffices to show that $\bigwedge_{i\in J}(y \vee x_i) \leqslant y \vee \bigwedge_{i\in J} x_i$. Pick the cover A from the definition of co-discreteness. For each $a \in A$ we have $a \wedge \bigwedge_{i\in J} x_i = \bigwedge_{i\in J}(a \wedge x_i) = a \wedge x_{i(a)}$ for a suitable $i(a) \in J$. Hence

$$\begin{aligned}\bigwedge_{i\in J}(y \vee x_i) &= \bigvee_{a\in A}(a \wedge \bigwedge_{i\in J}(y \vee x_i)) \leqslant \bigvee_{a\in A}(a \wedge (y \vee x_{i(a)})) = \\ &= \bigvee_{a\in A}((a \wedge y) \vee (a \wedge x_{i(a)})) = \bigvee_{a\in A}((a \wedge y) \vee (a \wedge \bigwedge_{i\in J} x_i)) = \\ &= \bigvee_{a\in A}(a \wedge (y \vee \bigwedge_{i\in J} x_i)) = y \vee \bigwedge_{i\in J} x_i.\end{aligned}$$

□

2.4.3 Proposition *A frame L is κ-collectionwise normal iff for any co-discrete $\{x_i\}_{i\in J}$ with $|J| \leqslant \kappa$, there is a discrete $\{u_i\}_{i\in J}$ such that $x_i \vee u_i = 1$ for all $i \in J$.*

Proof The implication "⇐" is obvious since any discrete system is disjoint.

Conversely, let $\{x_i\}_{i\in J}$ be a co-discrete system. Then there is a disjoint system $\{u_i\}_{i\in J}$ such that $u_i \vee x_i = 1$ for all i. Set

$$D = \{x \in L \mid x \wedge u_i \neq 0 \text{ for at most one } i\}$$

and $\overline{d} = \bigvee D$. Clearly $u_i \in D$, hence $u_i \leqslant \overline{d}$ for every i. Then, by 2.4.2, we have

$$\overline{d} \vee \bigwedge_{i\in J} x_i = \bigwedge_{i\in J}(\overline{d} \vee x_i) \geqslant \bigwedge_{i\in J}(u_i \vee x_i) = 1.$$

On the other hand, by the normality of L we have $u, v \in L$ such that $u \vee \bigwedge_{i\in J} x_i = 1 = v \vee \overline{d}$ and $u \wedge v = 0$. The system

$$\{y_i = u_i \wedge u\}_{i\in J}$$

is then the required discrete system. Indeed, $C = D \cup \{v\}$ is a cover of L (since $\bigvee C = \overline{d} \vee v = 1$), each $c \in C$ meets at most one y_i (since $y_i \wedge v \leqslant u \wedge v = 0$ for every i) and

$$y_i \vee x_i = (u_i \vee x_i) \wedge (u \vee x_i) = u \vee x_i \geqslant u \vee \bigwedge_{i\in J} x_i = 1$$

for every i. □

2.5 The next result is a counterpart of VII.1.4 for collectionwise normality.

Proposition *A frame L is κ-collectionwise normal iff for any co-discrete $\{x_i\}_{i\in J}$ with $|J| \leqslant \kappa$, there is a system $\{u_i^n\}_{i\in J}^{n\in\mathbb{N}}$ such that*

1. $\bigvee_{n=1}^{\infty}(x_i \vee u_i^n) = 1$ *and*
2. $\bigwedge_{n=1}^{\infty}(x_i \vee \bigwedge_{j\neq i}(u_j^n)^*) = 1$ *for every $i \in J$.*

Proof $\Rightarrow$: There is a disjoint $\{u_i\}_{i\in J}$ such that $u_i \vee x_i = 1$ for every i. Set $u_i^n = u_i$ for every $i \in J$ and $n \in \mathbb{N}$. 1 is obvious. Regarding 2, since $\{u_i\}_{i\in J}$ is a disjoint system, we have $u_i \leqslant u_j^*$ for every $i \neq j$. Hence

$$\bigwedge_{n=1}^{\infty}(x_i \vee \bigwedge_{j\neq i}(u_j^n)^*) = x_i \vee \bigwedge_{j\neq i} u_j^* \geqslant x_i \vee u_i = 1.$$

$\Leftarrow$: Let $\{x_i\}_{i\in J}$, $|J| \leqslant \kappa$, be a co-discrete subset of L and suppose $\{u_i^n\}_{i\in J}^{n\in\mathbb{N}} \subseteq L$ satisfies conditions 1 and 2. For each $i \in J$ set

$$u_i = \bigvee_{n=1}^{\infty}\big(u_i^n \wedge \bigwedge_{m\leqslant n}\bigwedge_{j\neq i}(u_j^m)^*\big).$$

Then

$$1 = \big(\bigvee_{n=1}^{\infty}(x_i \vee u_i^n)\big) \wedge \big(\bigwedge_{m=1}^{\infty}(x_i \vee \bigwedge_{j\neq i}(u_j^m)^*)\big) =$$

$$= \bigvee_{n=1}^{\infty}\big((x_i \vee u_i^n) \wedge \bigwedge_{n=1}^{\infty}(x_i \vee \bigwedge_{j\neq i}(u_j^m)^*)\big) \leqslant$$

$$\leqslant \bigvee_{n=1}^{\infty} \big((x_i \vee u_i^n) \wedge \bigwedge_{m\leqslant n} (x_i \vee \bigwedge_{j\neq i} (u_j^m)^*)\big) =$$

$$= \bigvee_{n=1}^{\infty} \big((x_i \vee u_i^n) \wedge \big(x_i \vee \bigwedge_{m\leqslant n} \bigwedge_{j\neq i} (u_j^m)^*\big)\big)$$

$$= \bigvee_{n=1}^{\infty} \big(x_i \vee \big(u_i^n \wedge \bigwedge_{m\leqslant n} \bigwedge_{j\neq i} (u_j^m)^*\big)\big) = x_i \vee u_i.$$

Moreover, the collection $\{u_i\}_{i\in J}$ is disjoint since

$$u_i \wedge u_j \leqslant \bigvee_{n=1}^{\infty} \bigvee_{n'=1}^{\infty} \big(u_i^n \wedge u_j^{n'} \wedge \bigwedge_{m\leqslant n} (u_j^m)^* \wedge \bigwedge_{m'\leqslant n'} (u_i^{m'})^*\big) = 0$$

for every $i \neq j$. Indeed, for $n \leqslant n'$,

$$u_i^n \wedge u_j^{n'} \wedge \bigwedge_{m\leqslant n} (u_j^m)^* \wedge \bigwedge_{m'\leqslant n'} (u_i^{m'})^* \leqslant u_i^n \wedge \bigwedge_{m'\leqslant n'} (u_i^n)^* = 0$$

and for $n' \leqslant n$,

$$u_i^n \wedge u_j^{n'} \wedge \bigwedge_{m\leqslant n} (u_j^m)^* \wedge \bigwedge_{m'\leqslant n'} (u_i^{m'})^* \leqslant u_j^{n'} \wedge (u_j^{n'})^* = 0.$$ □

2.6 Let $\mathcal{A} = \{A_1, A_2, \ldots, A_n, \ldots\}$ be a countable system of covers of a frame L. Recall II.7.2 and set

$$\alpha_n(x) = \bigvee\{y \in L \mid A_n y \leqslant x\}.$$

By the frame distributive law in L, $A_n\alpha_n(x) \leqslant x$ and hence $\bigvee_{n=1}^{\infty} \alpha_n(x) = x$ for all x whenever $\mathcal{A}$ is admissible.

2.6.1 Theorem *If a frame L admits an admissible countable system of covers, then it is collectionwise normal.*

Proof Let $\{x_i\}_{i\in J}$ be co-discrete in L with cover A witnessing it. Set

$$u_i = \bigvee\{a \in A \mid \forall j \neq i,\ a \leqslant x_j\}.$$

Obviously, $u_i \leqslant x_j$ for all $j \neq i$. Moreover $x_i \vee u_i = 1$ (since $1 = \bigvee A = \bigvee\{a \in A \mid a \leqslant x_i\} \vee \bigvee\{a \in A \mid a \nleqslant x_i\} \leqslant x_i \vee u_i$). Now, set

$$v_{in} = \alpha_n(u_i) \wedge \bigwedge_{k=1}^{n} (\alpha_k(x_i))^* \quad \text{and} \quad v_i = \bigvee_{n=1}^{\infty} v_{in}.$$

Let $i \neq j$ and $k \leqslant n$. Then

$$v_{in} \wedge \alpha_k(u_j) \leqslant (\alpha_k(x_i))^* \wedge \alpha_k(u_j) = 0$$

(because $u_j \leqslant x_i$ implies $\alpha_k(u_j) \leqslant \alpha_k(x_i)$ and then $\alpha_k(x_i)^* \leqslant \alpha_k(u_j)^*$). Hence $v_{in} \wedge v_{jk} = 0$ and, consequently, $v_i \wedge v_j = 0$. This shows that $\{v_i\}_{i\in J}$ is a disjoint family.

Finally, using 2.6,

$$x \vee v_i = \bigvee_{n=1}^{\infty} (x_i \vee v_{in}) = \bigvee_{n=1}^{\infty} ((x_i \vee \alpha_n(u_i)) \wedge \bigwedge_{k=1}^{n} (x_i \vee (\alpha_k(x_i))^*)) =$$

$$= \bigvee_{n=1}^{\infty} (x_i \vee \alpha_n(u_i)) = x_i \vee \bigvee_{n=1}^{\infty} \alpha_n(u_i) = x_i \vee u_i = 1. \qquad \square$$

2.6.2 Corollary *Any metrizable frame is collectionwise normal.*

2.7 Recall VII.2.3 (for a detailed treatment of paracompactness see [220]). Corollary 2.6.2 can be improved to

Theorem *Any paracompact (that is, fully normal) frame is collectionwise normal.*

To prove it we need to recall VII.2.3 and a result concerning *distributive families*, that is, subsets S of L such that

$$x \vee \bigwedge T = \bigwedge_{t\in T} (x \vee t) \text{ for all } x \in L \text{ and } T \subseteq S.$$

2.7.1 Lemma *Let $S \subseteq L$ and $t \in L$.*

1. *If S is locally finite, then $\{s^* \mid s \in S\}$ is distributive.*
2. *If $s \prec t$ for every $s \in S$ and $\{s^* \mid s \in S\}$ is distributive, then $\bigvee S \prec t$.*

Proof

1. It suffices to show that $\bigwedge_{t\in T}(x \vee t^*) \leqslant x \vee \bigwedge_{t\in T} t^*$ for all $x \in L$ and $T \subseteq S$.
   Pick the cover A from the definition of a locally finite family (VII.2.3). For each $a \in A$ there are only finitely many $s \in S$ such that $a \wedge s \neq 0$. Denote this finite subset of S by $F(a)$. Then, for any $a \in A$,

$$x \vee \bigwedge_{t\in T} t^* = x \vee (\bigwedge_{t\in T\cap F(a)} t^* \wedge \bigwedge_{t\in T\smallsetminus F(a)} t^*) \geqslant x \vee (\bigwedge_{t\in T\cap F(a)} t^* \wedge a) =$$

$$= \bigwedge_{t\in T\cap F(a)} (x \vee t^*) \wedge (x \vee a) \geqslant \bigwedge_{t\in T} (x \vee t^*) \wedge a,$$

   and since A is a cover, this implies that

$$\bigwedge_{t\in T} (x \vee t^*) \leqslant x \vee \bigwedge_{t\in T} t^*.$$

2. $t \vee (\bigvee S)^* = t \vee \bigwedge_{s\in S} s^* = \bigwedge_{s\in S}(t \vee s^*) = 1.$ $\square$

Proof of the theorem Let $\{x_i\}_{i\in J}$ be co-discrete in L and denote by C the cover of L such that for every $c \in C$, $c \nleqslant x_i$ for at most one $i = i(c) \in J$. It follows by regularity that

$$D = \{d \in L \mid \exists c \in C : d \prec c\}$$

is a cover of L. Then D has a locally finite refinement U by paracompactness. Further, for each $u \in U$ there is a $c \in C$ (that we shall denote by $c(u)$) such that $u \prec c(u)$.

For $c \in C$ let

$$u_c = \bigvee\{u \in U \mid u \prec c\} \quad \text{and} \quad U_0 = \{u_c \mid c \in C\}.$$

U_0 is a cover of L (since $u_{c(u)} \geqslant u$ for any u). Since U is locally finite we can conclude by the lemma that $u_c \prec c$ for all $c \in C$.

Now, for each $i \in J$ set $v_i = \bigvee\{u_c \in U_0 \mid c \leqslant x_i\}$. Again by the lemma, $v_i \prec x_i$ for all i. Then, it suffices to show that the system

$$\{v_i^* \mid i \in J\}$$

is disjoint.

To prove that, consider the cover U_0. For each $c \in C$, $c \leqslant x_i$ for all $i \neq i(c)$. Hence $u_c \leqslant v_i$ so that $u_c \wedge v_i^* = 0$ for all $i \neq i(c)$. In conclusion, $u_c \wedge v_i^* \wedge v_j^* = 0$ for every $c \in C$ and $i \neq j$. Since U_0 is a cover, then $v_i^* \wedge v_j^* = 0$ for every $i \neq j$. □

3 Real-Valued Functions

3.1 There are nontrivial spaces admitting only a few continuous real-valued functions; in fact there are regular spaces X such that the only real-valued continuous functions on X are the constant ones. On "more separated" spaces (starting with the completely regular ones, in particular $\mathbb{R}$ itself or $\mathbb{R}^2$) continuous real-valued functions abound.

The abundance of continuous real-valued functions in a space X can be assessed by the existence of functions that separate all subsets that can possibly be separated.[4] Urysohn's lemma, one of the fundamental classical results of point-set topology, characterizes such topological spaces. They are precisely the *normal* ones.

In terms of characteristic functions, Urysohn's lemma means precisely that in any normal space, whenever $\chi_F \leqslant \chi_A$ for a closed F and an open A, there exists a continuous function $h: X \to \mathbb{R}$ such that

$$\chi_F \leqslant h \leqslant \chi_A.$$

[4] Two subsets U and V of X are *separable* if there is a continuous map $f: X \to \mathbb{R}$ such that $f[U] = \{1\}$ and $f[V] = \{0\}$; of course, this is only possible if the closures of U and V are disjoint.

The (insertion) theorem of Katětov-Tong[5] extends this characterization by replacing the characteristic maps χ_F and χ_A, respectively, by an arbitrary upper semicontinuous real-valued function and an arbitrary lower semicontinuous one.

Such facts show the importance of normality among separation axioms and thus deserve the detailed treatment in this chapter.

3.2 Continuous real functions The point-free reals are presented in A.10 in two equivalent ways, as the frames $\mathfrak{L}(\mathbb{R})$ and $\mathfrak{L}_0(\mathbb{R})$. We will use the simpler symbol $\mathfrak{L}(\mathbb{R})$ but the technique will be mostly that of $\mathfrak{L}_0(\mathbb{R})$.

In a frame L, the *continuous real-valued functions* [30] are the frame homomorphisms

$$\mathfrak{L}(\mathbb{R}) \to L.$$

We denote by

$$\mathfrak{R}(L)$$

the set of all continuous real-valued functions on L.

3.3 Trails Recall VI.1.5. Now it will be more expedient to have general scales defined on the rationals—we will call them trails. Thus, a *trail* in a frame L is a map $\sigma\colon \mathbb{Q} \to L$ such that

(T1) $\sigma(s) \prec \sigma(r)$ whenever $r < s$ and
(T2) $\bigvee\{\sigma(r) \mid r \in \mathbb{Q}\} = 1 = \bigvee\{\sigma(r)^* \mid r \in \mathbb{Q}\}$.

3.3.1 Remark By (T1), a trail σ is necessarily an order-reversing map. Conversely, any order-reversing $\sigma\colon \mathbb{Q} \to L$ such that for each $p < q$ in $\mathbb{Q}$ there is some complemented element $a_{pq} \in L$ with $\sigma(q) \leqslant a_{pq} \leqslant \sigma(p)$ satisfies (T1). Indeed, $\sigma(p) \vee \sigma(q)^* \geqslant a_{pq} \vee a_{pq}^* = 1$. Note that in particular if $\sigma(p)$ is complemented for all p, then σ satisfies (T1) iff it is order-reversing.

3.4 Trails provide a general method for defining continuous real functions on a frame.

3.4.1 Lemma *For each trail σ in L the formulas*

$$f(r\cdot\!-) = \bigvee\{\sigma(s) \mid s > r\} \quad \textit{and} \quad f(-\!\cdot r) = \bigvee\{\sigma(s)^* \mid s < r\} \quad (r \in \mathbb{Q})$$

[5] Originally announced by Tong in 1948 (the proof was, however, not published until 1952 [273]), Katětov shares the name of the theorem because of his independent version of 1951, with a simpler proof [177]. Such results have roots in Baire (1905, [8]), Hahn (1917, [134]) and Dieudonné (1944, [78]) who proved it for the real line, metrizable spaces and paracompact spaces, respectively.

determine a frame homomorphism

$$f : \mathfrak{L}(\mathbb{R}) \to L$$

such that $f(r\cdot-) \leqslant \sigma(r) \leqslant f(-\cdot r)^*$ *for every* $r \in \mathbb{Q}$.

Proof The inequality $f(r\cdot-) \leqslant \sigma(r)$ is obvious; the other is also easy since $f(-\cdot r)^* = \bigwedge_{s<r} \sigma(s)^{**}$ and for each $s < r$, $\sigma(r) \leqslant \sigma(s) \leqslant \sigma(s)^{**}$.

To show that f is a frame homomorphism it suffices to check that it turns the defining relations (r1)–(r6) of $\mathfrak{L}(\mathbb{R})$ into identities in L (recall A.9.4.1). Relations (r3) and (r4) follow immediately from the fact that $\mathbb{Q}$ is dense in itself, while (r5) and (r6) follow from the identities $\bigvee\{\sigma(r) \mid r \in \mathbb{Q}\} = 1 = \bigvee\{\sigma(r)^* \mid r \in \mathbb{Q}\}$ on trails. Finally:

(r1): For any $r \geqslant s$, $f(r\cdot-) \wedge f(-\cdot s) = \bigvee_{s'<s\leqslant r<r'} \sigma(r') \wedge \sigma(s')^* = 0$.
(r2): Let $r < s$. Then

$$f(r\cdot-) \vee f(-\cdot s) = \bigvee_{r'>r} \sigma(r') \vee \bigvee_{s'<s} \sigma(s')^* \geqslant \bigvee_{r<r'<s'<s} \sigma(r') \vee \sigma(s')^* = 1$$

since $\sigma(s') \prec \sigma(r')$. □

3.5 Basic examples of continuous functions are the

Constant functions For $p \in \mathbb{Q}$ define σ_p by $\sigma_p(r) = 1$ if $r < p$ and $\sigma_p(r) = 0$ otherwise. This is clearly a trail. The corresponding function in $\mathfrak{R}(L)$ given by 3.4.1 is defined by

$$\mathbf{p}(r\cdot-) = \begin{cases} 1 & \text{if } r < p \\ 0 & \text{if } r \geqslant p \end{cases} \quad \text{and} \quad \mathbf{p}(-\cdot r) = \begin{cases} 0 & \text{if } r \leqslant p \\ 1 & \text{if } r > p. \end{cases}$$

Other ones are the

Characteristic functions of complemented elements For any complemented $a \in L$ define the trail σ_a by $\sigma_a(r) = 1$ if $r < 0$, $\sigma_a(r) = a^*$ if $0 \leqslant r < 1$ and $\sigma_a(r) = 0$ if $r \geqslant 1$. This trail generates the *characteristic function* $\chi_a \in \mathfrak{R}(L)$ given by

$$\chi_a(r\cdot-) = \begin{cases} 1 & \text{if } r < 0 \\ a^* & \text{if } 0 \leqslant r < 1 \\ 0 & \text{if } r \geqslant 1 \end{cases} \quad \text{and} \quad \chi_a(-\cdot r) = \begin{cases} 0 & \text{if } r \leqslant 0 \\ a & \text{if } 0 < r \leqslant 1 \\ 1 & \text{if } r > 1. \end{cases}$$

3.6 More general real functions Recall the structure of the coframe $\mathsf{S}(L)$ of sublocales of a locale L from A.5. Now it will be more expedient to work in its dual lattice, the frame

$$\mathsf{S}(L)^{\mathrm{op}} = (\mathsf{S}(L), \subseteq)^{\mathrm{op}}$$

(we shall denote "$\subseteq^{\mathrm{op}}$" by $\leqslant$ but joins and meets in $\mathsf{S}(L)^{\mathrm{op}}$ will still be denoted by $\bigvee$ and $\bigwedge$; note that joins are given by $\bigcap$ in $\mathsf{S}(L)$, $0_{\mathsf{S}(L)^{\mathrm{op}}} = L$ and $1_{\mathsf{S}(L)^{\mathrm{op}}} = \mathsf{O} = \{1\}$.) A general (not necessarily continuous) *real-valued function* in L is a frame homomorphism

$$f \colon \mathfrak{L}(\mathbb{R}) \to \mathsf{S}(L)^{\mathrm{op}}.$$

The set $\mathsf{F}(L)$ of real-valued functions in L is partially ordered by

$$f \leqslant g \equiv f(p\cdot\!-) \leqslant g(p\cdot\!-) \text{ for every } p \in \mathbb{Q}$$

$$\Leftrightarrow g(-\!\cdot q) \leqslant f(-\!\cdot q) \text{ for every } q \in \mathbb{Q}.$$

3.6.1 Notes

1. For any real function f, the families

$$\{f(r\cdot\!-)\}_{r\in\mathbb{Q}} \quad \text{and} \quad \{f(-\!\cdot r)^*\}_{r\in\mathbb{Q}}$$

are trails in $\mathsf{S}(L)^{\mathrm{op}}$. Both generate f.
2. Let $\{S_r\}_{r\in\mathbb{Q}}$ and $\{T_r\}_{r\in\mathbb{Q}}$ be trails of sublocales that generate real functions f and g, respectively. Then

$$f \leqslant g \quad \Leftrightarrow \quad S_q \leqslant T_p \quad \text{for every } p < q.$$

Indeed, for $p < q$ in $\mathbb{Q}$ pick some $r \in \mathbb{Q}$ with $p < r < q$. Then $S_q \leqslant f(r\cdot\!-) \leqslant g(r\cdot\!-) \leqslant T_p$. Conversely, for each $p \in \mathbb{Q}$,

$$f(p\cdot\!-) = \bigvee_{r>p} \bigvee_{s>r} S_s \leqslant \bigvee_{r>p} T_r = g(p\cdot\!-).$$

3.7 Semicontinuous real functions From A.6.3 we see that the system $\mathfrak{c}\mathsf{S}$ of closed sublocales of L is a subframe of $\mathsf{S}(L)^{\mathrm{op}}$ isomorphic to L. An *upper* resp. *lower semicontinuous real function* in L is a real function

$$f \colon \mathfrak{L}(\mathbb{R}) \to \mathsf{S}(L)^{\mathrm{op}}$$

such that $f(-\!\cdot r) \in \mathfrak{c}\mathsf{S}$ resp. $f(r\cdot\!-) \in \mathfrak{c}\mathsf{S}$ (for all $r \in \mathbb{Q}$). Denote the classes of upper resp. lower semicontinuous real functions in L by

$$\mathsf{U}(L) \quad \text{resp.} \quad \mathsf{L}(L).$$

Then

the continuous real functions are the functions in $\mathsf{U}(L) \cap \mathsf{L}(L)$, *that is, the* $f\colon \mathfrak{L}(\mathbb{R}) \to \mathsf{S}(L)^{\mathrm{op}}$ *for which* $f[\mathfrak{L}(\mathbb{R})] \subseteq \mathfrak{c}\mathsf{S}$.

Indeed, by the frame isomorphism $\mathfrak{c}\colon L \to \mathfrak{c}\mathsf{S}$ they correspond precisely to the class of continuous functions $\mathfrak{R}(L)$ defined in 3.2. We will denote $\mathsf{U}(L) \cap \mathsf{L}(L)$ by $\mathsf{C}(L)$ to distinguish the two classes of functions in this bijection.

3.7.1 Note Classically, when discussing general (not necessarily continuous) maps with a source space (X, τ) one replaces the topology by the discrete one, the power set $\mathfrak{P}(X)$. That is, one makes use of the trivial covering map

$$D(X, \tau)\colon (X, \mathfrak{P}(X)) \to (X, \tau).$$

This is mimicked in point-free topology by one-to-one homomorphic extensions

$$L \to DL,$$

with DL "much more discrete" than L. This is the role of the totally disconnected $\mathsf{S}(L)^{\mathrm{op}}$ and of the embedding

$$\mathfrak{c} = (a \mapsto \mathfrak{c}(a))\colon L \hookrightarrow \mathsf{S}(L)^{\mathrm{op}}.$$

For the purposes of this section it works satisfactorily, but for a more extensive theory one may wish for more. (Generalized) discreteness is in the point-free context modelled by Boolean frames, and $\mathsf{S}(L)^{\mathrm{op}}$, although nicely generated by complemented elements is not Boolean. A properly discrete extension (at least for subfit frames) will be presented as the Boolean $\mathsf{S}_{\mathfrak{c}}(L)$ in the next chapter. For a treatment of general real functions on L based on the extension $\mathsf{S}_{\mathfrak{c}}(L)$ see [225] (and compare it with [151]). Unlike that based on $\mathsf{S}(L)^{\mathrm{op}}$ which is not always conservative, this extension is so. On the other hand, the $\mathsf{S}(L)^{\mathrm{op}}$ approach may still have some advantages in cases where functoriality is of the essence.

3.7.2 Example For a complemented sublocale S consider the trail

$$\sigma_S\colon \mathbb{Q} \to \mathsf{S}(L)^{\mathrm{op}}$$

given by $\sigma_S(r) = L$ if $r < 0$, $\sigma_S(r) = S^*$ if $0 \leqslant r < 1$ and $\sigma_S(r) = \mathsf{O}$ otherwise. By 3.4.1, this trail determines the *characteristic function* χ_S defined by

$$\chi_S(r\cdot\!-) = \begin{cases} L & \text{if } r < 0 \\ S^* & \text{if } 0 \leqslant r < 1 \\ \mathsf{O} & \text{if } r \geqslant 1 \end{cases} \quad \text{and} \quad \chi_S(-\!\cdot r) = \begin{cases} \mathsf{O} & \text{if } r \leqslant 0 \\ S & \text{if } 0 < r \leqslant 1 \\ L & \text{if } r > 1. \end{cases}$$

Note that

$\chi_S \in \mathsf{U}(L)$ *iff* S *is closed*

and

$\chi_S \in \mathsf{L}(L)$ *iff* S *is open.*

Hence $\chi_S \in \mathsf{C}(L)$ iff S is clopen (compare with the characteristic functions in 3.5).

3.7.3 Proposition *An* $f\colon \mathfrak{L}(\mathbb{R}) \to \mathsf{S}(L)^{\mathrm{op}}$ *is*

1. *lower semicontinuous iff for each* $p < q$ *in* $\mathbb{Q}$ *there is an* $a_{pq} \in L$ *such that* $f(q\cdot -) \leqslant \mathfrak{c}(a_{pq}) \leqslant f(p\cdot -)$, *and*
2. *upper semicontinuous iff for each* $p < q$ *in* $\mathbb{Q}$ *there is an* $a_{pq} \in L$ *such that* $f(-\cdot p) \leqslant \mathfrak{c}(a_{pq}) \leqslant f(-\cdot q)$.

Proof It suffices to prove the first statement (the second follows in a similar way). The implication "⇒" is obvious since each $f(p\cdot -)$ is closed. Conversely, let $p \in \mathbb{Q}$. Then

$$f(p\cdot -) = \bigvee_{r>p} f(r\cdot -) \leqslant \bigvee_{r>p} \mathfrak{c}(a_{pr}) \leqslant f(p\cdot -)$$

and hence $f(p\cdot -) = \bigvee_{r>p} \mathfrak{c}(a_{pr}) = \mathfrak{c}(\bigvee_{r>p} a_{pr})$ is closed. □

4 Katětov–Tong Insertion Theorem

4.1 Katětov relations The basic ingredient of our treatment of the insertion theorem will be certain binary relations on a lattice M, here called Katětov relations.[6] A *Katětov relation* on M is a binary relation $\Subset$ on M such that

(K1) $a \Subset b \Rightarrow a \leqslant b$.
(K2) $a' \leqslant a \Subset b \leqslant b' \Rightarrow a' \Subset b'$.
(K3) $a \Subset b$ and $a' \Subset b \Rightarrow (a \vee a') \Subset b$.
(K4) $a \Subset b$ and $a \Subset b' \Rightarrow a \Subset (b \wedge b')$.
(K5) $a \Subset b \Rightarrow \exists c \in M, a \Subset c \Subset b$.

The following technical lemma about Katětov relations is straightforward. For a detailed proof see [188].

4.1.1 Lemma *Let* M *be a lattice,* $\Subset$ *a Katětov relation on* M *and* $\lhd$ *a transitive and irreflexive relation on a countable set* D. *Further, let* $\{a_d\}_{d\in D}$ *and* $\{b_d\}_{d\in D}$ *be two families of elements of* M *such that*

$$d_1 \lhd d_2 \quad\Rightarrow\quad (a_{d_2} \leqslant a_{d_1},\, b_{d_2} \leqslant b_{d_1} \text{ and } a_{d_2} \Subset b_{d_1})\,.$$

[6]Katětov relations appear in the literature under various other names such as, e.g. *quasi-proximities* or *subordinations*. They share some of the defining properties of the so-called *strong relations* that describe the compactifications of a frame [27].

Then there exists a system $\{c_d\}_{d\in D}$ *in* M *such that*

$$d_1 \lhd d_2 \;\Rightarrow\; (c_{d_2} \Subset c_{d_1},\, a_{d_2} \Subset c_{d_1} \text{ and } c_{d_2} \Subset b_{d_1})\,.$$

4.2 For a frame L define a relation $\Subset$ on $\mathsf{S}(L)^{\mathrm{op}}$ by

$$S \Subset T \;\equiv_{\mathrm{df}}\; \exists a, b \in L,\; S \leqslant \mathfrak{o}(b) \leqslant \mathfrak{c}(a) \leqslant T.$$

It is easy to see that $\Subset$ satisfies conditions (K1)–(K4).

4.2.1 Proposition *The relation from 4.2 interpolates iff L is normal.*

Proof Let $\Subset$ satisfy the interpolation axiom (K5). Consider $a, b \in L$ such that $a \vee b = 1$. Then $\mathfrak{o}(a) \leqslant \mathfrak{c}(b)$. This means that $\mathfrak{o}(a) \Subset \mathfrak{c}(b)$ and hence there is a $T \in \mathsf{S}(L)^{\mathrm{op}}$ such that $\mathfrak{o}(a) \Subset T \Subset \mathfrak{c}(b)$. Therefore there are $u_1, u_2, v_1, v_2 \in L$ such that

$$\mathfrak{o}(a) \leqslant \mathfrak{o}(v_1) \leqslant \mathfrak{c}(u_1) \leqslant T \leqslant \mathfrak{o}(v_2) \leqslant \mathfrak{c}(u_2) \leqslant \mathfrak{c}(b).$$

Then, immediately, $u_1 \wedge v_2 = 0$ and $a \vee u_1 = 1 = b \vee v_2$.

Conversely, let $S \leqslant \mathfrak{o}(b) \leqslant \mathfrak{c}(a) \leqslant T$. Then $a \vee b = 1$ and by normality there are $u, v \in L$ such that $u \wedge v = 0$ and $a \vee u = 1 = b \vee v$. It is then clear that

$$S \leqslant \mathfrak{o}(b) \leqslant \mathfrak{c}(v) \leqslant \mathfrak{o}(u) \leqslant \mathfrak{c}(a) \leqslant T,$$

which implies $S \Subset \mathfrak{c}(v) \Subset T$. □

4.2.2 Corollary *If L is normal, then the relation* $\Subset$ *from 4.2 is a Katětov relation.*

Theorem 4.3 (Katětov-Tong insertion theorem) *A frame L is normal iff for any* $f \in \mathsf{U}(L)$ *and* $g \in \mathsf{L}(L)$ *such that* $f \leqslant g$ *there is an* $h \in \mathsf{C}(L)$ *such that* $f \leqslant h \leqslant g$.

Proof Suppose L is a normal frame and $\Subset$ is the Katětov relation defined in 4.2. Let $f \in \mathsf{U}(L)$ and $g \in \mathsf{L}(L)$ with $f \leqslant g$. Furthermore set

$$A_r = f(-\cdot r)^* \quad \text{and} \quad b_r = g(r\cdot -) \quad \text{for} \quad r \in \mathbb{Q}.$$

Note that each A_r (resp. B_r) is open (resp. closed),

$$A_q \leqslant A_p \quad \text{and} \quad B_q \leqslant B_p \quad \text{whenever } p < q.$$

Then $A_q \Subset B_p$ for any $p < q$ since $A_q = f(-\cdot q)^* \leqslant f(p\cdot -) \leqslant g(p\cdot -) = B_p$. Now we can use Lemma 4.1.1 with

$$M = \mathsf{S}(L), \quad D = \mathbb{Q} \quad \text{and} \quad \lhd = <$$

and get a family $\{C_r\}_{r\in\mathbb{Q}}$ of sublocales of L such that

$$C_q \Subset C_p, \quad A_q \Subset C_p \quad \text{and} \quad C_q \Subset B_p \quad \text{whenever } p < q.$$

First note that $\bigvee_{p\in\mathbb{Q}} C_p \geqslant \bigvee_{p\in\mathbb{Q}} A_p = L$ and $\bigvee_{p\in\mathbb{Q}} C_p{}^* \geqslant \bigvee_{p\in\mathbb{Q}} B_p{}^* = L$. Also, if $p < q$, then $C_q \Subset C_p$ and hence there exists $u \in L$ such that $C_q \leqslant \mathfrak{c}(u) \leqslant C_p$. Since $\mathfrak{c}(u)$ is complemented it follows from 3.3.1 that $\{C_r\}_{r\in\mathbb{Q}}$ satisfies (T1). Hence $\{C_r\}_{r\in\mathbb{Q}}$ is a trail. Let

$$h\colon \mathfrak{L}(\mathbb{R}) \to \mathsf{S}(L)^{\mathrm{op}}$$

be the corresponding real function. Clearly, $f \leqslant h$ and $h \leqslant g$ (since $A_q \leqslant C_p$ and $C_q \leqslant B_p$ whenever $p < q$). Finally, let us check that h is continuous.

Let $p < r < q$. Since $C_q \Subset C_p$, there are $t_{rq}, t_{pr}, s_{rq}, s_{pr} \in L$ such that

$$C_q \leqslant \mathfrak{o}(t_{rq}) \leqslant \mathfrak{c}(s_{rq}) \leqslant C_r \leqslant \mathfrak{o}(t_{pr}) \leqslant \mathfrak{c}(s_{pr}) \leqslant C_p.$$

For any $p < q$

$$h(q\cdot\!-) = \bigvee_{q'>q} C_{q'} \leqslant C_q \leqslant \mathfrak{c}(s_{rq}) \leqslant C_r \leqslant \bigvee_{p'>p} C_{p'} = h(p\cdot\!-)$$

and

$$h(-\cdot p) = \bigvee_{p'<p} C_{p'}{}^* \leqslant C_p{}^* \leqslant \mathfrak{o}(t_{pr})^* = \mathfrak{c}(t_{pr}) \leqslant C_r{}^* \leqslant \bigvee_{q'<q} C_{q'}{}^* = h(-\cdot q).$$

By 3.7.3, this shows that h is continuous.

Conversely, suppose $a \vee b = 1$ in L. Then $\mathfrak{o}(b) \leqslant \mathfrak{c}(a)$, that is,

$$\chi_{\mathfrak{c}(a)} \leqslant \chi_{\mathfrak{o}(b)}.$$

Hence there is a continuous h such that $\chi_{\mathfrak{c}(a)} \leqslant h \leqslant \chi_{\mathfrak{o}(b)}$. Let $u, v \in L$ such that $\mathfrak{c}(u) = h(\frac{1}{2}\cdot\!-)$ and $\mathfrak{c}(v) = h(-\cdot\frac{1}{2})$. Then

$$\begin{aligned} \mathsf{O} = 1_{\mathsf{S}(L)^{\mathrm{op}}} &= \chi_{\mathfrak{c}(a)}(-\cdot\tfrac{3}{4}) \vee \chi_{\mathfrak{c}(a)}(\tfrac{1}{2}\cdot\!-) \leqslant \\ &\leqslant \chi_{\mathfrak{c}(a)}(-\cdot\tfrac{3}{4}) \vee h(\tfrac{1}{2}\cdot\!-) = \mathfrak{c}(a) \vee \mathfrak{c}(u) = \mathfrak{c}(a \vee u) \end{aligned}$$

and

$$\begin{aligned} \mathsf{O} = 1_{\mathsf{S}(L)^{\mathrm{op}}} &= \chi_{\mathfrak{o}(b)}(\tfrac{1}{4}\cdot\!-) \vee h\chi_{\mathfrak{o}(b)}(-\cdot\tfrac{1}{2}) \leqslant \\ &\leqslant \chi_{\mathfrak{o}(b)}(\tfrac{1}{4}\cdot\!-) \vee h(-\cdot\tfrac{1}{2}) = \mathfrak{c}(b) \vee \mathfrak{c}(v) = \mathfrak{c}(b \vee v). \end{aligned}$$

Thus, $a \vee u = 1 = b \vee v$. On the other hand, $L = 0_{\mathsf{S}(L)^{\mathrm{op}}} = \mathfrak{c}(u) \wedge \mathfrak{c}(v) = \mathfrak{c}(u \wedge v)$, that is, $u \wedge v = 0$. This proves the normality of L. □

4.4 Urysohn's separation lemma Let L be a normal frame. As we have just seen, if $a \vee b = 1$, then $\chi_{\mathfrak{c}(a)} \leqslant \chi_{\mathfrak{o}(b)}$ and the insertion theorem yields an $h \in \mathsf{C}(L)$ such that

$$\chi_{\mathfrak{c}(a)} \leqslant h \leqslant \chi_{\mathfrak{o}(b)}.$$

Let $\widetilde{h} = \mathfrak{c}^{-1} \circ h$ be the corresponding frame homomorphism $\mathfrak{L}(\mathbb{R}) \to L$. It is a straightforward exercise to check that the condition $\chi_{\mathfrak{c}(a)} \leqslant h$ is equivalent to $\widetilde{h}(-\cdot 0) = 0$ and $\widetilde{h}(-\cdot 1) \leqslant a$, whereas $h \leqslant \chi_{\mathfrak{o}(b)}$ is equivalent to $\widetilde{h}(1\cdot -) = 0$ and $\widetilde{h}(0\cdot -) \leqslant b$. This gives us the nontrivial implication of a well-known characterization of normality due to Dowker and Papert [80].

Corollary *A frame L is normal iff for every $a, b \in L$ such that $a \vee b = 1$ there exists an $h\colon \mathfrak{L}(\mathbb{R}) \to L$ such that $h(-\cdot 0 \vee 1\cdot -) = 0$, $h(-\cdot 1) \leqslant a$ and $h(0\cdot -) \leqslant b$.*

This is a point-free version of the celebrated Urysohn's separation lemma [275].

4.5 Continuous extensions Let S be a sublocale of L and consider the corresponding quotient homomorphism $\nu_S\colon L \to S$ given by

$$\nu_S(a) = \bigwedge\{s \in S \mid a \leqslant s\}.$$

A map $\overline{f}$ in $\mathfrak{R}(L)$ is a *continuous extension* of $f \in \mathfrak{R}(S)$ if the diagram

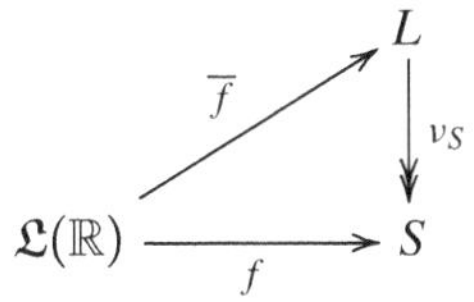

commutes.

4.5.1 Bounded functions An $f \in \mathfrak{R}(L)$ is *bounded* if $f(p\cdot q) = 1$ for some $p < q$ (or equivalently, if $f(-\cdot p) \vee f(q\cdot -) = 0$ for some $p < q$). Denote by

$$\mathfrak{R}^*(L)$$

the set of all bounded members f of $\mathfrak{R}(L)$ such that $f(-\cdot 0) \vee f(1\cdot -) = 0$.

4.5.2 Tietze's extension theorem Let L be a normal frame, $F = \mathfrak{c}(a)$ and $h \in \mathfrak{R}^*(F)$. Define $f\colon \mathfrak{L}(\mathbb{R}) \to \mathsf{S}(L)^{\mathrm{op}}$ by

$$f(-\cdot r) = \begin{cases} L & \text{if } r \leqslant 0 \\ \mathfrak{c}(h(-\cdot r)) & \text{if } 0 < r \leqslant 1 \\ \mathsf{O} & \text{if } r > 1 \end{cases}$$

and

$$f(r\cdot -) = \begin{cases} \mathsf{O} & \text{if } r < 0 \\ \bigvee_{s>r} \mathfrak{o}(h(-\cdot s)) & \text{if } 0 \leqslant r < 1 \\ L & \text{if } r \geqslant 1 \end{cases}$$

and $g\colon \mathfrak{L}(\mathbb{R}) \to \mathsf{S}(L)^{\mathrm{op}}$ by

$$g(r\cdot -) = \begin{cases} \mathsf{O} & \text{if } r < 0 \\ \mathfrak{c}(h(r\cdot -)) & \text{if } 0 \leqslant r < 1 \\ L & \text{if } r \geqslant 1. \end{cases}$$

and

$$g(-\cdot r) = \begin{cases} L & \text{if } r \leqslant 0 \\ \bigvee_{s<r} \mathfrak{o}(h(s\cdot -)) & \text{if } 0 < r \leqslant 1 \\ \mathsf{O} & \text{if } r > 1. \end{cases}$$

An easy computation shows that $f \in \mathsf{U}(L)$, $g \in \mathsf{L}(L)$ and $f \leqslant g$ ([220]). Hence by the Insertion Theorem there is a $k \in \mathsf{C}(L)$ such that $f \leqslant k \leqslant g$. Since

$$k(-\cdot 0 \vee 1\cdot -) \leqslant f(-\cdot 0) \vee g(1\cdot -) = L = 0_{\mathsf{S}(L)^{\mathrm{op}}}$$

k is bounded. Finally, the

$$\overline{h}\colon \mathfrak{L}(\mathbb{R}) \to L \quad \text{defined by} \quad \mathfrak{c}(\overline{h}(p\cdot q)) = k(p\cdot q)$$

is a continuous extension of the given h. Indeed, $\overline{h}(p\cdot q) \vee a = h(p\cdot q)$, that is, $k(p\cdot q) \vee \mathfrak{c}(a) = \mathfrak{c}(h(p\cdot q))$ (for all $p, q \in \mathbb{Q}$). To check this, observe first that $k(p\cdot q) \vee \mathfrak{c}(a) \leqslant (g(p\cdot -) \wedge f(-\cdot q)) \vee \mathfrak{c}(a) \leqslant \mathfrak{c}(h(p\cdot q))$. On the other hand, for each $r < q$ we have

$$\begin{aligned} h(-\cdot r) &= h(-\cdot r) \wedge (\overline{h}(-\cdot q) \vee \overline{h}(r\cdot -)) = \\ &= (h(-\cdot r) \wedge \overline{h}(-\cdot q)) \vee (h(-\cdot r) \wedge \overline{h}(r\cdot -)) \leqslant \\ &\leqslant \overline{h}(-\cdot q) \vee (h(-\cdot r) \wedge h(r\cdot -)) = \overline{h}(-\cdot q) \vee h(0) = \overline{h}(-\cdot q) \vee a. \end{aligned}$$

Likewise, $h(s\cdot -) \leqslant \overline{h}(p\cdot -) \vee a$ for every $s > p$. Hence, finally,

$$\begin{aligned} h(p\cdot q) &= (\bigvee_{r<q} h(-\cdot r)) \wedge (\bigvee_{s>p} h(s\cdot -)) \leqslant \\ &\leqslant (\overline{h}(-\cdot q) \vee a) \wedge (\overline{h}(p\cdot -) \vee a) = \overline{h}(p\cdot q) \vee a. \end{aligned}$$

This gives us the nontrivial implication of the point-free version of the (bounded) extension theorem of Tietze [272].

Corollary *A frame L is normal iff for any closed sublocale F of L and any $h \in \mathfrak{R}^*(F)$ there is a continuous extension $\overline{h} \in \mathfrak{R}^*(L)$.*

5 Dualizing Normality: Extremal Disconnectedness

5.1 Extemally disconnected frames Recall that a frame L is *extremally disconnected* (or *De Morgan*) if $a^* \vee a^{**} = 1$ for every $a \in L$ (in other words, if every regular element of L is complemented). This condition can be reformulated to

(edisc): *for any $a, b \in L$ with $a \wedge b = 0$ there are $u, v \in L$ such that $u \vee v = 1$ and $a \wedge u = 0 = b \wedge v$.*

Note that this is a condition dual to the (norm) in VII.1.1.

5.1.1 Proposition *The following statements about a frame L are equivalent.*

(1) *L satisfies (edisc).*
(2) *L is extremally disconnected.*
(3) *For every $a, b \in L$, $(a \vee b)^{**} = a^{**} \vee b^{**}$.*
(4) *The second De Morgan law holds in L.*
(5) *If $a \wedge b = 0$ with a and b regular, then there are $u, v \in L$ such that $u \vee v = 1$ and $a \wedge u = 0 = b \wedge v$.*
(6) *The interior of a closed sublocale of L is clopen.*
(7) *The closure of an open sublocale of L is clopen.*

Proof (1)⇒(2): Apply (edisc) to the identity $a \wedge a^* = 0$. Then $a^* \vee a^{**} \geqslant u \vee v = 1$.

(2)⇒(3): Since $(a \vee b)^{**} \geqslant a^{**} \vee b^{**}$ it suffices to show that $x \wedge (a \vee b)^* = 0$ implies $x \leqslant a^{**} \vee b^{**}$. Let $x \wedge a^* \wedge b^* = 0$. Then $x \wedge a^* \leqslant b^{**}$. Hence $x = x \wedge (a^* \vee a^{**}) = (x \wedge a^*) \vee (x \wedge a^{**}) \leqslant b^{**} \vee a^{**}$.

(3)⇒(4): Recall A.2.1.1. We have $(a \wedge b)^* = (a \wedge b)^{***} = (a^{**} \wedge b^{**})^* = (a^* \vee b^*)^{**}$. Then, by (3), $(a \wedge b)^* = a^* \vee b^*$.

(4)⇒(6): We have by De Morgan law $1 = (a \wedge a^*)^* = a^* \vee a^{**}$. Then $\mathfrak{c}(a^*) \cap \mathfrak{c}(a^{**}) = \mathsf{O}$ and $\mathfrak{c}(a^{**})$ is the complement of $\mathfrak{c}(a^*)$ (since $\mathfrak{c}(a^*) \vee \mathfrak{c}(a^{**}) = L$). Hence $\mathfrak{c}(a^{**}) = \mathfrak{o}(a^*)$ and $\operatorname{int} \mathfrak{c}(a) = \mathfrak{o}(a^*) = \mathfrak{c}(a^{**})$.

(6)⇒(7): $\operatorname{int}(\mathfrak{c}(a))$ is a clopen sublocale hence $\overline{\mathfrak{o}(a)} = \mathfrak{c}(a^*) = \mathfrak{o}(a^*)^* = (\operatorname{int} \mathfrak{c}(a))^*$ is also clopen.

(7)⇒(1): Since $\overline{\mathfrak{o}(a)} = \mathfrak{c}(a^*)$ is clopen it follows that any a^* is complemented in L. Let $a \wedge b = 0$ and pick $u = a^*$ and $v = a^{**}$. We have $u \vee v = 1$, $a \wedge u = 0$ and $b \wedge v = 0$.

(1)⇒(5) is trivial.

(5)⇒(2): Since $a^{**} \wedge a^* = 0$ and both the a^{**} and a^* are regular, there exist $u, v \in L$ such that $u \vee v = 1$ and $a^{**} \wedge u = 0 = v \wedge a^*$. This implies that $1 = u \vee v \leqslant a^{***} \vee a^{**} = a^* \vee a^{**}$. □

5.2 By passing to $\mathsf{S}(L)^{\mathrm{op}}$ via the isomorphism $L \cong \mathfrak{c}\mathsf{S}$ we can treat the notions of normality and extremal disconnectedness in parallel. Indeed, if we apply the condition (norm) to $\mathsf{S}(L)^{\mathrm{op}}$ we can conclude that a frame L is normal iff

(norm): *for any* $A, B \in \mathfrak{c}\mathsf{S}$ *such that* $A \vee B = \mathsf{O}$, *there are* $U, V \in \mathfrak{c}\mathsf{S}$ *such that* $U \wedge V = L$ *and* $A \vee U = \mathsf{O} = B \vee V$.

Similarly, a frame L is extremally disconnected iff

(edisc): *for any* $A, B \in \mathfrak{c}\mathsf{S}$ *such that* $A \wedge B = L$, *there are* $U, V \in \mathfrak{c}\mathsf{S}$ *such that* $U \vee V = \mathsf{O}$ *and* $A \wedge U = L = B \wedge V$.

5.2.1 We have proved in VII.1.6 that normality is invariant under closed localic maps. It will be instructive to see how the proof proceeds in the language of sublocales, using the above formulation of (norm) and the adjunction between images and preimages from A.7.2 (more precisely, its dual since we are now dealing with duals of sublocale lattices).

5.2.2 Proposition *Let* $f : L \to M$ *be a surjective closed localic map. If* L *is normal, then* M *is also normal.*

Proof Let A and B be closed sublocales of M with $A \vee B = \mathsf{O}_M$. Since

$$f_{-1}[-] \colon \mathsf{S}(M)^{\mathrm{op}} \to \mathsf{S}(L)^{\mathrm{op}}$$

is a frame homomorphism (see A.7.4.4), $f_{-1}[A]$ and $f_{-1}[B]$ are closed in L and $f_{-1}[A] \vee f_{-1}[B] = \mathsf{O}_L$. Then, by (norm), there are closed sublocales U_0 and V_0 of L such that $U_0 \wedge V_0 = L$ and $f_{-1}[A] \vee U_0 = \mathsf{O}_L = f_{-1}[B] \vee V_0$. Set

$$U = f[U_0] \quad \text{and} \quad V = f[V_0].$$

Clearly, U and V are closed sublocales of M such that $U \wedge V = f[U_0 \wedge V_0] = f[L] = M$. Moreover, since $f_{-1}[-]$ being a frame homomorphism preserves complements, $U_0 \geqslant f_{-1}[A]^* = f_{-1}[A^*]$ and thus

$$U = f[U_0] \geqslant ff_{-1}[A^*] \geqslant A^*,$$

that is, $A \vee U = \mathsf{O}_M$. Similarly, $B \vee V = \mathsf{O}_M$. □

5.2.3 Note Replacing "closed sublocale" by "open sublocale" and "closed map" by "open map" in the above proof makes it a valid reasoning still. Since changing $\mathfrak{c}\mathsf{S}$ (the lattice of *closed sublocales*) for $\mathfrak{o}\mathsf{S}$ (the lattice of *open sublocales*) in (norm) leads to a condition clearly equivalent to (edisc), we have an immediate proof of the following parallel result for extremal disconnectedness.

Proposition *Let* $f : L \to M$ *be a surjective open localic map. If* L *is extremally disconnected, then* M *is also extremally disconnected.*

It will be now our aim to explore this duality between the notions of normality and extremal disconnectedness.[7] We start by unifying several variants of normality under a single definition.

5.3 A relative notion of normality Let $\mathscr{C}$ be a *selection function* that assigns a system $\mathscr{C}L$ of complemented sublocales of L to L. A frame L is $\mathscr{C}$*-normal* if

($\mathscr{C}$-norm): *for any $A, B \in \mathscr{C}L$ such that $A \vee B = \mathsf{O}$, there are $U, V \in \mathscr{C}L$ such that $U \wedge V = L$ and $A \vee U = \mathsf{O} = B \vee V$.*

Denote by $\mathscr{C}^{\mathsf{c}}$ the selection function whose system of $\mathscr{C}^{\mathsf{c}}$-sublocales is the set $\{F^{\mathsf{c}} \mid F \in \mathscr{C}L\}$ of complements of all elements in $\mathscr{C}L$. A frame L is $\mathscr{C}$*-disconnected* if it is $\mathscr{C}^{\mathsf{c}}$-normal, that is,

($\mathscr{C}$-disc): *for any $A, B \in \mathscr{C}L$ such that $A \wedge B = L$, there are $U, V \in \mathscr{C}L$ such that $U \vee V = \mathsf{O}$ and $A \wedge U = L = B \wedge V$.*

5.3.1 Examples The following table lists the examples given by some of the standard selections.

	$\mathscr{C}L$	$\mathscr{C}$-normal frames	$\mathscr{C}$-disconnected frames
1.	$\mathfrak{c}(a)$ $(a \in L)$	Normal	Extremally disconnected
2.	$\mathfrak{c}(a)$ (a is regular)	Mildly normal	Extremally disconnected
3.	$\mathfrak{c}(a)$ $(a \in \mathrm{Coz}\,(L))$	All frames	F-frames[8]
4.	$\mathfrak{c}(a)$ (a is δ-regular)	δ-normal	extremally δ-disconnected

The first example is obvious. In Example 2, $\mathscr{C}$-disconnected frames are precisely the extremally disconnected ones by 5.1.1(5) (recall also (3) in A.2.1.1). Regarding Example 3, the fact that any frame is $\mathscr{C}$-normal is a consequence of Proposition 5 in [23], whereas the fact that $\mathscr{C}$-disconnected frames are precisely the F-frames follows from Proposition 8.4.10 in [18].

The last example yields the classes of frames that appear in the literature under the name of δ-normal and extremally δ-disconnected frames (see [131] for more information).

5.3.2 Note In examples 1, 3 and 4 in the table above, $\mathscr{C}L$ is a sublattice of $\mathsf{S}(L)^{\mathrm{op}}$. However, in example 2, the class $\mathscr{C}L$ of regular-closed sublocales is closed under finite meets only.

[7] T. Kubiak was the first to notice that several pairs of results in classical topology characterizing the concepts of normality and extremal disconnectedness show a "remarkable duality" [187] between the two concepts: each pair is identical in structure but prove facts about normal spaces on one side of the pair and about extremally disconnected spaces on the other [185]. The point-free study of this duality was undertaken in [131].

[8] *F-frames* are the frames in which the open quotient of any (dense) cozero element is a C^*-quotient.

5.4 Semicontinuities Recall 3.7.3. This is the right motivation for introducing general relative notions of continuity and semicontinuities.

A real function $f\colon \mathfrak{L}(\mathbb{R}) \to \mathsf{S}(L)^{\mathrm{op}}$ is

- *lower $\mathscr{C}$-semicontinuous* if for each $p < q$ in $\mathbb{Q}$, there is an $F_{pq} \in \mathscr{C}L$ such that $f(q\cdot\text{—}) \leqslant F_{pq} \leqslant f(p\cdot\text{—})$,
- *upper $\mathscr{C}$-semicontinuous* if for each $p < q$ in $\mathbb{Q}$, there is an $F_{pq} \in \mathscr{C}L$ such that $f(\text{—}\cdot p) \leqslant F_{pq} \leqslant f(\text{—}\cdot q)$, and
- *$\mathscr{C}$-continuous* if it is both lower and upper $\mathscr{C}$-semicontinuous.

Denote the classes of upper $\mathscr{C}$-semicontinuous, lower $\mathscr{C}$-semicontinuous and $\mathscr{C}$-continuous functions in L, respectively, by $\mathscr{C}$-$\mathsf{U}(L)$, $\mathscr{C}$-$\mathsf{L}(L)$ and $\mathscr{C}$-$\mathsf{C}(L)$.

5.4.1 Observation *An f is in $\mathscr{C}$-$\mathsf{U}(L)$ iff it is in $\mathscr{C}^{\mathsf{c}}$-$\mathsf{L}(L)$.*

(Indeed, if for $p < q$ in $\mathbb{Q}$ there is an $F_{pq} \in \mathscr{C}L$ such that $f(q\cdot\text{—}) \leqslant F_{pq} \leqslant f(p\cdot\text{—})$, pick a rational r such that $p < r < q$; since $f(r\cdot\text{—}) \leqslant F_{pr} \leqslant f(p\cdot\text{—})$, then $f(\text{—}\cdot p) \leqslant f(p\cdot\text{—})^* \leqslant F_{pr}{}^{\mathsf{c}}$ and, on the other hand, $f(\text{—}\cdot q) \vee F_{pr} \geqslant f(\text{—}\cdot q) \vee f(r\cdot\text{—}) = 1$ so that $F_{pr}{}^{\mathsf{c}} \leqslant f(\text{—}\cdot q)$.)

Hence

$$\mathscr{C}^{\mathsf{c}}\text{-}\mathsf{L}(L) = \mathscr{C}\text{-}\mathsf{U}(L) \quad \text{and} \quad \mathscr{C}^{\mathsf{c}}\text{-}\mathsf{C}(L) = \mathscr{C}\text{-}\mathsf{C}(L).$$

5.4.2 Examples For any $S \in \mathscr{C}L$, χ_S is upper $\mathscr{C}$-semicontinuous and $\chi_{S^{\mathsf{c}}}$ is lower $\mathscr{C}$-semicontinuous. The following table lists some other relevant examples (see [131] for more information).

	$\mathscr{C}L$	Lower $\mathscr{C}$-sc	Upper $\mathscr{C}$-sc	$\mathscr{C}$-continuous
1.	$\mathfrak{c}(a)$ ($a \in L$)	lsc	usc	Continuous
2.	$\mathfrak{c}(a)$ (a is regular)	Normal lsc	Normal usc	Normal continuous
3.	$\mathfrak{c}(a)$ ($a \in \mathrm{Coz}\,(L)$)	Zero lsc	Zero usc	Zero continuous
4.	$\mathfrak{c}(a)$ (a is δ-regular)	Regular lsc	Regular usc	Regular continuous

6 Dualizing Katětov–Tong Insertion Theorem

In this section we will formulate and prove an extended version of Katětov–Tong insertion theorem for $\mathscr{C}$-normality. A dual result for extremal $\mathscr{C}$-disconnectedness will follow in parallel.

6.1 Extended Katětov relation Recall the relation $\Subset$ from 4.2. It has an immediate formulation in terms of an arbitrary selection function $\mathscr{C}$. For $S, T \in \mathsf{S}(L)^{\mathrm{op}}$ set

$$S \Subset_{\mathscr{C}} T \equiv_{\mathrm{df}} \exists U \in \mathscr{C}L, \exists V \in \mathscr{C}^{\mathsf{c}}L\colon S \leqslant V \leqslant U \leqslant T.$$

Proposition 4.2.1 extends easily to

6.1.1 Proposition *The relation $\Subset_{\mathscr{C}}$ interpolates iff L is $\mathscr{C}$-normal.*

Proof Suppose that $\Subset_{\mathscr{C}}$ interpolates. Consider $A, B \in \mathscr{C}L$ such that $A \vee B = \mathsf{O}$. Then $A^{\mathsf{c}} \in \mathscr{C}^{\mathsf{c}}L$ satisfies $A^{\mathsf{c}} \leqslant B$. This means that $A^{\mathsf{c}} \Subset_{\mathscr{C}} B$ and hence there is a $T \in \mathsf{S}(L)$ such that $A^{\mathsf{c}} \Subset_{\mathscr{C}} T \Subset_{\mathscr{C}} B$. Hence there are $U_1, U_2 \in \mathscr{C}L$ and $V_1, V_2 \in \mathscr{C}^{\mathsf{c}}L$ such that

$$A^{\mathsf{c}} \leqslant V_1 \leqslant U_1 \leqslant T \leqslant V_2 \leqslant U_2 \leqslant B.$$

Then, immediately, $U_1 \wedge {V_2}^{\mathsf{c}} = L$ and $A \vee U_1 = \mathsf{O} = B \vee {V_2}^{\mathsf{c}}$.

Conversely, let $S \leqslant V \leqslant U \leqslant T$ for $S, T \in \mathsf{S}(L)$, $U \in \mathscr{C}L$ and $V \in \mathscr{C}^{\mathsf{c}}L$. Then $V^{\mathsf{c}} \vee U \geqslant V^{\mathsf{c}} \vee V = \mathsf{O}$ and by $\mathscr{C}$-normality there exist $A, B \in \mathscr{C}L$ such that $A \wedge B = L$ and $V^{\mathsf{c}} \vee A = \mathsf{O} = U \vee B$. This implies

$$S \leqslant V \leqslant A \leqslant B^{\mathsf{c}} \leqslant U \leqslant T,$$

that is, $S \Subset_{\mathscr{C}} A \Subset_{\mathscr{C}} T$. □

Note that in particular $\Subset_{\mathscr{C}^{\mathsf{c}}}$ interpolates iff L is $\mathscr{C}$-disconnected.

6.2 Katětov classes The relation $\Subset_{\mathscr{C}}$ from 6.1 above satisfies conditions (K1) and (K2). However (K3) and (K4) do not hold generally. Given a selection function $\mathscr{C}$, we say that the class of $\mathscr{C}$-sublocales of L is a *Katětov class* whenever $\Subset_{\mathscr{C}}$ also satisfies conditions (K3) and (K4). By 6.1.1,

> *if $\mathscr{C}L$ is a Katětov class in a $\mathscr{C}$-normal frame L, then $\Subset_{\mathscr{C}}$ is a Katětov relation.*

6.3 $\mathscr{C}$-Separable sublocales Two sublocales S and T of L are *completely $\mathscr{C}$-separable* if there is some $f \in \mathscr{C}\text{-}\mathsf{C}(L)$ such that $f(0 \cdot -) \leqslant S$ and $f(- \cdot 1) \leqslant T$.

6.4 Theorem *Let $\mathscr{C}$ be a selection function and let L be a frame such that $\mathscr{C}L$ is a Katětov class. The following statements about L are equivalent.*

(1) *L is $\mathscr{C}$-normal.*
(2) *For any $f \in \mathscr{C}\text{-}\mathsf{U}(L)$ and $g \in \mathscr{C}\text{-}\mathsf{L}(L)$ such that $f \leqslant g$ there is an $h \in \mathscr{C}\text{-}\mathsf{C}(L)$ such that $f \leqslant h \leqslant g$.*
(3) *Any $S, T \in \mathscr{C}L$ such that $S \vee T = \mathsf{O}$ are completely $\mathscr{C}$-separable.*

Proof (1)⇒(2): Let $\Subset_{\mathscr{C}}$ be the Katětov relation from 6.1. Set

$$A_r = f(- \cdot r)^* \quad \text{and} \quad B_r = g(r \cdot -) \quad \text{for} \quad r \in \mathbb{Q}.$$

It follows from Note 1 in 3.6.1 that $\{A_r\}_{r \in \mathbb{Q}}$ and $\{B_r\}_{r \in \mathbb{Q}}$ are trails generating f and g, respectively. In particular,

$$A_q \leqslant A_p \quad \text{and} \quad B_q \leqslant B_p \quad \text{whenever } p < q.$$

On the other hand, let $p, r, s, q \in \mathbb{Q}$ such that $p < r < s < q$. Since f is upper and g is lower $\mathscr{C}$-semicontinuous, there are $S_{sq}, T_{pr} \in \mathscr{C}L$ such that

$$f(-\cdot s) \leqslant S_{sq} \leqslant f(-\cdot q) \quad \text{and} \quad g(r\cdot -) \leqslant T_{pr} \leqslant g(p\cdot -).$$

Finally, since $f \leqslant g$, it follows from Note 2 in 3.6.1 that

$$A_q = f(-\cdot q)^* \leqslant S_{sq}{}^{\mathsf{c}} \leqslant f(-\cdot s)^* \leqslant f(r\cdot -) \leqslant g(r\cdot -) \leqslant T_{pr} \leqslant g(p\cdot -) = B_p,$$

that is, $A_q \Subset_{\mathscr{C}} B_p$.

Using Lemma 4.1.1 with

$$M = \mathsf{S}(L)^{\mathrm{op}}, \quad \Subset = \Subset_{\mathscr{C}}, \quad D = \mathbb{Q} \quad \text{and} \quad \lhd = <$$

we can pick a family $\{C_r\}_{r\in\mathbb{Q}}$ of sublocales of L such that

$$C_q \Subset_{\mathscr{C}} C_p, \quad A_q \Subset_{\mathscr{C}} C_p \quad \text{and} \quad C_q \Subset_{\mathscr{C}} B_p \quad \text{whenever } p < q.$$

Note that

$$\bigvee_{p\in\mathbb{Q}} C_p \geqslant \bigvee_{p\in\mathbb{Q}} A_p = \mathsf{O} \quad \text{and} \quad \bigvee_{p\in\mathbb{Q}} C_p{}^* \geqslant \bigvee_{p\in\mathbb{Q}} B_p{}^* = \mathsf{O}.$$

Further, if $p < q$, then $C_q \Subset_{\mathscr{C}} C_p$ and hence there is a $U \in \mathscr{C}L$ such that $C_q \leqslant U \leqslant C_p$. Since U is complemented, it follows from 3.3.1 that $\{C_r\}_{r\in\mathbb{Q}}$ satisfies (T1). Hence it is a trail and the associated generated function h has the following properties:

- $f \leqslant h$ (since $A_q \leqslant C_p$ whenever $p < q$, by 3.6.1.2) and
- $h \leqslant g$ (since $C_q \leqslant B_p$ whenever $p < q$, by 3.6.1.2).

It remains to be shown that h is $\mathscr{C}$-continuous. Let $p < r < q$. Since $C_q \Subset_{\mathscr{C}} C_p$, there exist $T_{rq}, T_{pr} \in \mathscr{C}^{\mathsf{c}}L$ and $S_{rq}, S_{pr} \in \mathscr{C}L$ such that $C_q \leqslant T_{rq} \leqslant S_{rq} \leqslant C_r$ and $C_r \leqslant T_{pr} \leqslant S_{pr} \leqslant C_p$. Then

$$h(q\cdot -) = \bigvee_{q'>q} C_{q'} \leqslant C_q \leqslant S_{rq} \leqslant C_r \leqslant \bigvee_{p'>p} C_{p'} = h(p\cdot -) \quad \text{and}$$

$$h(-\cdot p) = \bigvee_{p'<p} C_{p'}{}^* \leqslant C_p{}^* \leqslant T_{pr}{}^{\mathsf{c}} \leqslant C_r{}^* \leqslant \bigvee_{q'<q} C_{q'}{}^* = h(-\cdot q).$$

(2)⇒(3): Let $S, T \in \mathscr{C}L$ such that $S \vee T = \mathsf{O}$. Then χ_T is upper $\mathscr{C}$-semicontinuous, $\chi_{S^{\mathsf{c}}}$ is lower $\mathscr{C}$-semicontinuous and $\chi_T \leqslant \chi_{S^{\mathsf{c}}}$. By hypothesis, there is a $\mathscr{C}$-continuous h such that

$$\chi_T \leqslant h \leqslant \chi_{S^{\mathsf{c}}}.$$

This means that $h(-\cdot 1) \leqslant \chi_T(-\cdot 1) = T$ and $h(0\cdot -) \leqslant \chi_{S^c}(0\cdot -) = S$. Hence S and T are completely $\mathscr{C}$-separable.

(3)⇒(1): Let $A, B \in \mathscr{C}L$ with $A \vee B = \mathsf{O}$. Then there is a $\mathscr{C}$-continuous f such that $f(0\cdot -) \leqslant A$ and $f(-\cdot 1) \leqslant B$. Pick $U, V \in \mathscr{C}L$ such that

$$f(-\cdot \tfrac{1}{4}) \leqslant U \leqslant f(-\cdot \tfrac{1}{2}) \quad \text{and} \quad f(\tfrac{3}{4}\cdot -) \leqslant V \leqslant f(\tfrac{1}{2}\cdot -).$$

Clearly, $U \wedge V = L$ and $A \vee U = \mathsf{O} = B \vee V$. □

6.5 Stone-type insertion Using the duality

$$\mathscr{C}\text{-normality} \rightleftarrows \mathscr{C}\text{-disconnectedness}$$

the following result for $\mathscr{C}$-disconnectedness comes for free (by complementation).

Corollary *Let $\mathscr{C}$ be a selection function and let L be a frame such that $\mathscr{C}^{\mathsf{c}}L$ is a Katětov class. The following statements about L are equivalent.*

(1) *L is $\mathscr{C}$-disconnected.*
(2) *For any $f \in \mathscr{C}\text{-}\mathsf{U}(L)$ and $g \in \mathscr{C}\text{-}\mathsf{L}(L)$ such that $g \leqslant f$, there is an $h \in \mathscr{C}\text{-}\mathsf{C}(L)$ such that $g \leqslant h \leqslant g$.*
(3) *Any $S, T \in \mathscr{C}L$ such that $S \wedge T = L$ are completely $\mathscr{C}$-separable.*

The case $\mathscr{C}L = \mathfrak{c}\mathsf{S}$ is the point-free counterpart of Stone insertion theorem [268] presented in [121, 189].

6.6 Sufficient conditions for Katětov classes To see whether 6.4 and 6.5 cover all the examples of selection functions in 5.3.1, we need to analyse whether those examples yield Katětov classes. We have an obvious

6.6.1 Proposition *Every sublattice of complemented elements in $\mathsf{S}(L)^{\mathrm{op}}$ is a Katětov class.*

Hence the classes of closed sublocale (Example 1), coz-closed sublocales (Example 3) and δ-regular-closed sublocales (Example 4), as well as their corresponding classes of complements, are Katětov classes in any L. In order to conclude that the class of regular-closed sublocales (Example 2) is also a Katětov class we need the following result.

6.6.2 Proposition *If*

(a) $U_1, U_2 \in \mathscr{C}L \Rightarrow U_1 \vee U_2 \in \mathscr{C}L$ *and*
(b) $U_1, U_2 \in \mathscr{C}L,\ U_1 \wedge U_2 \geqslant V \in \mathscr{C}^{\mathsf{c}}L \Rightarrow \exists\, U \in \mathscr{C}L\colon\ U_1 \wedge U_2 \geqslant U \geqslant V$

then

(1) *$\mathscr{C}L$ is a Katětov class and*
(2) *$\mathscr{C}^{\mathsf{c}}L$ is a Katětov class whenever L is $\mathscr{C}^{\mathsf{c}}$-normal.*

Proof

(1) (K3): Let $S_i \leqslant U_i' \leqslant U_i \leqslant T$ with $U_i' \in \mathscr{C}^{\mathrm{c}}L$ and $U_i \in \mathscr{C}L$ $(i = 1, 2)$. Then

$$S_1 \vee S_2 \leqslant U_1' \vee U_2' \leqslant U_1 \vee U_2 \leqslant T,$$

where $U_1 \vee U_2 \in \mathscr{C}L$ by (a). Applying assumption (b) to the inequality

$$(U_1')^{\mathrm{c}} \wedge (U_2')^{\mathrm{c}} \geqslant (U_1)^{\mathrm{c}} \wedge (U_2)^{\mathrm{c}} \in \mathscr{C}^{\mathrm{c}}L$$

we get a $U \in \mathscr{C}L$ such that $U_1' \vee U_2' \leqslant U^{\mathrm{c}} \leqslant U_1 \vee U_2$. Then $S_1 \vee S_2 \leqslant U^{\mathrm{c}} \leqslant U_1 \vee U_2 \leqslant T$, that is, $S_1 \vee S_2 \Subset_{\mathscr{C}} T$ as required.

(K4): Let $S \leqslant U_i' \leqslant U_i \leqslant T_i$ with $U_i' \in \mathscr{C}^{\mathrm{c}}L$ and $U_i \in \mathscr{C}L$ $(i = 1, 2)$. Then

$$S \leqslant U_1' \wedge U_2' \leqslant U_1 \wedge U_2 \leqslant T_1 \wedge T_2,$$

where $U_1' \wedge U_2' \in \mathscr{C}^{\mathrm{c}}L$ by (a). Applying the assumption (b) to the inequality

$$U_1 \wedge U_2 \geqslant U_1' \wedge U_2' \in \mathscr{C}^{\mathrm{c}}L$$

we get a $U \in \mathscr{C}L$ such that $U_1' \wedge U_2' \leqslant U \leqslant U_1 \wedge U_2$. Then $S \leqslant U_1' \wedge U_2' \leqslant U \leqslant T_1 \wedge T_2$, that is, $S \Subset_{\mathscr{C}} T_1 \wedge T_2$.

(2) (K3): Let $S_i \leqslant U_i \leqslant U_i' \leqslant T$ with $U_i \in \mathscr{C}L$ and $U_i' \in \mathscr{C}^{\mathrm{c}}L$ $(i = 1, 2)$. Then $U_i \Subset_{\mathscr{C}^{\mathrm{c}}} U_i'$. By 6.1.1, $\Subset_{\mathscr{C}^{\mathrm{c}}}$ is interpolative hence

$$S_i \leqslant U_i \leqslant V_i' \leqslant V_i \leqslant U_i' \leqslant T$$

for $V_i \in \mathscr{C}L$ and $V_i' \in \mathscr{C}^{\mathrm{c}}L$. Then

$$S_1 \vee S_2 \leqslant U_1 \vee U_2 \leqslant V_1' \vee V_2' \leqslant V_1 \vee V_2 \leqslant U_1' \vee U_2' \leqslant T,$$

where $U_1 \vee U_2, V_1 \vee V_2 \in \mathscr{C}L$ (by (a)). Applying (b) to

$$(V_1')^{\mathrm{c}} \wedge (V_2')^{\mathrm{c}} \geqslant (V_1)^{\mathrm{c}} \wedge (V_2)^{\mathrm{c}} \in \mathscr{C}^{\mathrm{c}}L$$

we get $U \in \mathscr{C}L$ such that $V_1' \vee V_2' \leqslant U^{\mathrm{c}} \leqslant V_1 \vee V_2$. Hence

$$S_1 \vee S_2 \leqslant U_1 \vee U_2 \leqslant U^{\mathrm{c}} \leqslant T$$

and $S_1 \vee S_2 \Subset_{\mathscr{C}^{\mathrm{c}}} T$.

(K4): Let $S \leqslant U_i \leqslant U_i' \leqslant T_i$ with $U_i \in \mathscr{C}L$ and $U_i' \in \mathscr{C}^{\mathrm{c}}L$ $(i = 1, 2)$. Then $U_i \Subset_{\mathscr{C}^{\mathrm{c}}} U_i'$. By 6.1.1 we have

$$S \leqslant U_i \leqslant V_i' \leqslant V_i \leqslant U_i' \leqslant T_i$$

for a $V_i \in \mathscr{C}L$ and a $V_i' \in \mathscr{C}^{\mathsf{c}}L$. Then

$$S \leqslant U_1 \wedge U_2 \leqslant V_1' \wedge V_2' \leqslant V_1 \wedge V_2 \leqslant U_1' \wedge U_2' \leqslant T_1 \wedge T_2,$$

where $U_1' \wedge U_2', V_1' \wedge V_2' \in \mathscr{C}^{\mathsf{c}}L$. Finally, applying (b) to the inequality

$$V_1 \wedge V_2 \geqslant V_1' \wedge V_2' \in \mathscr{C}^{\mathsf{c}}L,$$

we get a $U \in \mathscr{C}L$ such that $V_1' \wedge V_2' \leqslant U \leqslant V_1 \wedge V_2$. Hence

$$S \leqslant U \leqslant U_1' \wedge U_2' \leqslant T_1 \wedge T_2$$

and $S \Subset_{\mathscr{C}^{\mathsf{c}}} T_1 \wedge T_2$ as required. □

The dual version of 6.6.2 follows immediately by complementation (note that $\mathscr{C}L$ satisfies properties (a) and (b) iff $\mathscr{C}^{\mathsf{c}}L$ satisfies (a') and (b') below).

6.6.3 Proposition *If*

(a') $U_1, U_2 \in \mathscr{C}L \;\Rightarrow\; U_1 \wedge U_2 \in \mathscr{C}L$ *and*
(b') $U_1, U_2 \in \mathscr{C}L,\; U_1 \vee U_2 \leqslant V \in \mathscr{C}^{\mathsf{c}}L \;\Rightarrow\; \exists\, U \in \mathscr{C}L:\; U_1 \vee U_2 \leqslant U \leqslant V$

then

(1) *$\mathscr{C}^{\mathsf{c}}L$ is a Katětov class and*
(2) *$\mathscr{C}L$ is a Katětov class whenever L is $\mathscr{C}$-normal.*

6.6.4 Corollary

1. *The class of regular-closed sublocales is a Katětov class in any mildly normal frame.*
2. *The class of regular-open sublocales is a Katětov class in any frame.*

Proof $\{\mathfrak{c}(a) \mid a \text{ is regular}\}$ has clearly the property (a'). To check property (b'), note just that for any regular elements a_1, a_2 and b in L, $\mathfrak{c}(a_1) \vee \mathfrak{c}(a_2) \leqslant \mathfrak{o}(b)$ implies

$$\mathfrak{c}(a_1) \vee \mathfrak{c}(a_2) \leqslant \mathfrak{c}((a_1 \vee a_2)^{**}) \leqslant \mathfrak{o}((a_1 \vee a_2)^{*}) \leqslant \mathfrak{o}(b)$$

and $(a_1 \vee a_2)^{**}$ is regular. □

Summing up, the extended versions of Katětov–Tong (6.4) and Stone (6.5) insertion results cover all examples in table 5.3.1. We refer the reader to [131] for more examples and information.

Chapter IX
Scatteredness: Joins of Closed Sublocales

The property to be discussed in this chapter delimits a rather special class of frames. About very restrictive properties one may sometimes hesitate to think about as of separation axioms (similarly like one does not usually think of discreteness as of a separation property: it says that distinct points can be separated by closed sets, but this is really somehow going too far). But

- this particular property is technically motivated by an analogy with a separation axiom about which there is not a slightest doubt,
- and, what is at least as relevant, the frames and spaces in question are in fact very important and useful.

Recall the characterizations of subfitness and fitness in terms of sublocales. We have seen in Chap. II that the formula for fitness stating that every closed sublocale is an intersection of open ones can be replaced by requiring that

every sublocale whatsoever *is an intersection of open ones.*

This comes without any restriction and hence, in particular, it still characterizes a property weaker than regularity. On the other hand, an analogous modification of subfitness, the requirement that

(ESJC): *Every Sublocale is a Join of Closed ones*

turns out to be very restrictive: we will see that it is equivalent with the property known (in spaces, but actually also in general frames) as *scatteredness*. This will be proved later in 2.4.1; so far let us use the abbreviation (ESJC).

But this is not all of the story. Joins of closed sublocales in general frames (unlike the ubiquitous meets of open ones) are fairly special objects [229]. It turns out that they constitute an interesting *frame* $\mathsf{S}_\mathfrak{c}(L)$ naturally embedded into the *coframe* $\mathsf{S}(L)$ of all sublocales of L. Studying this frame we gain a better insight into the structure of $\mathsf{S}(L)$, among other also into the role and status of the subspaces (induced sublocales) in among the general sublocales of a space (spatial frame).

J. Picado, A. Pultr, *Separation in Point-Free Topology*,
https://doi.org/10.1007/978-3-030-53479-0_9

Furthermore, for a subfit frame L the frame $\mathsf{S}_\mathfrak{c}(L)$ is a natural Boolean cover generalizing the (trivial) construction of the discrete cover of a space. This can be used e.g. for modelling discontinuity in the point-free context. It should be noted, however, that Boolean frames, although natural generalizations of discrete spaces, are much more richly structured than discrete spaces, and that such covers have many nontrivial features. The frames $\mathsf{S}_\mathfrak{c}(L)$ will be the topic of the second part of this chapter.

1 The Concept of Scatteredness in Frames

1.1 Recalling some basic properties of $\mathsf{S}(L)$ we will need Recall from A.6.4 that every sublocale S is a meet

$$S = \bigcap\{\mathfrak{o}(a) \vee \mathfrak{c}(b) \mid \mathfrak{o}(a) \vee \mathfrak{c}(b) \subseteq S\}.$$

Thus, the frame $\mathsf{S}(L)^{\mathrm{op}}$ is join-generated by complemented elements, that is,

- it is *zero-dimensional*,
- and, in particular, regular.

Further, recall that a sublocale S of a frame L is again a frame, and if $T \subseteq S \subseteq L$ then

- T is a sublocale of S iff it is a sublocale of L,

and (consequently)

- $\mathsf{S}(S)$ is embedded into $\mathsf{S}(L)$ preserving all meets (as the principal down-set $\downarrow_{\mathsf{S}(L)} S$).

1.2 A frame is said to be *scattered* [230, 232] if $\mathsf{S}(L)$ is a Boolean algebra.

1.2.1 Notes

1. In the original definition $\mathsf{S}(L)$ was required to be a frame. But of course, $\mathsf{S}(L)^{\mathrm{op}}$ is regular and hence being both a frame and a coframe makes it automatically Boolean (recall Theorem II.3.2.1).
2. Just assuming $\mathsf{S}(L)$ to be Boolean is not always a sufficiently strong condition. For instance, all finite frames are scattered in this definition. Therefore it makes sometimes a good sense to support it by a weak separation axiom. We will see in 2.4 that subfitness suits very well for this purpose.

1.3 First comparison with the classical concept It should be noted that the Boolean aspect appeared for a space X in the thorough analysis of $\mathsf{S}(\Omega(X))$ by Simmons in [256], and later in Niefield and Rosenthal [209]. We know that a sublocale of a space, that is, of its frame representation $\Omega(X)$, is not necessarily (induced by) a subspace. But a complemented sublocale of a space is always a subspace [220, VI.3.3]. Thus, the question whether $\mathsf{S}(\Omega(X))$ is Boolean is in fact equivalent to asking whether all sublocales of a given space are subspaces.

2 Joins of Closed Sublocales

2.1 Observation (ESJC) *is an hereditary property. Hence, in particular, it implies fitness.*

Proof Indeed, let S be a sublocale of an L with (ESJC), and let $T \subseteq S$ be a sublocale (of S and hence of L). Then a join of closed sublocales $T = \bigvee\{\mathfrak{c}(a) \mid \mathfrak{c}(a) \subseteq T\}$ in L is also such a join in S, because for $\mathfrak{c}(a) \subseteq S$ we have $\mathfrak{c}(a) = \mathfrak{c}(a) \cap S = \mathfrak{c}_S(a)$ (see, e.g., A.6.5), and the join in $\mathsf{S}(S)$ coincides with that in the $\mathsf{S}(L)$.

Since fitness is hereditary subfitness (II.5.3), the second statement follows. □

(We will see shortly that (ESCJ) is in fact much stronger than fitness.)

2.2 The reader probably remembers the beneficent role played in various situations by the down-set frame of a frame. For our purposes now it will be more expedient to use the analogous *up-set frame*

$$\mathfrak{U}(L) = \{A \subseteq L \mid \emptyset \neq A = \uparrow A\}$$

ordered, again, by inclusion so that the meets are the intersections and the joins are the unions, with the exception of the minimal $\bigvee \emptyset = \{1\}$.

Let S be a sublocale. Trivially, a join of up-sets $A_i \subseteq S$ is an up-set $A \subseteq S$ (which holds also for the non-standard $\bigvee \emptyset$) so that we have a largest up-set contained in S. It will be denoted by $\mathcal{U}(S)$. Since an up-set A is the union $A = \bigcup\{\uparrow a \mid a \in A\}$ we have

$$\mathcal{U}(S) = \bigcup\{\uparrow a \mid \uparrow a \subseteq S\},$$

the largest up-set contained in S (this join is always the standard union since we always have $\{1\} = \uparrow 1 \subseteq S$).

On the other hand, if A is a nonempty up-set and if $A \subseteq S$ for a sublocale S, then $\mathfrak{c}(a) = \uparrow a \subseteq S$ for each $a \in A$. Hence we have the smallest sublocale containing A, namely

$$\mathcal{J}(A) = \bigvee\{\mathfrak{c}(a) \mid \uparrow a \subseteq A\} = \bigvee\{\mathfrak{c}(a) \mid a \in A\} = \{\bigwedge B \mid B \subseteq A\}$$

with the joins computed in $\mathsf{S}(L)$.

Now we have monotone maps $\mathcal{U}\colon \mathsf{S}(L) \to \mathfrak{U}(L)$ and $\mathcal{J}\colon \mathfrak{U}(L) \to \mathsf{S}(L)$ and from the formulas we easily infer the following

2.2.1 Lemma *For any up-set $A \in \mathfrak{U}(L)$ and any sublocale $S \in \mathsf{S}(L)$ we have $\mathcal{J}(A) \subseteq S$ iff $A \subseteq \mathcal{U}(S)$, that is, we have a Galois adjunction*

$$\mathfrak{U}(L) \underset{\mathcal{U}}{\overset{\mathcal{J}}{\underset{\longleftarrow}{\overset{\longrightarrow}{\perp}}}} \mathsf{S}(L) \ .$$

Consequently, $\mathcal{J}$ preserves joins, and $\mathcal{U}$ preserves meets.

Now comes the crucial point of this section. Set

$$\sigma = \mathcal{U}\mathcal{J} \colon \mathfrak{U}(L) \to \mathfrak{U}(L).$$

We have

2.3 Proposition *σ is a nucleus in $\mathfrak{U}(L)$ so that $\sigma[\mathfrak{U}(L)]$ is a frame.*

Proof σ is monotone and we obviously have

$$A \subseteq \mathcal{U}\mathcal{J}(A) \quad \text{and} \quad \sigma\sigma = \mathcal{U}\mathcal{J}\mathcal{U}\mathcal{J} = \mathcal{U}\mathcal{J} = \sigma$$

so that we just have to prove that

$$\sigma(A) \cap \sigma(B) = \sigma(A \cap B).$$

The inclusion $\supseteq$ is trivial and the right Galois adjoint $\mathcal{U}$ preserves meets. Hence we have to prove the implication

$${\uparrow}x \subseteq \mathcal{J}(A) \cap \mathcal{J}(B) \quad \Rightarrow \quad {\uparrow}x \subseteq \mathcal{J}(A \cap B),$$

and since ${\uparrow}y \subseteq \mathcal{J}(X)$ whenever ${\uparrow}x \subseteq \mathcal{J}(X)$ and $y \geqslant x$ this reduces to the implication

$${\uparrow}x \subseteq \mathcal{J}(A) \cap \mathcal{J}(B) \quad \Rightarrow \quad x \in \mathcal{J}(A \cap B).$$

So let ${\uparrow}x \subseteq \mathcal{J}(A) \cap \mathcal{J}(B)$. Then, first, $x \in \mathcal{J}(A)$ and we have $x = \bigwedge_{i \in J} a_i$ for some $a_i \in A$. Now $a_i \in {\uparrow}x \subseteq \mathcal{J}(B)$ and hence we have $a_i = \bigwedge_{j \in K_i} b_{ij}$ with $b_{ij} \in B$. But A and B are up-sets and $b_{ij} \geqslant a_i$ so that b_{ij} are in $A \cap B$ and $x = \bigwedge_{i \in J} \bigwedge_{j \in K_j} b_{ij} \in \mathcal{J}(A \cap B)$. □

2.4 Let every sublocale of L be a join of closed sublocales. Then, in the notation above,

$$\mathcal{J}\mathcal{U}[\mathsf{S}(L)] = \mathsf{S}(L),$$

and we obtain

2.4.1 Theorem *The following statements about a frame are equivalent.*

(1) *Every sublocale of L is a join of closed sublocales.*
(2) *L is scattered and subfit.*
(3) *L is scattered and fit.*

Proof (1)⇒(3): The Galois adjunction from 2.2.1 provides an isomorphism between $\mathcal{JU}[\mathsf{S}(L)]$ and $\mathcal{UJ}[\mathfrak{U}(L)]$. The former is $\mathsf{S}(L)$ and the latter, $\sigma[\mathfrak{U}(L)]$, is by 2.3 a frame, and hence also $\mathsf{S}(L)$ is a frame. It is, then, both a frame and a coframe, and hence, by regularity, a Boolean algebra (recall 1.2.1).

L is fit by 2.1.

(3)⇒(2) is trivial.

(2)⇒(1): Consider a sublocale S of L. Since $\mathsf{S}(L)$ is Boolean, we have a complement $T = S^*$. In the standard representation we have

$$T = \bigwedge_{i \in J} \big(\mathfrak{o}(a_i) \vee \mathfrak{c}(b_i)\big)$$

and hence

$$S = T^* = \bigvee_{i \in J} \big(\mathfrak{c}(a_i) \wedge \mathfrak{o}(b_i)\big).$$

By virtue of subfitness, $\mathfrak{o}(b_i) = \bigvee_{j \in J_i} \mathfrak{c}(b_{ij})$ and hence (recall that we work in a frame, hence the distributivity is correct)

$$S = \bigvee_{i \in J} \big(\mathfrak{c}(a_i) \wedge \bigvee_{j \in J_i} \mathfrak{c}(b_{ij})\big) = \bigvee_{i \in J} \bigvee_{j \in J_i} \big(\mathfrak{c}(a_i) \wedge \mathfrak{c}(b_{ij})\big) = \bigvee_{i \in J} \bigvee_{j \in J_i} \mathfrak{c}(a_i \vee b_{ij}).$$

□

2.5 Note The subfitness in the theorem is essential. Indeed, a finite frame has a Boolean $\mathsf{S}(L)$. But in a finite $\mathsf{S}(L)$ *all* joins are finite and hence joins of closed sublocales are closed which does not necessarily make all of $\mathsf{S}(L)$.

Thus, one might be tempted to change the terminology and include the subfitness in the definition of scatteredness. But this would be imprudent: in the spatial case, we have to deal also with other matters (first of all, of course, with the classical property so termed) and as we will see in the following sections, the concept is conservative as it is.

3 Sublocales and Subspaces

3.1 Let us recall the two reasons for which sublocales of the frame $\Omega(X)$ are not always in a perfect correspondence with subspaces of a topological spaces X.

- First, we have the incorrect representation of subspaces by sublocales in spaces that are not T_D: it can happen that two distinct subspaces can have identical sublocale representations. This is not quite so important. We simply have to know that the spaces admitting correct point-free representation of subspaces should satisfy T_D. This leaves us with a pretty large class of spaces.
- The second reason, however, is really very important because this is what happens in all but rather exceptional cases: typically there are more sublocales than subspaces. This is, in fact, a useful phenomenon. To give an example: the system of regular open sets—the U's with $\operatorname{int}\overline{U} = U$—is a very important (and not at all esoteric) sublocale, and it is almost never a subspace. The status of subspaces in among sublocales is what we will be concerned with in the following pages.

The sublocales representing a subspace of a spatial frame $L = \Omega(X)$ are called *induced*, but without danger of confusion we can speak about them as of

$$\textit{subspaces of } L.$$

On the other hand, we will avoid speaking on them as *spatial sublocales* although this is sometimes done in literature. This expression may be confusing. Namely, it may be misunderstood for "sublocales that are themselves spatial frames" which is not the same.

Example In the naturally ordered unit interval $X = (\mathbb{I}, \leqslant)$ take the topology of all down-sets. Thus $L = \Omega(X)$ consists of the open sets

$$\downarrow x \quad \text{and} \quad < x = \{y \mid y < x\}.$$

L is linearly ordered, hence $u \to x$ in L is always either x or $\top$, so that the sublocales are precisely the subsets closed under meets. Such is, for instance

$$S = \{< x \mid x \in L\} \cup \{\top\}.$$

S is linearly ordered, hence spatial (in a linearly ordered frame all the elements but the top are prime), but it does not coincide with any $\Omega(Y)$ with $Y \subseteq X$ (which is an easy exercise).

3.2 The following fact (see [220, VI.3.3] for a proof) will play a most important role:

every complemented sublocale of $L = \Omega(X)$ *is a subspace.*

3.3 Spatial frames and points Recall from A.6.6 that the two-element sublocales of a frame L are precisely the pairs $P = \{p, 1\}$ with p prime. They are sometimes referred to as *one-point sublocales*,[1] or briefly as *points*.

In the construction based on primes (see, e.g., [220], or A.3.4) we have the spectrum ΣL represented as a sublocale of L, namely as

$$\Sigma L = \bigvee\{P \mid P \text{ point in } L\}.$$

Now if we take a T_D-space X we get

$$\Sigma\Omega(X) = \bigvee\{P \mid P \text{ point in } \Omega(X)\}$$

possibly larger than the original X. It is the *sober modification* of X (recall I.7.6); there, besides of the $x \in X$ represented as the points $\{X \smallsetminus \overline{\{x\}}, X\}$ one may also have $\{U, X\}$ with primes U not equal to any of the $X \smallsetminus \overline{\{x\}}$. Hence, in representing a space X we will restrict ourselves to the primes $X \smallsetminus \overline{\{x\}}$ and have an isomorphic representation

$$X \cong \bigvee\{\{X \smallsetminus \overline{\{x\}}, X\} \mid x \in X\}.$$

3.4 Representing subspaces We have

Proposition *Let $Y \subseteq X$. Then the corresponding subspace S of $\Omega(X)$ (that is, the sublocale S representing Y resp. $\Omega(Y)$) is*

$$S = \bigvee\{\{X \smallsetminus \overline{\{x\}}, X\} \mid x \in Y\}.$$

Proof This S comes from the adjunction

$$\Omega(X) \underset{V \mapsto k(V)}{\overset{U \mapsto U \cap Y}{\underset{\longleftarrow}{\overset{\longrightarrow}{\perp}}}} \Omega(Y)$$

where $k(V) = \bigcup\{U \mid U \cap Y = V\}$. That is, $S = \{U \mid k(U \cap Y) = U\}$.

We have always $(X \smallsetminus \overline{\{x\}}) \cap Y = Y \smallsetminus \overline{\{x\}}$. If $x \in Y$ the question is whether $X \smallsetminus \overline{\{x\}}$ is the largest U such that $U \cap Y = (X \smallsetminus \overline{\{x\}}) \cap Y$. Let U be larger; then we have a $y \in U \cap \overline{\{x\}}$, but then $U \ni x$, a contradiction.

If $x \notin Y$ use T_D and choose a $V \ni x$ such that $X \smallsetminus \{x\} = X \smallsetminus \overline{\{x\}}$. Then

$$(V \cup (X \smallsetminus \overline{\{x\}})) \cap Y = (Y \smallsetminus \{x\}) \cap (Y \smallsetminus \overline{\{x\}}) = Y \smallsetminus \overline{\{x\}}$$

while $X \smallsetminus \overline{\{x\}} \subsetneq V \cup (X \smallsetminus \overline{\{x\}}$. □

[1] They have *two elements* but this should not confuse us; the *void sublocale* O has *one element*.

3.4.1 Note Of course when we have a spatial L represented as the space

$$\bigvee\{P \mid P \text{ point in } L\}$$

then a sublocale $S \subseteq L$ is a subspace iff

$$S = \bigvee\{P \mid P \text{ point in } L,\ P \subseteq S\}.$$

4 Classical Scatteredness and Simmons Sublocale Theorem

In this section we will show that the point-free scatteredness agrees with the classical one (with the natural proviso that our spaces will be T_D). Furthermore, we will clarify the relation between sublocales and subspaces.

We will prove, first, some technical facts and then apply them to a property of classical spaces.

4.1 Notation For an element a in a frame L we will set

$$\mathsf{Pr}(a) = \{p \mid p \text{ prime},\ a \leqslant p\} \quad \text{and}$$

$$\mathsf{Pr}_{\min}(a) = \{p \mid p \text{ prime minimal with respect to } a \leqslant p\}.$$

By Zorn's lemma, $\bigwedge \mathsf{Pr}(a) = \bigwedge \mathsf{Pr}_{\min}(a)$ and hence

- *if L is spatial, then for every $a \in L$, $a = \bigwedge \mathsf{Pr}_{\min}(a)$.*

In a spatial L we will say that a minimal prime p is *essential* for a if

$$a \neq \bigwedge(\mathsf{Pr}_{\min}(a) \smallsetminus \{p\}).$$

An essential prime for a may not exist; if it does we say that a *has an essential prime*.

For $a \in L$ consider the Boolean sublocale

$$\mathfrak{b}(a) = \{x \to a \mid x \in L\} = \{x \mid x = (x \to a) \to a\}$$

(the smallest sublocale of L containing a, see A.6.7). Recall that always $x \leqslant (x \to a) \to a$.

4.2 Lemma *Each prime p such that $(p \to a) \to a = p$ is minimal.*

Proof Suppose $a \leqslant q < p$ for another $q \in \mathsf{Pr}(a)$. Since $p \wedge (p \to a) \leqslant a \leqslant q$ we have by primeness, $p \to a \leqslant q$ and hence $p \to a = q \wedge (p \to a) \leqslant p \wedge (p \to a) \leqslant a$ and $p = (p \to a) \to a = 1$, a contradiction. □

4.3 Lemma *A $p \in \mathsf{Pr}(a)$ is essential for a iff $p \in \mathfrak{b}(a)$.*

Proof

I. Set $b = \bigwedge(\mathsf{Pr}_{\min}(a) \smallsetminus \{p\})$. We have to prove that $b \neq a$. We have $p \wedge (p \to a) \leqslant a \leqslant q$ for any $q \in \mathsf{Pr}_{\min}(a)$, and if $q \neq p$, then by primeness $p \to a \leqslant q$ so that $p \to a \leqslant b$; since $p \wedge b = a$ we have $b \leqslant p \to a$ and conclude that

$$p \to a = b.$$

Thus $b \to a = (p \to a) \to a = p \neq 1$, and hence $b \nleqslant a$.

II. Let p be essential for a. For $b = \bigwedge(\mathsf{Pr}_{\min}(a) \smallsetminus \{p\})$ we have $b \wedge p = a$, and $b \nleqslant p$. Now $b \leqslant p \to a$ and

$$b \wedge ((p \to a) \to a) \leqslant (p \to a) \wedge ((p \to a) \to a) \leqslant a \leqslant p$$

and since $b \nleqslant p$, $(p \to a) \to a \leqslant p$, and finally $(p \to a) \to a = p$.

□

4.4 Theorem *Every sublocale S of a frame L is a subspace iff every $a \neq 1$ has an essential prime.*

Proof

I. If every sublocale of L is a subspace then, by 3.4.1, if $a \neq 1$,

$$\{1\} \neq \mathfrak{b}(a) = \bigvee\{\{p, 1\} \mid p \text{ prime}, \ \{p, 1\} \subseteq \mathfrak{b}(a)\}$$

and hence there is a prime $p \in \mathfrak{b}(a)$ essential for a by Lemma 4.3.

II. Now let S be a sublocale of L and let $a \in S$ be arbitrary. Set

$$\widetilde{a} = \bigwedge\{p \mid a \leqslant p \in S, \ p \text{ prime}\}.$$

If $a < \widetilde{a}$, $b = \widetilde{a} \to a \neq 1$ and hence it has an essential prime $p = (p \to b) \to b$. Now, however, we have

$$\begin{aligned} p &= (p \to (\widetilde{a} \to a)) \to (\widetilde{a} \to a) = \\ &= ((p \wedge \widetilde{a}) \to a) \to (\widetilde{a} \to a) = (\widetilde{a} \to a) \to (\widetilde{a} \to a) = 1, \end{aligned}$$

a contradiction.

□

4.5 Scatteredness This property appears in literature in various formulations. Perhaps the simplest is that

> *a space X is* scattered *if every nonempty subspace $A \subseteq X$ contains an isolated point.*

It is not necessary to check all the subspaces, though. We have

4.5.1 Proposition *A space X is scattered iff every nonempty closed subspace $A \subseteq X$ contains an isolated point.*

Proof Let $A \subseteq X$ be nonempty, let x be an isolated point of $\overline{A}$. Then there is an open $U \subseteq X$ such that $U \cap \overline{A} = \{x\}$ but since $x \in U$ and $x \in \overline{A}$, $U \cap A \neq \emptyset$ and hence $U \cap A = \{x\}$. □

4.6 Three lemmas

4.6.1 Lemma *Let x be isolated in $X \smallsetminus U$. Then $\{x\} \cup U$ is open.*

Proof Let W be an open set such that $(X \smallsetminus U) \cap W = \{x\}$. Then $\{x\} \cup U = ((X \smallsetminus U) \cap W) \cup U = W \cup U$. □

4.6.2 Lemma *Let x be isolated in $X \smallsetminus U$. Then*

$$((X \smallsetminus \overline{\{x\}}) \to U) \to U = X \smallsetminus \overline{\{x\}}.$$

Proof The Heyting operation in $\Omega(X)$ is given by

$$A \to B = \mathrm{int}((X \smallsetminus A) \cup B).$$

Thus, $V = (X \smallsetminus \overline{\{x\}}) \to U = \mathrm{int}(\overline{\{x\}} \cup U)$ and hence, by 4.6.1, $V \supseteq \{x\} \cup U$ and we have

$$V \to U \subseteq (\{x\} \cup U) \to U = \mathrm{int}(((X \smallsetminus \{x\}) \cap (X \smallsetminus U)) \cup U) =$$
$$= \mathrm{int}(X \smallsetminus \{x\}) = X \smallsetminus \overline{\{x\}}.$$

Since the other inclusion is trivial, the statement follows. □

4.6.3 Lemma *Let Q be essential for U and let $Q \subseteq X \smallsetminus \overline{\{x\}}$. Then*

$$U \subsetneq \bigwedge\{P \mid P \in \mathsf{Pr}_{\min}(U), U \subseteq P, P \nsubseteq X \smallsetminus \overline{\{x\}}\}.$$

Proof We have

$$U \subsetneq \bigwedge\{P \mid P \in \mathsf{Pr}_{\min}(U), U \subseteq P, P \neq Q\} \subseteq$$
$$\subseteq \bigwedge\{P \mid P \in \mathsf{Pr}_{\min}(U), U \subseteq P, P \nsubseteq X \smallsetminus \overline{\{x\}}\}.$$

□

4.7 Theorem *Every sublocale of a T_D-space X (that is, of its $\Omega(X)$) is a subspace iff X is scattered.*

Proof Let X be scattered, let $U \in \Omega(X)$, and $U \neq X$. Choose $x \in X \smallsetminus U$ isolated. Then we have by 4.2, 4.3 and 4.6.2, the essential

$$X \smallsetminus \overline{\{x\}} = ((X \smallsetminus \overline{\{x\}}) \to U) \to U$$

and, by 4.4, each sublocale is induced.

On the other hand, let every sublocale be induced and let $\emptyset \neq A \subseteq X$ be closed. Set $U = X \smallsetminus A$. By Theorem 4.4 and Lemma 4.6.3 there is an $x \in A$ such that

$$U \subsetneq V = \bigwedge\{P \mid P \in \mathsf{Pr}_{\min}(U), U \subseteq P, P \nsubseteq X \smallsetminus \overline{\{x\}}\}$$

(note that this meet is not necessarily, not even typically, the intersection).

We have $x \in V \smallsetminus U$ and, by T_D, an open $W \ni x$ such that $W \smallsetminus \{x\} = W \smallsetminus \overline{\{x\}}$. Now $x \in W \cap V \cap A$; let also $y \in W \cap V \cap A$ for some $y \neq x$. Then $y \in W \smallsetminus \overline{\{x\}}$ and hence $y \in W \cap X \smallsetminus \overline{\{x\}} = U$, a contradiction. Thus $W \cap V \cap A = \{x\}$ and we have an isolated $x \in A$. □

4.7.1 Note Since one needs the assumption T_D to have subspaces correctly represented as sublocales, Theorem 4.7 precisely describes the phenomenon of coincidence of subspaces and sublocales of spaces.

But one can also ask just whether $\mathsf{S}(\Omega(X))$ is Boolean or not, that is, whether $\Omega(X)$ is scattered as a frame. Then, by the general fact, we have still every sublocale represented as a subspace even if this subspace is in the non-T_D case not uniquely determined. A thorough analysis of this question was presented in Simmons [256], and later in Niefield and Rosenthal ([209], from which we borrowed Lemmas 4.2 and 4.3).

One needs modified definitions. A point $x \in X$ is *weakly isolated* in a closed set $A \subseteq X$ if there is an open U such that $\emptyset \neq A \cap U \subseteq \overline{\{x\}}$; a space is *weakly scattered* if every nonempty closed set has a weakly isolated point. For general spaces one then has that

> *a frame $\Omega(X)$ is scattered (that is, $\mathsf{S}(\Omega(X))$ is Boolean) iff X is weakly scattered.*

(Needless to say, in a T_D-space the concepts of *weakly isolated* and *isolated* coincide.)

4.8 So far we were interested in the question of complemented general sublocales. Such a sublocale is necessarily a subspace. On the other hand, a subspace is not necessarily complemented in $\mathsf{S}(\Omega(X))$ (typically, nontrivial dense subspaces are not). One knows that, for instance, open and closed subspaces are complemented. In 4.8.3 below we will present a characterization.

4.8.1 Observation *The join*

$$T = \bigvee\{\{X \smallsetminus \overline{\{x\}}, X\} \mid x \notin Y\}$$

is a supplement of the S from 3.4 iff $S \cap T = \mathsf{O}$.

4.8.2 Hereditary irresolvability and equi-dense sets In 1943, Hewitt introduced the concept of irresolvability [148]. Its hereditary variant can be formulated as follows.

Two subsets $A, B \subseteq X$ are *equi-dense* if $\overline{A} = \overline{B}$ and a space is *hereditarily irresolvable* if there are no nonempty disjoint equi-dense subsets.

4.8.3 Proposition *A subspace Y (that is, its representation S) is complemented in* $\mathsf{S}(\Omega(X))$ *iff there are no nonempty equi-dense sets A, B such that $A \subseteq Y$ and $B \subseteq X \smallsetminus Y$.*

Proof First, the sublocale T from 4.8.1 is a supplement of S unless $T \cap S \neq \mathsf{O}$. Indeed, let there be a $T' \subsetneq T$ with $T' \vee S = L$. Let, say, $X \smallsetminus \overline{\{z\}} \notin T'$ for a $z \notin Y$. Then $X \smallsetminus \overline{\{z\}} = U \cup V$ with $U \in T'$ and $V \in S$. By primeness either $X \smallsetminus \overline{\{z\}} = U$ which is impossible, or $X \smallsetminus \overline{\{z\}} = V \in S \cap T$.

Thus, S is complemented iff $S \cap T = \mathsf{O}$. Now we have $S \cap T \neq \mathsf{O}$ iff there are $\emptyset \neq A \subseteq Y$ and $B \subseteq X \smallsetminus Y$ such that

$$\bigwedge\{X \smallsetminus \overline{\{x\}} \mid x \in A\} = \bigwedge\{X \smallsetminus \overline{\{y\}} \mid y \in B\},$$

that is,

$$\operatorname{int}\bigcap\{X \smallsetminus \overline{\{x\}} \mid x \in A\} = \operatorname{int}\bigcap\{X \smallsetminus \overline{\{y\}} \mid y \in B\},$$

that is,

$$X \smallsetminus \overline{\bigcup_{x\in A} \overline{\{x\}}} = X \smallsetminus \overline{A} = X \smallsetminus \overline{B} = X \smallsetminus \overline{\bigcup_{y\in B} \overline{\{y\}}},$$

that is, iff $\overline{A} = \overline{B}$. □

As a consequence we immediately obtain

4.8.4 Theorem *Every subspace $Y \subseteq X$ (that is, its sublocale representation S) is complemented in* $\mathsf{S}(\Omega(X))$ *iff X is hereditarily irresolvable.*

4.8.5 Note We easily see that every scattered space is hereditarily irresolvable; this is simply because an isolated element of $\overline{A}$ is in A. The question whether scatteredness is strictly stronger than hereditary irresolvability was asked already in the original Hewitt's paper; in the same paper it was proved that the two concepts coincide for a broad class of spaces including all the metrizable spaces, and the locally compact Hausdorff ones. Later, in [7] the authors extended the coincidence class, and on the other hand presented examples of non-scattered hereditarily irresolvable spaces. Thus, there exist spaces in which all the subspaces are complemented while general sublocales are not always so. In other words, since a complemented sublocale of a space is always a subspace [220, VI.3.3], $\mathsf{S}(\Omega(X))$ of such a space is not Boolean, and its Boolean part consists precisely of the subspaces.

5 The Frame $\mathsf{S}_\mathfrak{c}(L)$

5.1 This very short section is just an extension of Sect. 2, a bridge to further consequences of the frame nature of the kernel of the adjunction of the maps $\mathcal{J}$ and $\mathcal{U}$.

Return to the subsections 2.2 and 2.3. On closer scrutiny one sees that the condition (ESCJ) did not yet play any role (it was crucial in 2.4, of course). Generally, we have the situation depicted in the following commutative diagram

$$\begin{array}{ccc} \mathfrak{U}(L) & \underset{\mathcal{U}}{\overset{\mathcal{J}}{\rightleftarrows}} & \mathsf{S}(L) \\ \downarrow{\scriptstyle \sigma'} & & \downarrow{\scriptstyle \kappa} \\ \mathcal{U}\mathcal{J}[\mathfrak{U}(L)] & \underset{\tilde{\mathcal{U}}}{\overset{\tilde{\mathcal{J}}}{\rightleftarrows}} & \mathsf{S}_\mathfrak{c}(L) = \mathcal{J}\mathcal{U}[\mathsf{S}(L)] \end{array}$$

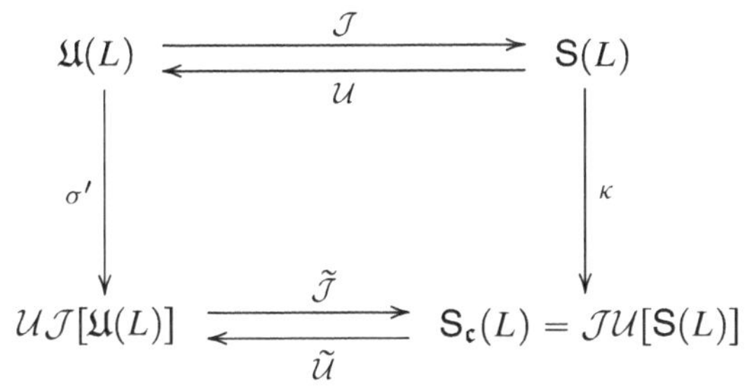

where $\tilde{\mathcal{J}}$ and $\tilde{\mathcal{U}}$ are restrictions of $\mathcal{J}$ and $\mathcal{U}$ (providing the isomorphisms between the kernels $\mathcal{U}\mathcal{J}[\mathfrak{U}(L)]$ and $\mathcal{J}\mathcal{U}[\mathsf{S}(L)]$) and κ is a restriction of $\mathcal{J}\mathcal{U}$. The diagram is commutative both in reading from left to right and from right to left because of the standard equalities $\mathcal{J}\mathcal{U}\mathcal{J} = \mathcal{J}$ and $\mathcal{U}\mathcal{J}\mathcal{U} = \mathcal{U}$. In the general case, unlike that of Sect. 2, $\mathcal{J}\mathcal{U}[\mathsf{S}(L)]$ does not have to coincide with $\mathsf{S}(L)$ and we have an onto map κ mapping $\mathsf{S}(L)$ onto the frame

$$\mathsf{S}_\mathfrak{c}(L) = \mathcal{J}\mathcal{U}[\mathsf{S}(L)].$$

The behaviour of this frame will be the main topic of the rest of this chapter.

5.2 Concentrate on the right-hand side of the diagram above

$$\begin{array}{c} \mathsf{S}(L) \\ {\scriptstyle j=\subseteq}\uparrow \;\; \downarrow{\scriptstyle \kappa} \\ \mathsf{S}_\mathfrak{c}(L) \end{array}$$

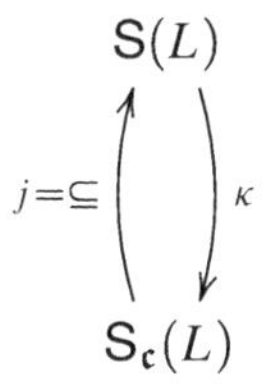

with $\mathsf{S}_\mathfrak{c}(L)$ isomorphic to $\mathcal{U}\mathcal{J}[\mathfrak{U}(L)]$ and hence a frame.

Thus the first fact that meets the eye is that we have here a somewhat peculiar embedding of a *frame* into a *coframe*.

But there is a (perhaps less conspicuous) feature of the embedding that deserves particular attention: the $\mathsf{S}_\mathfrak{c}(L)$ is a complete lattice closed under joins in $\mathsf{S}(L)$.

Recalling the definition of sublocale (closedness under meets) we cannot escape the question

isn't $\mathsf{S_c}(L)$*, besides of being a frame, also a sub-colocale of* $\mathsf{S}(L)$ *and hence also a coframe, moreover a very natural subobject of* $\mathsf{S}(L)$*?*

We will have a positive answer to that in the next section.

6 $\mathsf{S_c}(L)$ of a Subfit L (and of a T_1-Space)

6.1 Denote by $S \smallsetminus T$ the co-Heyting operation (the *difference*) in $\mathsf{S}(L)$.

Lemma *Let L be subfit. Then for any* $T \in \mathsf{S}(L)$ *and any* $x \in L$ *we have* $\mathfrak{c}(x) \smallsetminus T = {\uparrow}x \smallsetminus T \in \mathsf{S_c}(L)$.

Proof Let us use the standard representation

$$T = \bigcap_{i \in J} (\mathfrak{o}(a_i) \vee \mathfrak{c}(b_i)).$$

The adjointness of the difference (contravariant in the second variable and hence sending meets to joins) yields

$$\begin{aligned}\mathfrak{c}(x) \smallsetminus T = \mathfrak{c}(x) \smallsetminus (\bigcap_{i \in J} (\mathfrak{o}(a_i) \vee \mathfrak{c}(b_i))) = \bigvee_{i \in J} (\mathfrak{c}(x) \smallsetminus (\mathfrak{o}(a_i) \vee \mathfrak{c}(b_i))) =\\ = \bigvee_{i \in J} (\mathfrak{c}(x) \cap \mathfrak{c}(a_i) \cap \mathfrak{o}(b_i)) = \bigvee_{i \in J} (\mathfrak{c}(x \vee a_i) \cap \mathfrak{o}(b_i))\end{aligned}$$

(we have used the fact that for *complemented* B, $A \smallsetminus B = A \cap B^{\#}$ – recall A.5.6).

Now if L is subfit we can write $\mathfrak{o}(b_i)$ as $\bigvee_{j \in J_i} \mathfrak{c}(d_{ij})$ and conclude that

$$\mathfrak{c}(x) \smallsetminus T = \bigvee_{i \in J} (\mathfrak{c}(x \vee a_i) \cap \bigvee_{j \in J_i} \mathfrak{c}(d_{ij})) = \bigvee_{i \in J, j \in J_i} \mathfrak{c}(x \vee a_i \vee d_{ij})$$

(distributing $\mathfrak{c}(x \vee a) \cap \bigvee \mathfrak{c}(d_{ij})$ is correct since the $\mathfrak{c}(y)$ are complemented – recall A.6.2). □

6.2 Theorem *Let L be subfit. Then for every* $S \in \mathsf{S_c}(L)$ *and every* $T \in \mathsf{S}(L)$ *the difference* $S \smallsetminus T$ *is in* $\mathsf{S_c}(L)$.

Consequently,

- $\mathsf{S_c}(L)$ *is a sub-colocale of* $\mathsf{S}(L)$,
- *and it is a Boolean algebra.*

Proof S, an element of $\mathsf{S_c}(L)$, can be written as $\bigvee \mathfrak{c}(x_i)$. By Lemma 6.1 (and the standard distributivity of the difference following from the co-Heyting adjunction), $S \smallsetminus T = (\bigvee_{i \in J} \mathfrak{c}(x_i)) \smallsetminus T = \bigvee_{i \in J} (\mathfrak{c}(x_i) \smallsetminus T) \in \mathsf{S_c}(L)$ which is closed under joins.

Now $\mathsf{S}_\mathfrak{c}(L)$ is a sub-colocale recalling again that it is closed under joins. It is a Boolean algebra: $\mathsf{S}(L)^{\mathrm{op}}$ is regular, hence its sublocale $\mathsf{S}_\mathfrak{c}(L)^{\mathrm{op}}$ is regular, and since it is both a frame and a coframe, it is Boolean (recall II.3.2.1). □

6.3 Thus the mapping κ from the diagram above is a conucleus. But there is more to it. We have

Theorem *Let L be subfit. Then*

$$\kappa\colon \mathsf{S}(L) \to \mathsf{S}_\mathfrak{c}(L) \ \text{ is the (co-)Booleanization of } \mathsf{S}(L),$$

that is, if we denote by $S^\#$ *the supplement in* $\mathsf{S}(L)$,

$$\kappa(S) = S^{\#\#}.$$

Proof We already know that each $S^\# = L \smallsetminus S$ is in $\mathsf{S}_\mathfrak{c}(L)$ (use 6.2 for L which is, of course, in $\mathsf{S}_\mathfrak{c}(L)$) and since $\mathsf{S}_\mathfrak{c}(L)$ is a sub-colocale it is closed under the supplement. Moreover $\mathsf{S}_\mathfrak{c}(L)$ is a Boolean algebra and hence the restriction of the supplement is a complement, that is, for $S \in \mathsf{S}_\mathfrak{c}(L)$, $(S^\#)^\# = S$ and we conclude that

$$\mathsf{S}_\mathfrak{c}(L) = \{S^\# \mid S \in \mathsf{S}(L)\} = \mathfrak{B}(\mathsf{S}(L))^{\mathrm{op}},$$

the Booleanization of $\mathsf{S}(L)$.

Now it is immediate that κ, the restriction of $\mathcal{JU}$, is the left adjoint to the embedding j. But also $S \mapsto S^{\#\#}$ is obviously a left adjoint to j and hence $\kappa = (S \mapsto S^{\#\#})$. □

6.4 The condition of subfitness is essential. In fact, both the previous statements yield characterizations of subfitness.

Theorem *The following statements about a frame L are equivalent.*

(1) *L is subfit.*
(2) *$\kappa\colon \mathsf{S}(L) \to \mathsf{S}_\mathfrak{c}(L)$ is the Booleanization of $\mathsf{S}(L)$.*
(3) *$\mathsf{S}_\mathfrak{c}(L)$ is a sub-colocale of $\mathsf{S}(L)$.*
(4) *$\mathsf{S}_\mathfrak{c}(L)$ is Boolean.*

Proof (1)⇒(2) is in 6.3, and (2)⇒(3)&(4) is trivial.

(3)⇒(1): If $\mathsf{S}_\mathfrak{c}(L)$ is a sub-colocale, then $\mathfrak{o}(a) = L \smallsetminus \mathfrak{c}(a)$ is in $\mathsf{S}_\mathfrak{c}(L)$ and hence it is a join of closed sublocales.

(4)⇒(1): Generally, the meets $S \sqcap T$ in $\mathsf{S}_\mathfrak{c}(L)$ need not coincide with the meets $S \cap T$ in $\mathsf{S}(L)$ but we do have

$$\mathfrak{c}(a) \cap \mathfrak{c}(b) = \mathfrak{c}(a \vee b) = \mathfrak{c}(a) \sqcap \mathfrak{c}(b)$$

because the intersection is in $\mathsf{S}_\mathfrak{c}(L)$.

Now let $S = \bigvee_{i \in J} \mathfrak{c}(b_i)$ be the complement of $\mathfrak{c}(a)$ in $\mathsf{S}_\mathfrak{c}(L)$. We have

$$\mathsf{O} = \mathfrak{c}(a) \sqcap (\bigvee_{i \in J} \mathfrak{c}(b_i)) = \bigvee_{i \in J} (\mathfrak{c}(a) \sqcap \mathfrak{c}(b_i)) = \bigvee_{i \in J} (\mathfrak{c}(a) \cap \mathfrak{c}(b_i)) = \mathfrak{c}(a) \cap S$$

in $\mathsf{S}(L)$ and hence $S \subseteq \mathfrak{o}(a)$. On the other hand, $S \vee \mathfrak{c}(a) = L$ and hence $\mathfrak{o}(a) = \mathfrak{o}(a) \cap (S \vee \mathfrak{c}(a)) = \mathfrak{o}(a) \cap S$, that is, $\mathfrak{o}(a) \subseteq S$, and $\mathfrak{o}(a) = S \in \mathsf{S}_\mathfrak{c}(L)$. Hence each $\mathfrak{o}(a)$ is a join of closed sublocales. □

6.4.1 Note Hence, if L is not subfit, then $\mathsf{S}_\mathfrak{c}(L)$ is neither Boolean nor a subcolocale of $\mathsf{S}(L)$. But it does not mean that it cannot be a coframe. For instance, it is easy to prove that if L is both a frame and a coframe, then so is also $\mathsf{S}_\mathfrak{c}(L)$. The existence of an L such that $\mathsf{S}_\mathfrak{c}(L)$ is not a coframe is an open problem.

6.5 For T_1-spaces, $\mathsf{S}_\mathfrak{c}(\Omega(X))$ picks out precisely the subspaces among sublocales.

Theorem *Let X be a T_1-space, $L = \Omega(X)$. Then $\mathsf{S}_\mathfrak{c}(L) = \mathfrak{B}(\mathsf{S}(\Omega(X)))^{\mathrm{op}}$ is precisely the system of all subspaces of $\Omega(X)$.*

Proof Recall 3.3. We will represent, again, L as $\bigvee\{\{p, 1\} \mid p \in M\}$ with M the system of all the $p = X \smallsetminus \{x\}$, $x \in X$. The subspaces will be then precisely the sublocales of the form $S = \bigvee\{\{p, 1\} \mid p \in A,\ A \subseteq M\}$. Now since our $p = X \smallsetminus \{x\}$ are obviously maximal, we have the subspaces represented as

$$S = \bigvee\{\mathfrak{c}(p) = \uparrow p \mid p \in A,\ A \subseteq M\}$$

and hence we see that all of them are in $\mathsf{S}_\mathfrak{c}(L)$.

Thus it suffices to prove that every $S \in \mathsf{S}_\mathfrak{c}(L)$, that is, $S = \bigvee_{i \in J} \mathfrak{c}(a_i)$, is a subspace and this will immediately follow if we prove that for any a, $\mathfrak{c}(a) = \bigvee\{\mathfrak{c}(p) \mid p \in M, a \leqslant p\}$.

Set $M(x) = \{p \in M \mid x \leqslant p\}$ (so that, in particular, for every x, $x = \bigwedge M(x)$). Since $p \geqslant a$ yields $\mathfrak{c}(p) \leqslant \mathfrak{c}(a)$ we immediately have that

$$\mathfrak{c}(a) \supseteq \bigvee\{\mathfrak{c}(p) \mid p \in M(a)\}$$

and we just have to prove the opposite inclusion. Let $b \geqslant a$. Then every $q \in M(b)$ is in $M(a)$ and hence $M(b) \subseteq \bigcup\{M(p) \mid p \in M(a)\}$ and since $M(p) \subseteq \uparrow p$ we see that $b \in \bigvee\{\mathfrak{c}(p) \mid p \in M(a)\}$. □

6.6 Another important feature of subfit locales in this context is that we have a frame embedding

$$\mathfrak{o} = (a \mapsto \mathfrak{o}(a))\colon L \hookrightarrow \mathsf{S}_\mathfrak{c}(L)$$

(although the embedding $\mathsf{S}_\mathfrak{c}(L) \subseteq \mathsf{S}(L)$ in general does not preserve finite meets, since the intersection $\mathfrak{o}(a) \cap \mathfrak{o}(b) = \mathfrak{o}(a \wedge b)$ in $\mathsf{S}(L)$ is open, it is in $\mathsf{S}_\mathfrak{c}(L)$ and hence coincides with the meet in this frame).

Although we have this embedding only for the subfit frames (by the sublocale definition of subfitness), it is a part of a more general situation. This will be discussed in the next section.

6.7 The Boolean ("discrete") cover For a subfit L, the embedding

$$\mathfrak{o} = (a \mapsto \mathfrak{o}(a)) \colon L \hookrightarrow \mathsf{S}_\mathfrak{c}(L)$$

can be viewed as an extension of the discrete cover of a space, the identity carried $\delta_X \colon DX = (X, \mathfrak{P}(X)) \to X = (X, \tau)$. Even in spaces, where this mapping is trivial, it is useful in dealing with mappings $X \to Y$ that are not continuous.

What happens in T_1-spaces: In the proof of 6.5 we saw that the mapping

$$\phi = (A \mapsto \bigvee\{\mathfrak{c}(X \smallsetminus \{x\}) \mid x \in A\}) \colon \mathfrak{P}(X) = \Omega(DX) \to \mathsf{S}_\mathfrak{c}(\Omega(X))$$

is one-to-one onto, and since it is obviously monotone, it is a Boolean isomorphism. Now consider the embedding

$$\mathfrak{o}_{\Omega(X)} \colon \Omega(X) \to \mathsf{S}_\mathfrak{c}(\Omega(X)).$$

We have

$$\mathfrak{o}(U) = \bigvee\{\mathfrak{c}(V) \mid \mathfrak{c}(V) \subseteq \mathfrak{o}(U)\} = \bigvee\{\mathfrak{c}(V) \mid V \cup U = X\}.$$

In a T_1-space, $V = \bigcap\{X \smallsetminus \{x\} \mid x \notin V\}$, and

$$\bigcap\{X \smallsetminus \{x\} \mid x \notin V\} \cup U = \bigcap\{(X \smallsetminus \{x\}) \cup U \mid x \notin V\} = X$$

iff for all $x \notin V$, $(X \smallsetminus \{x\}) \cup U = X$, that is, for all $x \notin V$, $x \in U$, and we see that

$$\mathfrak{o}(U) = \bigvee\{\mathfrak{c}(X \smallsetminus \{x\}) \mid x \in U\}.$$

Since obviously $\Omega(\delta)(U) = U$, we have the commutative diagram

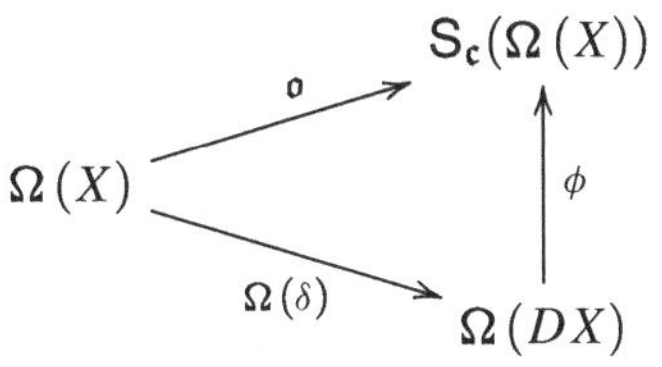

with ϕ an isomorphism.

6.7.1 Note The reader might be interested in an explicit formula for the counterpart of the discrete cover $\delta : DX \to X$, that is, for the localic map $(\mathfrak{o}_L)_* : \mathsf{S}_\mathfrak{c}(L) \to L$

adjoint to the $\mathfrak{o}_L$. Here it is: since $\mathsf{S}_\mathfrak{c}(L)$ is Boolean we have complements $S^\#$, and observe that $\mathfrak{o}(a) \subseteq S$ iff $\uparrow a \vee S = L$ iff $S^\# \subseteq \uparrow a$ iff $a \leqslant \bigwedge S^\#$ so that

$$(\mathfrak{o}_L)_*(S) = \bigwedge S^\#.$$

7 The Nature of the $\mathsf{S}_\mathfrak{c}(L)$ as an Extension of the Frame L

7.1 Essential extensions A monomorphism $m\colon A \to B$ in a category is an *essential extension* if every morphism $f\colon B \to C$ such that $f \cdot m\colon A \to C$ is a monomorphism is itself a monomorphism. If $m\colon A \to B$ is an essential extension, we say that A *is essential* in B.

The category we are interested in is a class of algebras in which the monomorphisms are the one-to-one homomorphisms. Then, an algebra B is an essential extension of a subalgebra A if there is no nontrivial congruence on B restricting to a trivial one on A (in other words: if a factorization on B identifies two distinct elements of B, then it has to identify two distinct elements of A as well).

We will be interested in essential extensions in the category of frames, and for comparison also in the category **DLat** of distributive lattices.

7.1.1 Maximal essential extensions A *maximal essential extension* of L is an essential extension $m\colon L \to M$ such that for every essential extension $n\colon L \to N$ there is precisely one morphism $g\colon M \to N$ such that $g \cdot m = n$.

While we will be explicit in the questions of general essential extensions, to save space we will resort to quoting literature [9, 17, 33] in the following standard facts concerning the maximal ones:

- the maximal essential extensions in **DLat** and **Frm** are precisely the Boolean ones and
- every frame has a maximal essential extension (in **Frm**).

7.2 Computing in $\mathsf{S}_\mathfrak{c}(L)$

1. The embedding of $\mathsf{S}_\mathfrak{c}(L)$ into $\mathsf{S}(L)$ preserves joins but not meets. However, if an element of $\mathsf{S}_\mathfrak{c}(L)$ is complemented in $\mathsf{S}(L)$ and if the complement sits in $\mathsf{S}_\mathfrak{c}(L)$, then it is complemented in $\mathsf{S}_\mathfrak{c}(L)$. In particular, this holds for the closed sublocales in subfit L.
2. By 1 we have in the subfit case $(\bigvee \uparrow b_i) \cap \uparrow a = \bigvee(\uparrow b_i \cap \uparrow a) = \bigvee \uparrow(b_i \vee a) \in \mathsf{S}_\mathfrak{c}(L)$ so that for an $S \in \mathsf{S}_\mathfrak{c}(L)$ the meet $S \cap \uparrow a$ from $\mathsf{S}(L)$ is also the meet in $\mathsf{S}_\mathfrak{c}(L)$.
3. We have the join $\uparrow a \vee \uparrow b = \uparrow(a \wedge b)$ both in $\mathsf{S}(L)$ and $\mathsf{S}_\mathfrak{c}(L)$. Indeed: $\uparrow a \vee \uparrow b = \{u \wedge v \mid u \geqslant a, v \geqslant b\}$ and for $x \geqslant a \wedge b$ we have $x = (x \vee a) \wedge (x \vee b)$.

7.3 Proposition *Let L and M be distributive lattices or frames. A one-to-one lattice or frame homomorphism $\iota\colon L \to M$ is essential iff for each pair $x < y$ in M there is a pair $a < b$ in L such that $x \wedge \iota(b) \leqslant \iota(a)$ and $y \vee \iota(a) \geqslant \iota(b)$.*

Proof $\Leftarrow$: Consider a morphism $h\colon M \to N$ that is not one-to-one, say $h(x) = h(y)$ for $x < y$ in M. Take $a < b$ in L such that $x \wedge \iota(b) \leqslant \iota(a)$ and $y \vee \iota(a) \geqslant \iota(b)$. Then we have

$$\begin{aligned} h(\iota(a)) &= h((x \wedge \iota(b)) \vee \iota(a)) = (h(x) \wedge h(\iota(b))) \vee h(\iota(a)) = \\ &= (h(y) \wedge h(\iota(b))) \vee h(\iota(a)) = h((y \wedge \iota(b)) \vee \iota(a)) = \\ &= h((y \vee \iota(a)) \wedge \iota(b)) \geqslant h(\iota(b)), \end{aligned}$$

so that $h(\iota(a)) = h(\iota(b))$, and $h\iota$ is not one-to-one.

$\Rightarrow$: Suppose $\iota\colon L \to M$ is essential and $x < y$ in M. Consider the congruence

$$C = \{(u, v) \mid x \wedge u = x \wedge v \text{ and } y \vee u = y \vee v\}$$

with quotient homomorphism $h\colon M \to M/C$. It is not a monomorphism because $h(x) = h(y)$, and neither is $h\iota$ because ι is essential. Consequently, there exist $a < b$ in L for which $h(a) = h(b)$, which is to say that

$$x \wedge \iota(b) = x \wedge \iota(a) \leqslant \iota(a) \quad \text{and} \quad y \vee \iota(a) = y \vee \iota(b) \geqslant \iota(b). \qquad \square$$

7.4 Lemma *Let L be an arbitrary frame. Then in the context of distributive lattices,*

$$\lambda = (a \mapsto {\uparrow}a)\colon L \to \mathsf{S}_\mathfrak{c}(L)^{\mathrm{op}} = M$$

is an essential extension preserving, moreover, all joins.

Proof λ is a lattice embedding preserving all joins:

$${\uparrow}\bigvee a_i = \bigcap {\uparrow}a_i = \bigwedge^{\mathsf{S}_\mathfrak{c}(L)} {\uparrow}a_i = \overset{M}{\bigvee} {\uparrow}a_i,$$

$${\uparrow}(a \wedge b) = {\uparrow}a \overset{\mathsf{S}_\mathfrak{c}(L)}{\vee} {\uparrow}b = {\uparrow}a \overset{M}{\wedge} {\uparrow}b.$$

Now let $S <^M T$, that is, $T \subsetneq S$. Hence, there is an x such that ${\uparrow}x \subseteq S$ and ${\uparrow}x \nsubseteq T$ so that there exists an $a > x$ such that $a \notin T$ and as ${\uparrow}a \subseteq {\uparrow}x$, ${\uparrow}a \subseteq S$. Set $b = \bigwedge\{y \in T \mid y \geqslant a\}$. Then $b \in T$, and as $a \notin T$, $a < b$. Now we have, by the definition,

$${\uparrow}a \cap T \supseteq {\uparrow}b, \quad \text{that is,} \quad {\uparrow}a \overset{M}{\vee} T \geqslant^M {\uparrow}b,$$

and since ${\uparrow}a \subseteq S$ we also have that

$$S \vee {\uparrow}b \supseteq {\uparrow}a, \quad \text{that is,} \quad {\uparrow}b \overset{M}{\wedge} S \leqslant^M {\uparrow}a,$$

and we can use 7.3. $\square$

7.4.1 Theorem *For an arbitrary frame L, L and $\mathsf{S}_\mathfrak{c}(L)$ have the same maximal essential extension, up to isomorphism. Hence in particular, for any L, the maximal essential extension of* $\mathsf{S}_\mathfrak{c}(L)$ *is isomorphic to* $\mathfrak{B}(\mathsf{S}(L))^{\mathrm{op}}$.

Proof We will use the facts from 7.1.1.

Consider a maximal essential extension B of $\mathsf{S}_\mathfrak{c}(L)$, $m\colon \mathsf{S}_\mathfrak{c}(L) \to B$, and the composed embedding

$$L \xrightarrow{(a\mapsto\uparrow a)} \mathsf{S}_\mathfrak{c}(L)^{\mathrm{op}} \xrightarrow{\quad m \quad} B^{\mathrm{op}}.$$

By 7.3 it is essential and since B^{op} is Boolean, it is a maximal essential extension. As a Boolean algebra, B is isomorphic to B^{op}. □

7.4.2 More precisely Let us describe the situation in more detail. Consider the following diagram in which the dotted arrows make sense only in the subfit case.

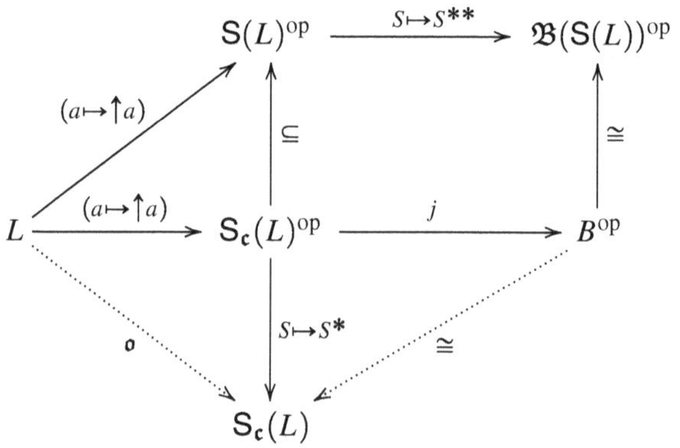

We see that the maximal essential extensions of L and of $\mathsf{S}_\mathfrak{c}(L)$ are isomorphic only because a Boolean algebra is isomorphic, by complementation, with its opposite. The embeddings of L into $\mathfrak{B}(\mathsf{S}(L))^{\mathrm{op}}$ and of $\mathsf{S}_\mathfrak{c}(L)$ into B are in fact connected by a *contravariant* isomorphism.

If L is subfit (and only in that case), j is the identity.

8 Lifting Subfit Frames by $\mathsf{S}_\mathfrak{c}$: Boolean Target

8.1 Since Booleanization is not functorial, Theorem 6.3 raises doubts that the construction $\mathsf{S}_\mathfrak{c}$ be such. And it really generally is not (see [16]). Therefore it is of interest whether it naturally extends a homomorphisms at least in some particular cases. We say that $\mathsf{S}_\mathfrak{c}$ *lifts* a homomorphism $h\colon L \to M$ if there is an $\widetilde{h}$ such that the diagram

$$\begin{array}{ccc} \mathsf{S}_\mathfrak{c}(L) & \xrightarrow{\widetilde{h}} & \mathsf{S}_\mathfrak{c}(M) \\ \uparrow{\scriptstyle \mathfrak{o}_L} & & \uparrow{\scriptstyle \mathfrak{o}_M} \\ L & \xrightarrow{h} & M \end{array} \qquad \text{(lift)}$$

commutes.

Since $\mathsf{S}_\mathfrak{c}$ does not do anything with a Boolean L, each $h\colon L \to M$ with a Boolean source lifts. In this section we will discuss the case of a Boolean *target* M.

8.2 Let us start with an explicit formula for the lift mapping.

8.2.1 Proposition *The unique candidate for the $\widetilde{h}$ in the diagram (lift) above is given by the formula*

$$\widetilde{h}(S) = \bigvee\{\uparrow h(a) \mid \uparrow a \subseteq S\}. \tag{1}$$

Proof If the diagram commutes and $\widetilde{h}$ is a homomorphism, we have

$$\widetilde{h}(\uparrow a) = \widetilde{h}(\mathfrak{o}(a)^*) = \widetilde{h}(\mathfrak{o}(a))^* = \mathfrak{o}(h(a))^* = \uparrow h(a)$$

and hence

$$\widetilde{h}(S) = \widetilde{h}(\bigvee\{\uparrow a \mid \uparrow a \subseteq S\}) = \bigvee\{\uparrow h(a) \mid \uparrow a \subseteq S\}. \qquad \square$$

8.2.2 Proposition *The mapping $\widetilde{h}$ from (1) preserves meets.*

Proof Since $\widetilde{h}$ is obviously monotone, we have $\widetilde{h}(S) \wedge \widetilde{h}(T) \geqslant \widetilde{h}(S \wedge T)$.

Now, although the meet in $\mathsf{S}_\mathfrak{c}(L)$ does not generally coincide with the meet in $\mathsf{S}(L)$ (which is the intersection), we have in $\mathsf{S}_\mathfrak{c}(L)$, $\uparrow a \wedge \uparrow b = \uparrow a \cap \uparrow b = \uparrow(a \vee b)$ simply because the intersection is in $\mathsf{S}_\mathfrak{c}(L)$. Hence we have, since $\mathsf{S}_\mathfrak{c}(L)$ is a frame with the same joins as in $\mathsf{S}(L)$,

$$\begin{aligned} \widetilde{h}(S) \wedge \widetilde{h}(T) &= \bigvee\{\uparrow h(a) \mid \uparrow a \subseteq S\} \wedge \bigvee\{\uparrow h(b) \mid \uparrow b \subseteq T\} = \\ &= \bigvee\{\uparrow h(a) \wedge \uparrow h(b) \mid \uparrow a \subseteq S, \uparrow b \subseteq T\} = \\ &= \bigvee\{\uparrow (h(a) \vee h(b)) \mid \uparrow a \subseteq S, \uparrow b \subseteq T\} = \\ &= \bigvee\{\uparrow h(a \vee b) \mid \uparrow a \subseteq S, \uparrow b \subseteq T\} \leqslant \\ &\leqslant \bigvee\{\uparrow h(c) \mid \uparrow c \subseteq S \wedge T\} \leqslant \widetilde{h}(S \wedge T). \end{aligned} \qquad \square$$

8.2.3 Thus, one has to deal with two questions. For a particular h,

- does the $\widetilde{h}$ from (1) commute in the diagram (lift)? And
- does it preserve arbitrary joins?

8.2.4 Observation *The diagram (lift) commutes iff*

$$\uparrow h(a)^* = \bigvee\{\uparrow h(x) \mid x \vee a = 1\}. \tag{2}$$

(Indeed, we have

$$\widetilde{h}(\mathfrak{o}(a)) = \bigvee\{\uparrow h(x) \mid \uparrow x \subseteq \mathfrak{o}(a)\} = \bigvee\{\uparrow h(x) \mid x \vee a = 1\}.)$$

8.3 The Boolean target If M is Boolean we will consider, to simplify the computing, the g in the diagram

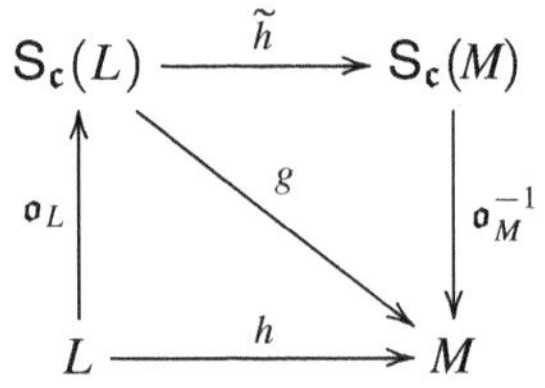

We have $\mathfrak{o}_M(a^*) = \mathfrak{o}_M(a)^* = \uparrow a$ and hence $\mathfrak{o}_M^{-1}(\uparrow a) = a^*$. Thus,

$$g(\uparrow a) = \mathfrak{o}_M^{-1}(\widetilde{h}(\uparrow a)) = \mathfrak{o}_M^{-1}(\uparrow h(a)) = h(a)^*,$$

$$g(S) = \bigvee\{h(x)^* \mid \uparrow x \subseteq S\},$$

and hence, in particular, $g(\mathfrak{o}(a)) = \bigvee\{h(x)^* \mid x \vee a = 1\}$ and the condition (2) from 8.2.4 transforms to

$$h(a) = \bigvee\{h(x)^* \mid x \vee a = 1\}. \tag{3}$$

8.3.1 Proposition *Let M be Boolean and let $h\colon L \to M$ be a complete*[2] *homomorphism. Then $\widetilde{h}$ is a frame homomorphism.*

Proof $\widetilde{h}$ preserves finite meets generally by 8.2.2.

Now let us prove that g preserves joins. Obviously it suffices to prove that it preserves joins of closed sublocales, that is, that

$$g(\bigvee_{i\in J} \uparrow a_i) = \bigvee\{h(x)^* \mid \uparrow x \subseteq \bigvee_{i\in J} \uparrow a_i\} = \bigvee_{i\in J} g(\uparrow a_i) = \bigvee_{i\in J} h(a_i)^* = (\bigwedge_{i\in J} h(a_i))^*.$$

Since $g(\bigvee_{i\in J} \uparrow a_i) \geqslant \bigvee_{i\in J} g(\uparrow a_i)$ is trivial, it suffices to prove that

$$\forall x \text{ such that } \uparrow x \subseteq \bigvee_{i\in J} \uparrow a_i,\ \ h(x)^* \leqslant (\bigwedge_{i\in J} h(a_i))^*.$$

[2] Which is the same as assuming h to be open—see X, Sect. 3.

Hence let $\uparrow x \subseteq \bigvee_{i \in J} \uparrow a_i$ so that $x = \bigwedge_{i \in J} y_i$ for some $y_i \geqslant a_i$. Thus, by completeness of h, $h(x) = \bigwedge_{i \in J} h(y_i) \geqslant \bigwedge_{i \in J} h(a_i)$, and $h(x)^* \leqslant (\bigwedge_{i \in J} h(a_i))^*$. □

8.3.2 Theorem *Let L be regular and let M be Boolean. Then every complete $h\colon L \to M$ lifts.*

Proof We know that $\widetilde{h}$ is a homomorphism so that it remains to prove the commutativity of the diagram. Thus, by 8.2.4 it suffices to prove the equation (3) in 8.3.

If $x \vee a = 1$, then $h(x)^* = h(x)^* \wedge (h(x) \vee h(a)) = h(x)^* \wedge h(a)$, hence $h(x)^* \leqslant h(a)$, and we have $h(a) \geqslant \bigvee\{h(x)^* \mid x \vee a = 1\}$. On the other hand, if L is regular, we have

$$\begin{aligned} h(a) = h(\bigvee\{y \mid y \prec a\}) &= \bigvee\{h(y) \mid y^* \vee a = 1\} \leqslant \\ &\leqslant \bigvee\{h(y^{**}) \mid y^* \vee a = 1\} \leqslant \bigvee\{h(x^*) \mid x \vee a = 1\} \leqslant \\ &\leqslant \bigvee\{h(x)^* \mid x \vee a = 1\}. \end{aligned}$$

□

8.3.3 Note In our paper [16] we have mistakenly claimed the results for general frame homomorphisms, not just for complete ones. We are indebted to I. Arrieta Torres for alerting us to the error.

9 Aside: Exact Filters in $\mathfrak{U}(L)$

We have studied the properties of the frame $\mathsf{S}_\mathfrak{c}(L)$ as a natural sublattice of the coframe $\mathsf{S}(L)$. Let us return to the diagram in 5.1. The question naturally arises of what is the nature of the frame $\mathcal{U}\mathcal{J}[\mathfrak{U}(L)]$ (isomorphic with $\mathsf{S}_\mathfrak{c}(L)$) as a subset of $\mathfrak{U}(L)$. The answer is of some interest ([12], see also [205]).

9.1 Exact filters A meet of a subset A of a complete distributive lattice L is said to be *exact* if we have

$$(\bigwedge A) \vee b = \bigwedge\{a \vee b \mid a \in A\}$$

for any $b \in L$.

9.2 Notes

1. Thus, e.g. every finite meet is exact. Other examples are provided by the meets in $\Omega(X)$ that happen to be set-theoretic intersections. In fact, under a very weak separation condition on X, these are precisely the exact meets in $\Omega(X)$ (see, e.g., [14]).
2. The concept of an exact meet is due to MacNeille, who introduced it under a different name in his dissertation [193] and published it in [194]. It reappeared

in a crucial role (the authors referred to *admissible meets*) in a beautiful result of Bruns and Lakser characterizing frames as the injective meet-semilattices [66]. Ball rediscovered the idea in [10] and coined the term "exact", contributing, inter alia, a first-order characterization of the complete distributive lattices in which every join and meet is exact.

3. Since the other inequality is trivial, the exactness amounts to requiring that

$$\forall b \in L, \quad \left(\bigwedge A\right) \vee b \geqslant \bigwedge\{a \vee b \mid a \in A\}.$$

9.3 Exact filters An *exact filter* in a complete distributive lattice L is an up-set F closed under all exact meets. Note that every finite meet is exact and hence exact filters are indeed a special type of filters.

The set of all exact filters in a frame L ordered by inclusion is obviously a complete lattice. It will be denoted by

$$\mathcal{F}_{\mathsf{ex}}(L).$$

9.4.1 Lemma *Let $A \subseteq L$ be a (nonempty) up-set. Then $b \in \mathcal{UJ}(A)$ iff*

$$\forall y \in L, \quad \bigwedge\{a \vee y \mid a \in A\} \leqslant b \vee y.$$

Proof Let $b \in \mathcal{UJ}(A)$, that is, $\uparrow b \subseteq \bigvee\{\uparrow a \mid a \in A\}$. Since $b \vee y \geqslant b$ we have $b \vee y = \bigwedge\{a_y \mid a \in A\}$ for some $a_y \geqslant a$. Now for each individual a_y we have $a_y \geqslant a \vee y$ and hence $b \vee y \geqslant \bigwedge\{a \vee y \mid a \in A\}$.

On the other hand, let the formula hold. Then, if $y \geqslant b$, we have $y = b \vee y \geqslant \bigwedge\{a \vee y \mid a \in A\} \geqslant y$, that is, $y = \bigwedge\{a \vee y \mid a \in A\} \in \mathcal{J}(A)$. Hence $\uparrow b \subseteq \bigwedge\{a \vee y \mid a \in A\}$, and $b \in \mathcal{UJ}(A)$. □

9.4.2 Theorem *The set $\mathcal{UJ}[\mathfrak{U}(L)]$ is precisely the set $\mathcal{F}_{\mathsf{ex}}(L)$ of all exact filters in L.*

Proof

I. Each $\mathcal{UJ}(A)$ is exact:

Let $b = \bigwedge\{c \mid c \in B\}$, with $B \subseteq \mathcal{UJ}(A)$, be an exact meet. Let $y \in L$ be arbitrary. Then for every $c \in B$ ($\subseteq \mathcal{UJ}(A)$) we have by Lemma 9.4.1, $c \vee y \geqslant \bigwedge\{a \vee y \mid a \in A\}$ and consequently, by exactness,

$$b \vee y = \left(\bigwedge_{c \in B} c\right) \vee y = \bigwedge\{c \vee y \mid c \in B\} \geqslant \bigwedge\{a \vee y \mid a \in A\}.$$

Hence, by Lemma 9.4.1 again, $b \in \mathcal{UJ}(A)$.

II. If A is an exact filter, $A = \mathcal{UJ}(A)$:

That is, we want to show that $\mathcal{UJ}(A) \subseteq A$. Hence, let $b \in \mathcal{UJ}(A)$. By Lemma 9.4.1 we have in particular for $y = b$,

$$b = b \vee b \geqslant \bigwedge_{a \in A}(a \vee b) \geqslant b,$$

and since the $a \vee b$ are in A it suffices to show that the meet $b = \bigwedge_{a \in A}(a \vee b)$ is exact. Let $x \in L$ be arbitrary. We have by Lemma 9.4.1

$$\Big(\bigwedge_{a \in A}(a \vee b)\Big) \vee x = b \vee x = b \vee (b \vee x) \geqslant$$

$$\geqslant \bigwedge\{a \vee b \vee x \mid a \in A\} = \bigwedge\{(a \vee b) \vee x \mid a \in A\}$$

as desired. □

9.4.3 Corollary *The system $\mathcal{F}_{\mathsf{ex}}(L)$ of all exact filters L, ordered by inclusion, is a frame.*

9.4.4 Note This is closely related to results of Bruns and Lakser [66], who not only characterize frames as injective meet-semilattices, but also show that the injective hull of a meet-semilattice is obtained by taking *exact ideals*, the duals of exact filters. In particular, $\mathcal{F}_{\mathsf{ex}}(L)$ is precisely the injective hull of L^{op} regarded as a meet-semilattice.

Chapter X
Subfit, Fit, Open and Complete

In this final chapter, after briefly summarizing some of the already discussed facts concerning subfitness and fitness, we will tackle an aspect of these properties we have not examined yet, the role they play in the links of the phenomena of completeness, openness and the Heyting structure. Let us explain the main topic we will be interested in.

Special types of sublocales (open, closed, one-point) typically have obvious geometric motivation. On the other hand, sublocales are closely connected with frame congruences, and there are natural properties of congruences of algebraic nature that do not seem to have anything to do with topology. Thus in particular there is the property of completeness, namely preserving (besides all joins) all the meets (not only the finite ones), the property of the congruences connected with the complete lattice homomorphisms (that are, of course, a special kind of frame homomorphisms). Note in particular that the congruences associated with open sublocales, namely

$$\Delta_a = \{(x, y) \mid x \wedge a = y \wedge a\}$$

are obviously complete. Are there others, and what is the nature (preferably topologically expressed) of the *complete sublocales*, that is, sublocales associated with them? For many frames, but not for all, the only complete congruences are the open ones. What is behind this phenomenon? Then there is the famous Joyal–Tierney Theorem stating that the open frame homomorphisms $h \colon L \to M$ (the homomorphisms corresponding to open continuous maps) are precisely the *complete Heyting homomorphisms*. Here we have the completeness again; and what actually is the role of the Heyting operation in this respect?

Subfitness and fitness play a fundamental role in answering these questions. In subfit frames open and complete homomorphisms coincide (although one could, for good reason, expect this to be connected rather with fitness). On the other hand,

J. Picado, A. Pultr, *Separation in Point-Free Topology*,
https://doi.org/10.1007/978-3-030-53479-0_10

however, for analysing the nature of complete sublocales, fitness (and the associated operation of fitting) is essential.

In the first two sections of this chapter we will summarize some facts about subfitness and fitness, on the one hand aiming to show generally how important concepts they are, on the other hand, however, with emphasis on some specific new aspects. Then we will analyse the Joyal–Tierney Theorem, in particular asking the question when and how the Heyting part is of importance. The next section is then concerned with the operator of fitting, the basic technical means for the study of complete sublocales that follows. Subfitness seems to be the border property for which openness and closedness coincide; formally we can formulate a necessary and sufficient condition, but the problem is still open whether this is not, after all, still equivalent with subfitness. This is discussed in the penultimate section. The last section then contains a few remarks on the role of the T_D-axiom when interpreting the point-free results in classical spaces.

1 Some Merits of Subfitness

1.1 Although subfitness ("conjunctivity", "disjunctivity") was around since ages, it was the regularity (extremely important, extremely expedient—but for some purposes too strong) that became in the first decades of point-free topology central as a basic separation axiom. But it is, at least when looking at the formulas, closely connected with fitness and subfitness. We wonder since when people realized that we have here, in a (remote) parallel with the spatial T_k's, a rudimentary sequence

$$a \nleq b \;\Rightarrow \exists c,\; a \vee c = 1 \neq b \vee c \qquad \text{(sfit)}$$

$$a \nleq b \;\Rightarrow \exists c,\; a \vee c = 1 \neq \text{ and } c \to b \nleq b \qquad \text{(fit)}$$

$$a \nleq b \;\Rightarrow \exists c,\; a \vee c = 1 \neq \text{ and } c \to 0 \nleq b \qquad \text{(reg)}$$

as already mentioned in V.5.1.

1.2 After being neglected for decades, subfitness reappeared in 1974 [152] in a different, equivalent form, namely as the condition that

L is subfit iff each open sublocale of L is a join of closed ones

(recall II.4.2). It is certainly one of the most important equivalents of the property.

1.3 Recall the classical notion of density (Y is *dense* in X if for every nonempty open U, $U \cap Y \neq \emptyset$). In a dual analogy one says that Y is *codense* (or, *replete*) if for every nonempty closed A, $A \cap Y \neq \emptyset$; in the point-free context we say that a sublocale $S \subseteq L$ is *codense* (or, *replete*) if for every nonempty closed sublocale $\mathfrak{c}(a)$, $\mathfrak{c}(a) \cap S \neq \mathsf{O}$. More generally, if $S \subseteq T \subseteq L$, we say that S is *codense* (or, *replete*) in T if for every closed sublocale such that $\mathfrak{c}(a) \cap T \neq \mathsf{O}$ also $\mathfrak{c}(a) \cap S \neq \mathsf{O}$ (this creates no confusion—recall A.6.5: the closed sublocales in T are the intersections $\mathfrak{c}(a) \cap T$).

Subfitness is closely connected with codensity. Namely, we have (II.4.1, II.4.2(2)) that

L is subfit iff for every sublocale $S \subsetneq L$ there is a nonempty closed sublocale $\mathfrak{c}(a)$ such that $S \cap \mathfrak{o}(a) = \mathsf{O}$,

and hence,

L is subfit iff the only codense (replete) sublocale $S \subseteq L$ is L itself.

1.4 Subfitness plays a fundamental role when working with congruences. One has the technically expedient fact that

1.4.1 *L is subfit iff for a sublocale S, $S \smallsetminus \{1\}$ is cofinal in $L \smallsetminus \{1\}$ only if $S = L$* which leads to the following equivalent condition, a very useful one (recall II.6.2):

1.4.2 *L is subfit iff a congruence E with $E1 = \{1\}$ has to be trivial.*

In the language of homomorphisms:

1.4.3 *L is subfit iff every codense homomorphism $h\colon L \to M$ is one-to-one* (note the link with 1.3: codense homomorphisms h are those for which $h(a) = 1$ implies $a = 1$, which is the same as $f[M]$ being codense (replete) for the localic map f adjoint to h).

1.5 A rather surprising equivalent of subfitness is the validity of a useful formula for the Heyting arrow, namely

L is subfit iff the Heyting operation in L is given by

$$a \to b = \bigwedge\{x \mid x \vee a = 1 \text{ and } x \geqslant b\}$$

(recall II.3). It has very important consequences. In particular recall from II.3.5.1 that

if a frame L is subfit then every complete homomorphism $h\colon L \to M$ preserves the Heyting operation.

1.6 Another important characterization was proved in II.7, namely that

L is subfit iff it admits a generalized nearness.

1.7 Then one has the behaviour of the frame $\mathsf{S}_\mathfrak{c}(L)$ in the subfit case (recall Chap. IX). Since each open sublocale is a join of closed ones we have a natural frame embedding

$$\mathfrak{o} = (a \mapsto \mathfrak{o}(a))\colon L \to \mathsf{S}_\mathfrak{c}(L).$$

Furthermore, we have that

the subfitness of L is equivalent with any of the following statements:

- $\mathsf{S}_\mathfrak{c}(L)$ *is Boolean,*
- $\mathsf{S}_\mathfrak{c}(L)$ *is the (co)Booleanization of* $\mathsf{S}(L)$*,*
- $\mathsf{S}_\mathfrak{c}(L)$ *is a sub-colocale of* $\mathsf{S}(L)$*.*

1.8 Let us add a few consequences (more will come in the sequel).
Under subfitness

- *a compact frame is spatial* (Isbell's spatiality theorem),
- by 1.4 we can, in particular, compute the pseudocomplement as

$$a^* = \bigwedge\{x \mid x \vee a = 1\}$$

(already holding for weak subfitness), and hence
- *a frame that is also a coframe is Boolean.*

1.9 And we should also mention the role of subfitness in combination with other properties. Thus for instance,[1] *supported by subfitness,*

- *normality implies regularity (and hence also complete regularity),*
- *the weak Hausdorff property (and all its variants, see III.2.3.1) implies the Hausdorff property (H),*
- *the weak Hausdorff property becomes conservative (as observed in III.2.3),*
- *and T_D implies T_1.*

2 Some Merits of Fitness

2.1 First of all, for the category minded let us recall from [152] (see V.7.4.3 above) that

the subcategory of fit locales is reflective in ***Loc****.*

2.2 Fitness is the weakest property making for a (categorically) well-behaved extension of the subcategory of subfit locales. In particular we have the characterization of fit locales as those for which

each sublocale is subfit

or those for which

each sublocale is weakly subfit.

Similarly, fit locales are characterized by *prefitness of all sublocales.*

[1] To add a further example, not treated here, see [132], where the role of subfitness in combination with the property of *monotone normality* is studied.

2.3 An interesting (and technically expedient already when applied just for the stronger regular case) equivalent formula is the (somewhat surprising) variation of the original definition

L is fit iff every sublocale $S \subseteq L$ is an intersection of open ones.

Intersections of open sublocales (and of fit sublocales) will play a fundamental role in Sect. 4 below.

2.4 Finally, let us recall a useful and interesting fact that (see II.6)

L is fit iff each congruence on L is determined by the class of the top, that is, if $E1 = E'1$ implies that $E = E'$.

Compare this statement with the weaker one characterizing subfitness in 1.4.2 (if $E1$ is trivial, then E is trivial).

3 Joyal–Tierney Theorem

3.1 Open continuous maps are naturally modelled as *open localic maps*, that is, localic maps $f: L \to M$ such that the image $f[U]$ of every open sublocale is open. The associated frame homomorphisms $h = f^*$ are usually referred to as *open frame homomorphisms*.

3.2 It will be technically of advantage to represent open sublocales $\mathfrak{o}(a)$ by isomorphic frames $\downarrow a$ as explained in A.6.1.2.3 and seen in detail in the following diagram where j is the embedding $\mathfrak{o}(a) \subseteq L$ and the dotted isomorphism consists of the mappings $(x \mapsto a \wedge x): \mathfrak{o}(a) \to \downarrow a$ and $(x \mapsto a \to x): \downarrow a \to \mathfrak{o}(a)$.

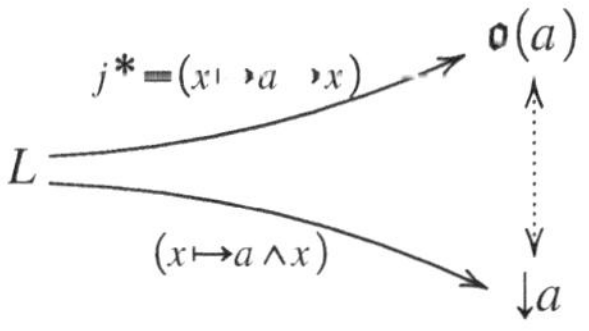

3.3 Theorem *A localic map $f: M \to L$ is open iff the adjoint frame homomorphism $h = f^*: L \to M$ is a complete Heyting homomorphism, that is, if it preserves (also) all meets and the Heyting operation.*

Proof For each $a \in M$ we have a uniquely defined $\phi(a)$ such that $f[\mathfrak{o}(a)] = \mathfrak{o}(\phi(a))$ resulting in the decomposition $f \cdot j_a = j_{\phi(a)} \cdot g$ (where $j_a: \mathfrak{o}(a) \subseteq M$ and $j_{\phi(a)}: \mathfrak{o}(\phi(a)) \subseteq L$ are the embeddings). Obviously this map $\phi: M \to L$ is monotone. In terms of the adjoining frame homomorphism we thus have $j_a^* \cdot h = g^* \cdot j_{\phi(a)}^*$. Replacing the j^*'s isomorphically as in 3.2 we obtain a commutative diagram

$$\begin{array}{ccc} M & \xrightarrow{\;h\;} & L \\ {\scriptstyle (x\mapsto\phi(a)\wedge x)}\big\downarrow & & \big\downarrow{\scriptstyle (x\mapsto a\wedge x)} \\ {\downarrow}\phi(a) & \longrightarrow & {\downarrow}a \end{array}$$

Thus the openness of f is characterized by the existence of a monotone $\phi\colon M \to L$ such that

$$x \wedge \phi(a) = y \wedge \phi(a) \quad \text{iff} \quad h(x) \wedge a = h(y) \wedge a \qquad ⚠$$

or, equivalently,

$$x \wedge \phi(a) \leqslant y \quad \text{iff} \quad h(x) \wedge a \leqslant h(y). \qquad (*)$$

For $x = 1$, in particular, $\phi(a) \leqslant y$ iff $a \leqslant h$ so that ϕ is a left adjoint of h and hence h preserves all meets. Further, we have by $(*)$, for arbitrary a,

$$\begin{aligned} a \leqslant h(x) \to h(y) &\text{ iff } a \wedge h(x) \leqslant h(y) \\ &\text{ iff } x \wedge \phi(a) \leqslant y \text{ iff } \phi(a) \leqslant x \to y \text{ iff } a \leqslant h(x \to y) \end{aligned}$$

and hence $h(x) \to h(y) = h(x \to y)$.

On the other hand, if h preserves the Heyting operation, we have

$$\begin{aligned} x \wedge \phi(a) \leqslant y &\text{ iff } \phi(a) \leqslant x \to y \\ &\text{ iff } a \leqslant h(x \to y) = h(x) \to h(y) \text{ iff } h(x) \wedge a \leqslant h(y), \end{aligned}$$

and hence $(*)$. □

3.4 The subfit case Recall II.3.5.1 (and 1.5 above). For a subfit frame the complete homomorphism is automatically Heyting and hence we immediately infer the following

Theorem *Let L be subfit. Then a localic map $f\colon M \to L$ is open iff the adjoint frame homomorphism $h = f^*\colon L \to M$ is a complete lattice homomorphism.*

3.5 The fit case: notes Thus we have in particular that

> *for a fit frame L the open frame homomorphisms $L \to M$ are precisely the complete ones.*

This is trivial: fitness is stronger than subfitness.

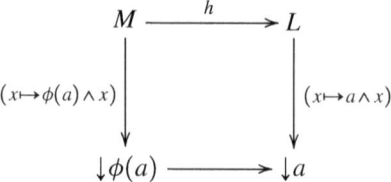

But there is an interesting aspect in comparison of the two facts. Return to the proof of 3.3 and stop at the tag ⚠. Realize that it says precisely that $E = E'$ for the congruences

$$E = \{(x, y) \mid x \wedge \phi(a) = y \wedge \phi(a)\} \text{ and } E' = \{(x, y) \mid h(x) \wedge a = h(y) \wedge a\}.$$

Now if M is fit this is equivalent with $E1 = E'1$, that is with

$$x \wedge \phi(a) = \phi(a) \quad \text{iff} \quad h(x) \wedge a = a$$

which is the same as saying that

$$\phi(a) \leqslant x \quad \text{iff} \quad x \leqslant h(x),$$

or that ϕ is a left adjoint to h. The proof is already finished at this moment. Hence, if L is fit, we obtain the statement just by comparing two equivalences and observing that they coincide because their tops do. The equality of congruences E, E' resulting from $E1 = E'1$ is a characteristic property of fitness and does not hold for the weaker (much weaker) subfitness. We see that there is a subtle difference between the reasons for the result: In the fit case, preserving of the Heyting operation did not play any role; in the subfit case it did, but after everything was finished we got it for free.

Thus, the question naturally arises whether one may have the implication

$$\text{(completeness)} \Rightarrow \text{(openness)}$$

under a still weaker assumption, for yet another reason. This will be discussed in Sect. 6.

4 Fitting, "Another Closure"

4.1 The main point of this section is a brief discussion of a certain operator of a closure type. It will have a technical use in the sequel, but may be found, as we hope, interesting in itself.

A sublocale S of L is said to be *fitted* if

$$S = \bigcap\{\mathfrak{o}(a) \mid S \subseteq \mathfrak{o}(a)\}$$

(thus, L is fit iff each of its sublocales is fitted). For a general sublocale S define the *fitting*

$$S^\circ = \bigcap\{\mathfrak{o}(a) \mid S \subseteq \mathfrak{o}(a)\} \;\; (= \bigcap\{T \mid S \subseteq T,\ T \text{ fitted}\})$$

(thus, S is fitted iff $S^\circ = S$ and L is fit iff the fitting operates on L trivially).

4.1.1 Proposition *$S \mapsto S^\circ$ is an operator of closure type. That is, it is monotone, and we have*

$$\mathsf{O}^\circ = \mathsf{O}, \quad S \subseteq S^\circ, \quad (S^\circ)^\circ = S^\circ, \quad \textit{and} \quad (S \vee T)^\circ = S^\circ \vee T^\circ.$$

Proof Trivially it is monotone, $\mathsf{O}^\circ = \mathsf{O}$, $S \subseteq S^\circ$, $(S^\circ)^\circ = S^\circ$ and $(S \vee T)^\circ \supseteq S^\circ \vee T^\circ$. Finally, by the coframe distributivity we have

$$\begin{aligned} S^\circ \vee T^\circ &= \bigcap\{\mathfrak{o}(a) \mid S \subseteq \mathfrak{o}(a)\} \vee \bigcap\{\mathfrak{o}(b) \mid T \subseteq \mathfrak{o}(b)\} = \\ &= \bigcap\{\mathfrak{o}(a) \vee \mathfrak{o}(b) \mid S \subseteq \mathfrak{o}(a), T \subseteq \mathfrak{o}(b)\} = \\ &= \bigcap\{\mathfrak{o}(a \vee b) \mid S \subseteq \mathfrak{o}(a), T \subseteq \mathfrak{o}(b)\} \supseteq \\ &\supseteq \bigcap\{\mathfrak{o}(c) \mid S \vee T \subseteq \mathfrak{o}(c)\} = (S \vee T)^\circ. \end{aligned}$$

□

4.1.2 Notes

1. The operator $A^\circ = \bigcap\{U \mid A \subseteq U,\ U \text{ open}\}$ makes sense also in classical topology, but it is not of much interest there. One obviously has $A^\circ = A$ for all $A \subseteq X$ iff X is T_1, and $A^\circ = A$ for all closed $A \subseteq X$ iff X is symmetric.
2. We have not found many instances of using the fitting operator in the point-free context. Simpson uses it in the context of σ-frames, in his point-free approach to measure theory [260]. More recently, Dube used a variant of it, defined for cozero elements only, in [88].

4.2 Recall the images and preimages of localic maps (see A.7). We have

Proposition *Let $f\colon L \to M$ be a localic map, let S be a sublocale of L and T a sublocale of M. Then:*

(1) *for images, $f[S^\circ] \subseteq f[S]^\circ$,*
(2) *for preimages, $f_{-1}[T]^\circ \subseteq f_{-1}[T^\circ]$ and*
(3) *if f is one-to-one then $f_{-1}[f[S]^\circ] = S^\circ$.*

Proof

(1) Denote by $h\colon M \to L$ the frame homomorphism adjoint to f and recall from A.4.4 that $f(h(b) \to x) = b \to f(x)$. From this and the formula $\mathfrak{o}(a) = \{a \to x \mid x \in L\}$ we immediately infer that

$$f[\mathfrak{o}(h(b))] \subseteq \mathfrak{o}(b)$$

and then compute

$$\begin{aligned} f[S^\circ] &\subseteq \bigcap\{f[\mathfrak{o}(a)] \mid S \subseteq \mathfrak{o}(a)\} \subseteq \bigcap\{f[\mathfrak{o}(f^*(b))] \mid S \subseteq \mathfrak{o}(f^*(b))\} = \\ &= \bigcap\{f[\mathfrak{o}(f^*(b))] \mid f[S] \subseteq \mathfrak{o}(b)\} \subseteq \bigcap\{\mathfrak{o}(b) \mid f[S] \subseteq \mathfrak{o}(b)\} = f[S]^\circ. \end{aligned}$$

(2) Use the adjunction $f[S] \subseteq T$ iff $S \subseteq f_{-1}[T]$ and set in (1) $S = f_{-1}[T]$. Then $f[f_{-1}[T]^\circ] \subseteq f[f_{-1}[T]]^\circ \subseteq T^\circ$ and the inclusion $f_{-1}[T]^\circ \subseteq f_{-1}[T^\circ]$ follows.
(3) By (2) one has

$$S^\circ \subseteq f_{-1}[f[S]]^\circ \subseteq f_{-1}[f[S]^\circ].$$

Furthermore, writing h for the homomorphism adjoint to f again,

$$f_{-1}[f[S]^\circ] = \bigcap\{\mathfrak{o}(h(b)) \mid f[S] \subseteq \mathfrak{o}(b)\} = \bigcap\{\mathfrak{o}(h(b)) \mid S \subseteq \mathfrak{o}(h(b))\}$$

and this is S° whenever h is onto, that is, whenever f is one-to-one. □

4.3 Codensity (although related with subfitness rather than with fitness, recall 1.2) can be expressed in the language of fitting. In fact, it relates to the operator S° precisely as density does to the standard closure $\overline{S}$. We have

4.3.1 Proposition *S is codense in T iff $T \subseteq S^\circ$.*
Consequently, L is subfit iff $S^\circ = L$ implies that $S = L$.

Proof If $T \subseteq S^\circ$ and $S \cap \mathfrak{c}(a) = \mathsf{O}$, then $S \subseteq \mathfrak{o}(a)$, hence $T \subseteq S^\circ \subseteq \mathfrak{o}(a)$, and $T \cap \mathfrak{c}(a) = \mathsf{O}$. Thus, S is codense in T. Conversely, if S is codense in T and $S \subseteq \mathfrak{o}(a)$, then $S \cap \mathfrak{c}(a) = \mathsf{O}$, hence $T \cap \mathfrak{c}(a) = \mathsf{O}$, and $T \subseteq \mathfrak{o}(a)$.
The second statement follows from 1.3. □

Then, using 4.2, we immediately obtain

4.3.2 Corollary *Let $f : L \to M$ be a localic map and let S, T be sublocales of L. If S is codense in T, then $f[S]$ is codense in $f[T]$. If f is one-to-one, we have also the converse implication.*

5 Complete and Weakly Complete Sublocales

A (frame) congruence on a frame L is *complete* if it respects all the joins and *all* the meets (not only the finite ones). Complete congruences are (of course) associated with complete frame homomorphisms.

In particular, all the open congruences $\Delta_a = \{(x, y) \mid a \wedge x = a \wedge y\}$ are complete. In fact they are more special, being associated with the complete *Heyting* homomorphisms ([173, 220]; see II.3.5); however, under subfitness they coincide (II. 3.5.1, 3.4).

Although it is very easy to specify complete congruences in between frame congruences, a specification of *complete sublocales* (the associated sublocales) is not quite so straightforward. This will be the main topic of this section.

5.1 The congruence E_S associated with a sublocale in more detail Recall the translations from A.5.5. We have

$$E_S = \{(a,b) \mid \nu_S(a) = \nu_S(b)\} = \{(a,b) \mid \bigwedge(\uparrow a \cap S) = \bigwedge(\uparrow b \cap S)\}.$$

Now if $\bigwedge(\uparrow a \cap S) = \bigwedge(\uparrow b \cap S)$ and $x \in \uparrow a \cap S$, then $x \geqslant \bigwedge(\uparrow b \cap S) \geqslant b$, and since it is in S, it is in $\uparrow b \cap S$.

5.1.1 Thus we have obtained a well-known handier formula

$$\begin{aligned} E_S &= \{(a,b) \mid \uparrow a \cap S = \uparrow b \cap S\} = \\ &= \{(a,b) \mid \forall s \in S,\ a \leqslant s \text{ iff } b \leqslant s\} = \{(a,b) \mid \mathfrak{o}(a) \cap S = \mathfrak{o}(b) \cap S\} \end{aligned}$$

(the last equality[2] follows from $\mathfrak{o}(x)$ being the complement of $\uparrow x = \mathfrak{c}(x)$).

5.1.2 For a localic map f and the associated frame homomorphism h we obtain

$$aE_{f[S]}b \quad \text{iff} \quad h(a)E_S h(b).$$

(Indeed, $aE_{f[S]}b$ iff $(\forall s \in S, a \leqslant f(s) \equiv b \leqslant f(s))$ iff $(\forall s \in S, h(a) \leqslant s \equiv h(b) \leqslant s)$.)

5.2 Proposition *A congruence $E \subseteq L \times L$ is complete iff each congruence class Ea contains a least element.*

Proof The implication "$\Rightarrow$" is obvious.

Now consider the quotient map $q : L \to L/E$ as in A.5.5. From the saturation formula

$$xEy \Rightarrow (x \leqslant s \equiv y \leqslant s)$$

we immediately see that $q(a)$ is the *maximum* element of the equivalence class Ea. For $x \in L/E = q[L]$ define $\phi(x)$ as the *minimum* element of Ex. Then, as $\phi(x)Ex$, we have

$$q(\phi(x)) = x.$$

We claim that ϕ is monotone: indeed, if $x \leqslant y$ for $x, y \in q[L]$, we have

$$(\phi(x) \wedge \phi(y))E(x \wedge y) \equiv xE\phi(x)$$

[2]Compare it with formula $E_Y = \{(U,V) \mid U \cap Y = V \cap Y\}$ from I.5.3 representing a subspace Y of a topological space X as a congruence in $\Omega(X)$.

and hence $\phi(x) \wedge \phi(y) = \phi(x)$ by the minimality of $\phi(x)$. Now from $q(\phi(x)) = x$ and from the obvious inequality $\phi(q(x)) \leqslant x$ we infer that ϕ is a left adjoint of q, hence q preserves all meets, and E is complete. □

5.2.1 A congruence E is said to be *weakly complete* if $E1$ contains a least element.

This property seems to be very much weaker than completeness, and in general it is. But in special cases it is not always so: see 5.4.2 below.

5.3 For $a \in L$ and each sublocale S of L set

$$u_{aS} = \bigwedge\{x \mid \mathfrak{o}(x) \cap S = \mathfrak{o}(a) \cap S\} = \bigwedge\{x \mid \mathfrak{o}(a) \cap S \subseteq \mathfrak{o}(x)\}$$

(the second equality because if $\mathfrak{o}(a) \cap S \subseteq \mathfrak{o}(x)$, then $\mathfrak{o}(a) \cap S = \mathfrak{o}(x) \cap \mathfrak{o}(a) \cap S = \mathfrak{o}(x \wedge a) \cap S$ so that $a \wedge x \in \{y \mid \mathfrak{o}(y) \cap S = \mathfrak{o}(a) \cap S\}$ and $x \geqslant a \wedge x$ while on the other hand $\{x \mid \mathfrak{o}(x) \cap S = \mathfrak{o}(a) \cap S\} \subseteq \{x \mid \mathfrak{o}(a) \cap S \subseteq \mathfrak{o}(x)\}$).

This element u_{aS} is defined for any S. Note, however, that by 5.1.1,

u_{aS} is the least element of $E_S a$ whenever this equivalence class has one.

5.3.1 Denote by $T^{\#}$, as usual, the supplement of T in $\mathsf{S}(L)$. We have

Lemma *For any S and $a \in L$,*

$$\mathfrak{o}(u_{aS}) \subseteq ((\mathfrak{o}(a) \cap S)^{\circ})^{\#\#} \subseteq (\mathfrak{o}(a) \cap S)^{\circ}.$$

Proof We have

$$\uparrow u_{aS} = \uparrow(\bigwedge\{x \mid \mathfrak{o}(x) \cap S = \mathfrak{o}(a) \cap S\}) \supseteq \bigvee\{\uparrow x \mid \mathfrak{o}(a) \cap S \subseteq \mathfrak{o}(x)\}$$

in L and hence

$$\begin{aligned}\uparrow u_{aS} &\supseteq \bigvee\{\mathfrak{o}(x)^{\#} \mid \mathfrak{o}(a) \cap S \subseteq \mathfrak{o}(x)\} = \\ &= (\bigcap\{\mathfrak{o}(x) \mid \mathfrak{o}(a) \cap S \subseteq \mathfrak{o}(x)\})^{\#} = ((\mathfrak{o}(a) \cap S)^{\circ})^{\#}\end{aligned}$$

so that $\mathfrak{o}(u_{aS}) = (\uparrow u_{aS})^{\#} \subseteq ((\mathfrak{o}(a) \cap S)^{\circ})^{\#\#}$. □

5.3.2 Lemma *If $E_S a$ has a least element, then $(\mathfrak{o}(a) \cap S)^{\circ} \subseteq \mathfrak{o}(u_{aS})$.*

Proof The least element is u_{aS} and hence $\mathfrak{o}(u_{aS}) \cap S = \mathfrak{o}(a) \cap S$ so that $\mathfrak{o}(a) \cap S \subseteq \mathfrak{o}(u_{aS})$ and since $\mathfrak{o}(u_{aS})$ is open, $(\mathfrak{o}(a) \cap S)^{\circ} \subseteq \mathfrak{o}(u_{aS})$. □

5.3.3 Proposition *$E_S a$ has a least element iff $(\mathfrak{o}(a) \cap S)^{\circ}$ is open.*

Proof Let $(\mathfrak{o}(a) \cap S)^{\circ} = U$ be open. Since $\mathfrak{o}(a) \cap S$ is codense in U, it is also codense in $\mathfrak{o}(a) \cap U$ and we see that $U = \mathfrak{o}(b)$ with $\mathfrak{o}(b) \subseteq \mathfrak{o}(a)$ and hence $b \leqslant a$. We have

$$\uparrow b \cap S = \uparrow b \cap S \cap (\mathfrak{o}(a) \vee \uparrow a) = (\uparrow b \cap \mathfrak{o}(a) \cap S) \vee (\uparrow a \cap \uparrow b \cap S) = \uparrow a \cap S$$

so that $bE_S a$. Now if $cE_S a$, we have $\mathfrak{o}(a) \cap S = \mathfrak{o}(c) \cap S \subseteq \mathfrak{o}(c)$ and hence $\mathfrak{o}(b) = (\mathfrak{o}(a) \cap S)^\circ \subseteq \mathfrak{o}(c)$ and $b \leqslant c$. Thus, b is the least element of $E_S a$.

On the other hand, if $E_S a$ has a least element it is u_{aS} and we have, by 5.3.1 and 5.3.2, $\mathfrak{o}(u_{aS}) \subseteq (\mathfrak{o}(a) \cap S)^\circ \subseteq \mathfrak{o}(u_{aS})$, that is, $(\mathfrak{o}(a) \cap S)^\circ = \mathfrak{o}(u_{aS})$. □

5.4 A sublocale S is said to be *complete* resp. *weakly complete* if the congruence E_S is complete resp. weakly complete.

5.4.1 Theorem *A sublocale $S \subseteq L$ is complete iff for every open U, $(S \cap U)^\circ$ is open.*

It is weakly complete iff S° is open.

Proof The first statement follows immediately from 5.3.3 and 5.2; for the second one apply $\mathfrak{o}(1) = L$. □

5.4.2 Corollary *Open localic maps preserve weakly complete sublocales.*

Proof Let $f : L \to M$ be an open localic map and S a weakly complete sublocale of L. By 4.2, $f[S]^\circ \supseteq f[S^\circ] \supseteq f[S]$. Since S° is open, $f[S^\circ]$ is also open and hence $f[S]^\circ = f[S^\circ]$. □

5.4.3 Notes

1. Here is a simple example of a weakly complete but not complete sublocale.

   In the unit interval $[0, 1]$ with the natural order we have $\mathfrak{o}(a) = [0, a) \cup \{1\}$; if we consider
   $$S = \{s_1 < s_2 < \cdots < s_n < \cdots\}$$
   where s_n converges to $\frac{1}{2}$, we have $E_S 1 = [\frac{1}{2}, 1]$ with the least element $\frac{1}{2}$ so that S is weakly complete, while the other congruence classes $E_S a$ with $0 < a < \frac{1}{2}$ are not. Then any subset containing 0 makes for a weakly complete sublocale, but many of these are not complete: for $S = [0, \frac{1}{2}]$ and $U = (0, 1)$, $S \cap U = (0, \frac{1}{2}]$ is closed (in U and X) for the closure operator $(-)^\circ$ but $S \cap U$ is not open in U.
2. A closer scrutiny of the proofs shows that we can be more specific, that is, in the case of weak completeness we know that the open set is $\mathfrak{o}(u_{1S})$, while in the case of completeness it is $\mathfrak{o}(u_{aS})$. Hence, we have observed that a sublocale $S \subseteq L$ is

   - weakly complete iff $S^\circ = \mathfrak{o}(u_{1S})$ and
   - complete iff for each $a \in L$ $(S \cap \mathfrak{o}(a))^\circ = \mathfrak{o}(u_{aS})$.

5.4.4 By 5.1.1, $E_S 1 = \{x \mid \mathfrak{o}(x) \cap S = S\} = \{x \mid S \subseteq \mathfrak{o}(x)\}$. Since $S \subseteq \mathfrak{o}(x)$ iff $S^\circ \subseteq \mathfrak{o}(x)$ we have
$$E_{S^\circ} 1 = E_S 1.$$

5.5 Lemma *If S is weakly complete, then*

$$\downarrow(S \smallsetminus \{1\}) = \downarrow(\mathfrak{o}(u_{1S}) \smallsetminus \{1\}).$$

Proof The equality states that $x(\neg E_S)1$ iff $x(\neg E_{\mathfrak{o}(u_{1S})})1$, that is, xE_S1 iff $xE_{\mathfrak{o}(u_{1S})}1$. If S is weakly complete, then $S^\circ = \mathfrak{o}(u_{1S})$. Use 5.4.4. □

Here is another characterization of subfitness:

5.5.1 Theorem *A frame L is subfit iff each weakly complete sublocale of L is open.*

Proof $\Rightarrow$: If S is weakly complete, then it is codense in an open $U \subseteq L$. Every open sublocale of a subfit L is subfit (in fact, every complemented one is—see II.5), and hence we can apply 1.3 and conclude that $S = U$.

$\Leftarrow$: Let S be codense in L. Since L is open, S is weakly complete and hence open, and an open codense $S \subseteq L$ is equal to L (consider the complement of S). Use 1.3. □

5.5.2 Corollary *Let L be subfit. Then the following statements about a sublocale $S \subseteq L$ are equivalent.*

(1) *S is weakly complete.*
(2) *S is complete.*
(3) *S is open.*

6 A Formal Relaxation of Subfitness

6.1 Since an open sublocale of a subfit locale is subfit we can replace the second statement in 4.3.1 by (only a formally stronger) claim concerning an arbitrary open $U \subseteq L$ instead of L. Thus we can characterize subfitness by stating that

(sfit'): *for every open $U \subseteq L$ and every sublocale $S \subseteq L$,*
$S^\circ = U \;\Rightarrow\; S = U.$

Now we will relax this condition to

(c-subfit): *for every open $U \subseteq L$ and every* **complete** *sublocale $S \subseteq L$,*
$S^\circ = U \;\Rightarrow\; S = U.$

We will present two necessary and sufficient conditions for c-subfitness, one of them technical, another stating that it is precisely the borderline of the coincidence of completeness and openness.

6.2 Theorem *A frame L is c-subfit iff for every complete sublocale S and every $a \in L$,*

$$\downarrow(S \smallsetminus \{1\}) = \downarrow(\mathfrak{o}(a) \smallsetminus \{1\}) \quad \Leftrightarrow \quad S = \mathfrak{o}(a).$$

Proof Since $S^{\circ} = \mathfrak{o}(a)$ is obviously the same as claiming that $S \subseteq \mathfrak{o}(a)$ and $\downarrow(S \smallsetminus \{1\})$ is cofinal in $\downarrow(\mathfrak{o}(a) \smallsetminus \{1\})$, and since "$\Leftarrow$" is trivial, it suffices to prove that $S \subseteq \mathfrak{o}(a)$.

Let

$$\downarrow(S \smallsetminus \{1\}) = \downarrow(\mathfrak{o}(a) \smallsetminus \{1\}).$$

We shall show that $\mathfrak{c}(a) \cap S = \mathsf{O}$. Indeed, if $s \in S$ and $s \neq 1$ and if $a \leqslant s$ then $a \in \downarrow(\mathfrak{o}(a) \smallsetminus \{1\})$, hence $a \leqslant a \to x \neq 1$ for some x, but then $a = a \wedge a \leqslant x$, and $a \to x = 1$, a contradiction. □

6.3 Theorem *A frame L is c-subfit iff each complete sublocale $S \subseteq L$ is open.*

Proof $\Rightarrow$: Let L be c-subfit and let $S \subseteq L$ be complete. Then it is weakly complete and hence by 5.5, $\downarrow(S \smallsetminus \{1\}) = \downarrow(\mathfrak{o}(u_{1S}) \smallsetminus \{1\})$. Thus, by 6.2, $S = \mathfrak{o}(u_{1S})$ is open.

$\Leftarrow$: By 6.2 it suffices to prove that

$$\downarrow(\mathfrak{o}(b) \smallsetminus \{1\}) = \downarrow(\mathfrak{o}(a) \smallsetminus \{1\}) \quad \Rightarrow \quad b = a.$$

Let the first equality hold and let $b \nleqslant a$. Then $1 \neq b \to a \in \mathfrak{o}(b) \smallsetminus \{1\}$. Then $b \to a \in \downarrow(\mathfrak{o}(a) \smallsetminus \{1\})$ and $b \to a \leqslant a \to x$ for some x such that $a \to x \neq 1$. Then $a \wedge (b \to a) \leqslant x$ but since $a \leqslant b \to a$ this yields $a \leqslant x$ and $a \to x = 1$, a contradiction. □

6.4 Complete Heyting homomorphisms We have recalled in 5.2 that complete homomorphisms $h\colon M \to L$ with subfit M automatically preserve the Heyting operation. Now we will show that this is true precisely for c-subfit frames.

6.4.1 Lemma *Let $h\colon M \to L$ be a frame homomorphism. Then we have for congruences associated with sublocales*

$$(h \times h)^{-1}[E_S] = E_{h_*[S]}.$$

Proof Use A.5.5. From the adjunction $h \dashv h_*$ we obtain that

$$(x, y) \in (h \times h)^{-1}[E_S], \quad \text{that is,} \quad \forall s \in S,\ \ h(x) \leqslant s \text{ iff } h(y) \leqslant s$$

if and only if

$$\forall s \in S,\ \ x \leqslant h_*(s) \text{ iff } y \leqslant h_*(s), \quad \text{that is,} \quad (x, y) \in E_{h_*[S]}.$$ □

6.4.2 Theorem *Every complete frame homomorphism $h\colon M \to L$ is a complete Heyting homomorphism if and only if M is c-subfit.*

Proof $\Leftarrow$: Let M be c-subfit and let $h\colon M \to L$ be a complete frame homomorphism. Let S be an open sublocale of L. Then in particular it is complete, and hence E_S is complete. From the completeness of h and from 6.4.1 we now immediately infer that $E_{h_*[S]}$ is complete. Thus, $h_*[S]$ is complete and by 6.3 it is open. We conclude that h is an open homomorphism and apply the Joyal–Tierney theorem.

$\Rightarrow$: Let S be a complete sublocale of M. Consider the quotient homomorphism $h\colon M \to S$ adjoint to the embedding $j\colon S \subseteq M$. Then h is a complete homomorphism and hence it is a Heyting one. Thus, it is an open homomorphism and in particular $S = j[S]$ is an open sublocale. □

6.5 Note Restricting in (c-sfit) the condition from general sublocales to (very special) complete ones seems to be a very radical reduction. Yet, it is still an open problem whether it is not, after all, just another formula for subfitness. It has defied solution for years, even in modified contexts like that of Heyting meet-semilattices [228].

Any answer to this problem would be of interest. The first step might be replacing (c-subfit) by a first order formula. As it is presented in 5.3 it is (like (sfit'), and of course like the Isbell's sublocale axiom) a second order condition. We miss a first order equivalent akin to the conjunctivity implication. It would probably make the problem much more transparent.

7 Remarks on the Role of T_D

7.1 Suppose $L = \Omega(X)$ and $M = \Omega(Y)$ and let $g\colon X \to Y$ be a continuous map. Can we interpret in such a case the Joyal–Tierney theorem by stating that g is open iff $\Omega(g)$ is a complete Heyting homomorphism? Almost, but not quite.

Recall the formula from 5.1.2. For $h = \Omega(g)$ and f the adjoint localic map and an open $S = \mathfrak{o}(U)$ we obtain

$$AE_{f[\mathfrak{o}(U)]}B \quad \text{iff} \quad g^{-1}[A]E_{\mathfrak{o}(U)}g^{-1}[B].$$

Applying 5.1.2 we see that the latter says that $g^{-1}[A] \cap U = g^{-1}[B] \cap U$ which in turn is easy to see to be equivalent with $A \cap g[U] = B \cap g[U]$. Now from the Joyal–Tierney theorem we have learned that each $f[\mathfrak{o}(U)]$ is an open $\mathfrak{o}(V)$, that is we always have

$$A \cap V = B \cap V \quad \text{iff} \quad AE_{\mathfrak{o}(V)}B \quad \text{iff} \quad A \cap g[U] = B \cap g[U],$$

that is, $E_{g[U]} = E_V$ with an open V iff $\Omega(g)$ is complete Heyting. Now if Y is T_D this yields $g[U] = V$ (recall I.6.3) and g is indeed an open map.

If Y is not T_D, however, there is a point x such that no $U \smallsetminus \{x\}$ with open $U \ni x$ is open. Then the embedding $j\colon Y \smallsetminus \{x\} \subseteq Y$ is not open while the $\Omega(j)$ is an isomorphism. Thus we have

7.1.1 Theorem *If Y is a T_D-space, then a continuous map $g\colon X \to Y$ is open iff $\Omega(g)$ is a complete Heyting homomorphism. If Y is not T_D, then this statement does not hold.*

7.2 Equally met subspaces Recall from I.6.2.1 that there is an open set $U \ni x$ such that $U \smallsetminus \{x\}$ is open iff $(X \smallsetminus \overline{\{x\}}) \cup \{x\}$ is open. The points x such that

$$(X \smallsetminus \overline{\{x\}}) \cup \{x\} \text{ is open}$$

are called T_D*-points*. Thus, in a T_D-space each point is a T_D-point.

Two subsets A, B of a space are said to be *equally met* if $E_A = E_B$. (Thus, a continuous map $g\colon X \to Y$ with open—that is, complete Heyting—$\Omega(g)$ may not be open, but a $g[U]$ with open U is always at least equally met with an open set.)

7.2.1 Proposition *If an open set U is equally met with A, then $A \subseteq U$. Hence in particular two open sets are equally met only if they coincide.*

Proof Suppose $x \in A \smallsetminus U$. We have $U \cap U = U \cap X$ while $A \cap U \not\ni x \in A \cap X$. □

7.2.2 Proposition *Let A and B be equally met and let $x \in A \smallsetminus B$. Then x is not a T_D-point.*

Proof Suppose it is. Then $U_x = (X \smallsetminus \overline{\{x\}}) \cup \{x\}$ is open. We have $B \cap U_x = B \cap (X \smallsetminus \overline{\{x\}})$ while $A \cap U_x \ni x \notin A \cap (X \smallsetminus \overline{\{x\}})$ so that A and B are not equally met. □

Appendix

1 Partial Order

1.1 If there is no danger of confusion we denote a *partial order* on a set by $\leqslant$, even if we work in fact with distinct posets. Whenever the situation calls for distinction, we use, e.g., indices like in $\leqslant_1$ or ad hoc symbols like, for instance, $\sqsubseteq$. In case of the order of inclusion on a system of subsets of a set, we use as a rule the standard $\subseteq$.

For subsets A resp. elements a of a poset $(X, \leqslant)$ we use the standard notation

$$\downarrow A = \{x \in X \mid x \leqslant a \text{ for an } a \in A\}, \quad \downarrow a = \downarrow\{a\}, \quad \text{and}$$

$$\uparrow A = \{x \in X \mid x \geqslant a \text{ for an } a \in A\}, \quad \uparrow a = \uparrow\{a\}.$$

Suprema resp. infima of pairs of elements (if they exist) will be denoted by

$$a \vee b \quad \text{resp.} \quad a \wedge b,$$

for suprema resp. infima of subsets A (if they exist) we typically use

$$\bigvee A \text{ or } \bigvee\{a \mid a \in A\}, \text{ or } \bigvee_{i \in J} a_i \text{ (for } \bigvee\{a_i \mid i \in J\}),$$

$$\bigwedge A \text{ or } \bigwedge\{a \mid a \in A\}, \text{ or } \bigwedge_{i \in J} a_i \text{ (for } \bigvee\{a_i \mid i \in J\}).$$

But we may also write sup A resp. inf A (typically in cases when the existence is not assumed generally), and in cases when there is a danger of confusion we use ad hoc symbols like $a \sqcup b$, $a \sqcap b$, $\bigsqcup_{i \in J} a_i$, etc. For unions and intersections of sets we use

J. Picado, A. Pultr, *Separation in Point-Free Topology*,
https://doi.org/10.1007/978-3-030-53479-0

the standard

$$A \cup B, \quad \bigcup_{i \in J} A_i, \quad A \cap B, \quad \bigcap_{i \in J} A_i,$$

whether they happen to be suprema resp. infima in the particular system of subsets discussed or not.

The least element of a poset (the bottom) is usually denoted by 0 or $\perp$, the largest one by 1 or $\top$, often specified, like 0_L for the bottom of L; sometimes we have quite specific symbols, like $\emptyset$, of course.

For the *dual*, or *opposite*, of a poset $(X, \leqslant)$ we write

$$(X, \leqslant)^{\mathrm{op}} = (X, \leqslant^{\mathrm{op}}) \quad \text{where} \quad x \leqslant^{\mathrm{op}} y \text{ iff } y \leqslant x.$$

1.2 Lattices, and complete lattices A (bounded) *lattice* is a poset where every finite subset has both a supremum and an infimum. With very few exceptions we will work with *complete* lattices where this holds for *any* subset. Recall the well-known fact that for the latter it suffices to check the existence of just one of the two (since, say, $\bigvee A$ can be obtained as $\bigwedge\{x \mid A \subseteq {\downarrow}x\}$).

A lattice L is *distributive* if one has

$$a \vee (b \wedge c) = (a \vee b) \wedge (a \vee c)$$

for all $a, b, c \in L$. This is equivalent with the requirement that

$$a \wedge (b \vee c) = (a \wedge b) \vee (a \wedge c)$$

(thus, L is distributive iff L^{op} is).

1.3 Monotone mappings A mapping

$$f\colon (X, \leqslant) \to (Y, \leqslant)$$

(the two $\leqslant$'s may designate distinct orders) is *monotone* if

$$\forall x_1, x_2, \quad x_1 \leqslant x_2 \;\Rightarrow\; f(x_1) \leqslant f(x_2).$$

1.4 (Galois) adjunctions Two monotone maps

$$(X, \leqslant) \underset{r}{\overset{\ell}{\rightleftarrows}} (Y, \leqslant)$$

are said to be (*Galois*) *adjoint* [95], ℓ the *left adjoint* of r, r the *right adjoint* of ℓ, if

$$\forall x \in X\ \forall y \in Y \quad \ell(x) \leqslant y \text{ iff } x \leqslant r(y),$$

equivalently if

$$\forall x \in X,\ x \leqslant r(\ell(x)) \quad \text{and} \quad \forall y \in Y,\ \ell(r(y)) \leqslant y. \tag{$*$}$$

A left resp. right adjoint of a given monotone map does not have to exist, but if it does, it is uniquely determined.

A left (resp. right) adjoint of a monotone map f is usually denoted by f^* (resp. f_*).

1.4.1 From $(*)$ one immediately infers that

$$\ell r \ell = \ell \quad \text{and} \quad r \ell r = r.$$

As a consequence we immediately see that

a left adjoint is onto (resp. one-to-one) iff the associated right adjoint is one-to-one (resp. onto).

1.5 Preserving suprema and infima The following is a well-known, easy, and very useful fact.

Theorem *Left adjoints preserve all the existing suprema and right adjoints preserve all the existing infima.*

On the other hand, if $(X, \leqslant)$ and $(Y, \leqslant)$ are complete lattices and if $f: (X, \leqslant) \to (Y, \leqslant)$ preserves all suprema (resp. infima), then it is a left (resp. right) adjoint.

Proof Let f_* exist and let $s = \sup M$ in X. Then, trivially, $f(s) \geqslant f(m)$ for all $m \in M$. Now if $y \geqslant f(m)$ for all $m \in M$ we have $f_*(y) \geqslant m$ for all $m \in M$, hence $s \leqslant f_*(y)$, and $f(s) \leqslant y$ using the adjunction again.

On the other hand, if the posets are complete lattices and f preserves all suprema, it is easy to check that for $g: (Y, \leqslant) \to (X, \leqslant)$ defined by $g(y) = \bigvee\{x \mid f(x) \leqslant y\}$ we have $f(x) \leqslant y$ iff $x \leqslant g(y)$. □

2 Pseudocomplements, Supplements and Complements; (Co)linearity

2.1 Pseudocomplements and supplements Let L be a lattice with 0 and 1. A *pseudocomplement* (resp. *supplement*) of an element a in L is an element b such that

$$a \wedge x = 0 \text{ iff } x \leqslant b \quad (\text{resp.} \quad a \vee x = 1 \text{ iff } x \geqslant b). \tag{psc}$$

Obviously a pseudocomplement resp. supplement does not have to exist, but if it exists it is uniquely determined. If it exists it will be denoted by

$$a^* \quad \text{resp.} \quad a^{\#}.$$

2.1.1 Basic facts about pseudocomplements

(1) $a \leqslant a^{**}$.
(2) $a \leqslant b \Rightarrow a^* \geqslant b^*$; *hence* $a \mapsto a^{**}$ *is monotone.*
(3) $a^* = a^{***}$; *hence* $x \wedge a = 0$ *iff* $x \wedge a^{**} = 0$.
(4) $(a \wedge b)^{**} = a^{**} \wedge b^{**}$.
(5) $(\bigvee_{i \in J} a_i)^* = \bigwedge_{i \in J} a_i^*$ *(De Morgan law).*

Proof

(1) Apply $a \wedge a^*$ in (psc) for a^* instead of for a.
(2) If $x \wedge b = 0$, then $x \wedge a = 0$.
(3) Use (1) and (2) to compare $(a^*)^{**}$ with $(a^{**})^*$.
(4) Trivially $(a \wedge b)^{**} \leqslant a^{**} \wedge b^{**}$ and $a \wedge b \leqslant (a \wedge b)^{**}$. From the latter, (psc), and (3) used twice we obtain

$$0 = a \wedge b \wedge (a \wedge b)^* = a^{**} \wedge b \wedge (a \wedge b)^* = a^{**} \wedge b^{**} \wedge (a \wedge b)^*$$

so that $a^{**} \wedge b^{**} \leqslant (a \wedge b)^{**}$.
(5) $x \leqslant \bigwedge_i a_i^*$ iff for all i, $x \leqslant a_i^*$ iff for all i, $x \wedge a_i = 0$ iff for all i, $a_i \leqslant x^*$ iff $\bigvee_i a_i \leqslant x^*$ iff $(\bigvee_i a_i) \wedge x = 0$ iff $x \leqslant (\bigvee_i a_i)^*$. □

Remark For the supplement one has the obvious dual formulas; in particular we have the De Morgan law

$$(\bigwedge_{i \in J} a_i)^{\#} = \bigvee_{i \in J} a_i^{\#}.$$

Note that in both cases we have only one general De Morgan law. The two De Morgan laws in Boolean algebras result from the complement being there both a pseudocomplement and a supplement (see below).

2.2 Complements A *complement* of an element a in a lattice L is an element b such that

$$a \wedge b = 0 \quad \text{and} \quad a \vee b = 1. \tag{$*$}$$

A complement of a, of course, does not have to exist; if it does we say that a is *complemented.*

Unlike a pseudocomplement or supplement, a complement is not in general even uniquely determined. But we have

2.2.1 Proposition *In a distributive lattice, a complement is both a pseudocomplement and a supplement (and hence, in particular, it is uniquely determined).*

Proof Let b be a complement of a and let $a \wedge x = 0$. Then $x = (a \vee b) \wedge x = (a \wedge x) \vee (b \wedge x) = b \wedge x$, that is, $x \leqslant b$. The supplement case follows by symmetry. □

2.2.2 Convention In a distributive lattice we will use, mostly, the symbol a^* also for the complement (keeping in mind, however, that not every pseudocomplement a^* is a complement).

2.2.3 Boolean algebras A (*complete*) *Boolean algebra* is a (complete) distributive lattice in which every element is complemented.

2.3 Linear and co-linear elements An element a in a distributive lattice L is said to be *linear* resp. *co-linear* [152] if for all $B \subseteq L$,

$$a \wedge \bigvee B = \bigvee\{a \wedge b \mid b \in B\} \quad \text{resp.} \quad a \vee \bigwedge B = \bigwedge\{a \vee b \mid b \in B\}$$

whenever $\bigvee B$ resp. $\bigwedge B$ exists. The following theorem, usually presented with a more complicated proof, is well known.

2.3.1 Theorem *Each complemented element in a distributive lattice L is both linear and co-linear.*

Proof We will prove the linearity, the co-linearity follows considering L^{op}.

Let $\overline{a}$ be the complement of a. We have

$$a \wedge x \leqslant y \quad \text{iff} \quad x \leqslant \overline{a} \vee y$$

(if $y \geqslant a \wedge x$, then $\overline{a} \vee y \geqslant \overline{a} \vee (a \wedge x) = (\overline{a} \vee a) \wedge (\overline{a} \vee x) \geqslant 1 \wedge x = x$; if $x \leqslant \overline{a} \vee y$, then $a \wedge x \leqslant a \wedge (\overline{a} \vee y) = (a \wedge \overline{a}) \vee (a \wedge y) \leqslant y$). Thus, $a \wedge (-) \colon L \to L$ is a left Galois adjoint and hence by 1.5 above it preserves all existing joins. □

3 Frames, Coframes, and Locales

3.1 Frames and coframes A *frame* is a complete lattice L satisfying the distributivity law

$$a \wedge \bigvee B = \bigvee\{a \wedge b \mid b \in B\} \qquad \text{(frm)}$$

for every $a \in L$ and $B \subseteq L$ (in other words, it is a complete lattice in which every element is linear).

Dually, a *coframe* is a complete lattice L satisfying the distributivity law

$$a \vee \bigwedge B = \bigwedge\{a \vee b \mid b \in B\} \qquad \text{(cofrm)}$$

for every $a \in L$ and $B \subseteq L$.

A typical frame is the lattice $\Omega(X)$ of open sets of a topological space; for good reasons (recall Sect. 2 in Chap. I) frames can be viewed as generalized spaces. But also coframes have an important geometric connotation: we will see that the system of all generalized subspaces is a coframe (and it can be argued that, actually, the system of classical subspaces of a space should be viewed as a coframe rather than as a frame—see 5.4.2 below). Also the associated notions are important. To avoid tedious repetitive dualizations of definitions we will summarize the most important ones in Remarks 5.6 below.

3.1.1 Observation In view of 2.3.1, every complete Boolean algebra is both a frame and a coframe.

3.2 Frame homomorphisms and the functor Ω Let L, M be frames. A mapping $h\colon L \to M$ is said to be a *frame homomorphism* if it preserves all joins and all finite meets (including 1). The resulting category will be denoted by

$$\mathbf{Frm}.$$

3.2.1 Notes

1. Typically, a frame homomorphism *does not* preserve general meets.
2. In case of Boolean algebras B_1, B_2, however, a frame homomorphism $h\colon B_1 \to B_2$ is automatically a complete Boolean homomorphism: from the formulas $(*)$ in 2.2 above we immediately infer that such an h preserves complements, and then we can also prove that it preserves general meets using De Morgan formulas.

3.2.2 For a continuous map $f\colon X \to Y$ define

$$\Omega(f)\colon \Omega(Y) \to \Omega(X) \quad \text{by setting} \quad \Omega(f)(U) = f^{-1}[U]$$

(recall Chap. I, Sect. 2). Thus, if we denote by **Top** the category of topological spaces, we obtain a contravariant functor

$$\Omega\colon \mathbf{Top} \to \mathbf{Frm}.$$

3.3 The category of locales; localic maps The fact 3.4 in Chap. I can be interpreted as that for an important subcategory of **Top**, the category of sober spaces (Sect. I.7), the functor Ω is a full embedding, only contravariant. Thus, if we consider the dual $\mathbf{Frm}^{\mathrm{op}}$, the *category of locales*, which we will denote by

$$\mathbf{Loc}$$

we have a *covariant*

$$\Omega\colon \mathbf{Top} \to \mathbf{Loc},$$

and **Loc** can be viewed as a natural extension of an important category of spaces.

3.3.1 Localic maps It is of advantage to view the category **Loc** as a concrete category, that is, a category in which the "generalized continuous maps" $L \to M$ are not just represented as formally inverted homomorphism arrows $h\colon L \leftarrow M$ but as well defined mappings $L \to M$. Since a frame homomorphism $h\colon M \to L$ preserves all suprema (joins) it has a uniquely defined right adjoint $f\colon L \to M$ which preserves all meets. Thus, we can represent the morphisms of **Loc** as

> *meet-preserving $f\colon L \to M$ such that their left adjoints f^* preserve finite meets.*

Such f's are referred to as *localic maps*.
A more expedient characterization will be presented in 4.4 below.

3.4 Spectra We have seen in Sect. 3 of Chap. I how the original space X can be reconstructed from the lattice (frame) $\Omega(X)$. Applying the procedure to a general frame we obtain a construction of the so-called *spectrum*

$$\Sigma(L) = (\{F \mid F \text{ completely prime filters on } L\}, \{\Sigma_a \mid a \in L\}),$$

where $\Sigma_a = \{F \mid a \in F\}$ and $\{\Sigma_a \mid a \in L\}$ is easily seen to constitute a topology on the set of completely prime filters.

Moreover, for localic $f\colon L \to M$ we have continuous maps

$$\Sigma(f) = (F \mapsto (f^*)^{-1}[F])\colon \Sigma(L) \to \Sigma(M)$$

and obtain a functor

$$\Sigma\colon \mathbf{Loc} \to \mathbf{Top}.$$

3.4.1 Notes

1. Thus, in the sober case, the spectrum reconstructs not only spaces from frames $\Omega(X)$, but also continuous maps from frame homomorphisms.
2. For the category minded, Σ is the right adjoint to Ω. See [220] or [161].

3.4.2 Spectra in terms of primes Denote by $\mathsf{Pr}(L)$ the set of all *prime* elements in L (that is, $p \in L$, $p \neq 1$, such that $a \wedge b \leqslant p$ implies that either $a \leqslant p$ or $b \leqslant p$). The spectrum functor can be, equivalently, described as

$$\Sigma(L) = (\mathsf{Pr}(L), \{\Sigma_a \mid a \in L\}) \quad \text{with} \quad \Sigma_a = \{p \mid a \nleqslant p\},$$

$$\Sigma(f)(p) = p$$

(localic maps send prime elements to primes, another advantage of the concrete representation of the category **Loc**). See [219, 220].

4 The Heyting Structure on a Frame

4.1 The Heyting arrow Recall 1.4 again. By the distribution law (frm) from 3.1 each mapping $(-) \wedge b$ has a right adjoint $b \to (-)$ resulting in the Heyting equivalence

$$a \wedge b \leqslant c \quad \text{iff} \quad a \leqslant b \to c. \tag{hey}$$

Thus, a frame is a (complete) Heyting algebra, and we will use its structure extensively.

Note that, on the other hand, every complete Heyting algebra is by 1.4 a frame.

Caution This is not to say that we can reduce the theory of frames to that of complete Heyting algebras. The respective categories have the same objects, but the morphisms are different. A frame homomorphism does not preserve general meets and, what is more important, $h(a \to b)$ is not generally equal to $h(a) \to h(b)$.

But the Heyting arrow is nevertheless very expedient when working in individual frames.

4.1.1 Pseudocomplements Note that, in particular,

$$a \wedge b \leqslant 0 \quad \text{iff} \quad b \leqslant a \to 0$$

so that we have pseudocomplements given by the formula

$$a^* = a \to 0.$$

4.1.2 Complements and Heyting arrows By the proof of 2.3.1, *if an element a has a complement a^*, then*

$$a \to b = a^* \vee b.$$

Note, however, that this formula holds only for complemented a.

4.2 Basic formulas resulting from the Heyting adjunction First, we have order-reversing mappings

$$(-) \to a \colon L \to L.$$

Indeed, if $x \leqslant y$ then, since $y \to a \leqslant y \to a$ gives $(y \to a) \wedge y \leqslant a$, we obtain $(y \to a) \wedge x \leqslant a$, and hence $y \to a \leqslant x \to a$. Thus, since $a \wedge b \leqslant c$ iff $b \wedge a \leqslant c$, the equivalence (hey) also yields the adjunction

$$a \to c \leqslant^{\mathrm{op}} c \quad \text{iff} \quad a \leqslant b \to c. \tag{op-hey}$$

Now from 1.4, (hey) and (op-hey) we obtain the rules

$$a \to \bigwedge_{i\in J} b_i = \bigwedge_{i\in J}(a \to b_i), \quad \text{and} \tag{distr}$$

$$(\bigvee_{i\in J} a_i) \to b = \bigwedge_{i\in J}(a_i \to b). \tag{op-distr}$$

Trivially, because $a \wedge b = b \wedge a$ we have the exchange rule

$$a \leqslant b \to c \quad \text{iff} \quad b \leqslant a \to c. \tag{exch}$$

4.3 Several simple computation rules in a Heyting lattice The Heyting lattice L in this subsection does not have to be complete; moreover, the existence of the top 1 does not have to be assumed, but this is not important for our purposes.

Proposition *We have*

(H1) $1 \to a = a$,
(H2) $a \leqslant b$ *iff* $a \to b = 1$,
(H3) $a \leqslant b \to a$,
(H4) $a \to b = a \to (a \wedge b)$,
(H5) $a \wedge (a \to b) \leqslant b$ *(modus ponens), consequently* $a \wedge (a \to b) = a \wedge b$,
(H6) $a \wedge b = a \wedge c$ *iff* $a \to b = a \to c$,
(H7) $(a \wedge b) \to c = a \to (b \to c) = b \to (a \to c)$,
(H8) $a = (a \vee b) \wedge (b \to a)$,
(H9) $a \leqslant (a \to b) \to b$, *and*
(H10) $((a \to b) \to b) \to b = a \to b$.

Proof

(H1) We have $x \leqslant 1 \to a$ iff $x = x \wedge 1 \leqslant a$.
(H2) $1 \leqslant a \to b$ iff $a = 1 \wedge a \leqslant b$.
(H3) $a \wedge b \leqslant a$, hence $a \leqslant b \to a$.
(H4) $a \to (a \wedge b) = (a \to a) \wedge (a \to b) = 1 \wedge (a \to b) = a \to b$.
(H5) $a \to b \leqslant a \to b$, hence $a \wedge (a \to b) \leqslant b$ so that $a \wedge (a \to b) \leqslant a \wedge b$. By (H3), $a \wedge b \leqslant a \wedge (a \to b)$.
(H6) immediately follows from (H4) and (H5).
(H7) We have $x \leqslant (a \wedge b) \to c$ iff $x \wedge a \wedge b \leqslant c$ iff $x \wedge a \leqslant b \to c$ iff $x \leqslant a \to (b \to c)$.
(H8) $a \leqslant (a \vee b) \wedge (b \to a)$ by (H3). On the other hand, $(a \vee b) \wedge (b \to a) = (a \wedge (b \to a)) \vee (b \wedge (b \to a) \leqslant a \vee (b \wedge a) = a$ by (H5).
(H9) Use $a \wedge (a \to b) \leqslant b$ from (H5) and (hey).

(H10) $((a \to b) \to b) \to b \geqslant a \to b$ by (H9) for $a \to b$, while applying the antitone $(-) \to b$ on (H9) yields the other inequality. □

4.4 Theorem *Let L, M be frames and let a mapping $f\colon L \to M$ preserve all meets. Let $h = f^*$ be its left Galois adjoint. Then h is a frame homomorphism (hence, f is a localic map) iff*

(a) $f(x) = 1$ *only if* $x = 1$, *and*
(b) $f(h(a) \to b) = a \to f(b)$.[1]

Proof $\Rightarrow$: Let h be a frame homomorphism. Then for (a), if $f(x) = 1$ we have $1 \leqslant f(x)$ and hence $1 = h(1) \leqslant x$. For (b), $x \leqslant f(h(a) \to b)$ iff $h(x) \leqslant h(a) \to b$ iff $h(x \wedge a) = h(x) \wedge h(a) \leqslant b$ iff $x \wedge a \leqslant f(b)$ iff $x \leqslant a \to f(b)$.

$\Leftarrow$: Let the conditions hold. Then, first, $h(1) = 1$ because $f(h(1)) \geqslant 1$. Second, by 1.4 $(*)$, $a \wedge b \leqslant f(h(a \wedge b))$ and hence, using (b), $a \leqslant b \to f(h(a \wedge b)) = f(h(b) \to h(a \wedge b))$. By the adjunction, $h(a) \leqslant h(b) \to h(a \wedge b)$ and we conclude that $h(a) \wedge h(b) \leqslant h(a \wedge b)$; $h(a) \wedge h(b) \geqslant h(a \wedge b)$ is trivial. □

5 The Coframe of Sublocales. Nuclei and Congruences

5.1 Motivation Subobjects in categories can be defined in various ways. Perhaps the most natural approach is using the concept of an *extremal monomorphism*, that is, a monomorphism $m\colon S \to A$ such that there is no decomposition $m = f \circ e$ with an epimorphism e which is not an isomorphism. Not very roughly speaking, if one thinks of concrete categories of structured objects, this amounts to one-to-one embedding of a smaller structured object S such that the structure on S is a border case: it cannot be made any stronger with the embedding map still carrying a morphism (think, for instance, of a subspace formed on a subset by the induced topology as opposed to just a continuous one-to-one map, or of an induced subgraph as opposed by just a subgraph with some of the edges between chosen vertices possibly omitted).

Recall the well-known (and very easy) fact that

- in the category of frames, the extremal epimorphisms are precisely the onto frame homomorphisms (while the plain epimorphisms are somewhat wild),
- and hence (recall A.1.4.1) the extremal monomorphisms in **Loc** are precisely the one-to-one localic maps.

5.2 Sublocales A subset $S \subseteq L$ is called a *sublocale* if

(S1) for every $M \subseteq S$, $\bigwedge M \in S$, and
(S2) for every $a \in L$ and every $s \in S$, $a \to s \in S$.

[1] This is often referred to as the *Frobenius identity*.

5.3 Theorem *A subset S of a frame L is a sublocale iff it is a frame and the embedding $j\colon S \subseteq L$ is a localic map.*

Proof $\Rightarrow$: If S is a sublocale it is a frame because by (S1) it is a complete lattice and by (S2), in particular, it has a Heyting operation, which takes care for the frame distributivity.

By (S1), $j\colon S \to L$ preserves all meets, and by 1.4 we have a left adjoint $h\colon L \to S$. We have $a \leqslant jh(a)$ for all $a \in L$ and $hj(s) \leqslant s$ for all $s \in S$. Since, by A.1.4.1, h is onto and $hjh = h$ we see that $hj = \mathrm{id}$. Summing up, we have

$$\forall a \in L,\ a \leqslant h(a) \quad \text{and} \quad \forall s \in S,\ h(s) = s.$$

Now let us use 4.4. We have $j(x) = 1$ only for $x = 1$ because $j(x) = x$. Further, if $a \in L$ and $s \in S$, then $x \leqslant j(h(a) \to s) = h(a) \to s$ iff $h(a) \leqslant x \to s$ iff, by adjunction, $a \leqslant j(x \to s) = x \to s$ iff $x \leqslant a \to s = a \to j(s)$. Thus, $j(h(a) \to s) = a \to j(s)$.

$\Leftarrow$: If S is a frame and $j\colon S \subseteq L$ is a localic map, we have immediately (S1) because j preserves meets. Now let a be arbitrary and $s \in S$. Then by 4.4, $a \to s = a \to j(s) = j(h(a) \to s)$ is in S. □

Now, as we have observed that a sublocale is a frame we can make a further

5.3.1 Observation *A sublocale of a sublocale is a sublocale.*

5.4 The coframe of sublocales Denote by

$$\mathsf{S}(L)$$

the system of all sublocales of L ordered by inclusion.

5.4.1 Theorem $\mathsf{S}(L)$ *is a coframe with the meets and joins given by*

$$\bigwedge_{i\in J} S_i = \bigcap_{i\in J} S_i, \quad \bigvee_{i\in J} S_i = \{\bigwedge M \mid M \subseteq \bigcup_{i\in J} S_i\},$$

and the least element $\mathsf{O} = \{1\}$.

Proof By (S1) each sublocale contains 1 and $\mathsf{O} = \{1\}$ is a sublocale, hence it is the smallest one.

Obviously, intersections of sublocales are sublocales (with the proviso that we set $\bigwedge \emptyset = L$). Hence $\mathsf{S}(L)$ is a complete lattice and $\bigwedge_{i\in J} S_i = \bigcap_{i\in J} S_i$.

Now if T contains all the S_i, $i \in J$, then by (S1) it has to contain $\{\bigwedge M \mid M \subseteq \bigcup_{i\in J} S_i\}$. On the other hand, if a is arbitrary and $M \subseteq \bigcup_{i\in J} S_i$, then

$$a \to \bigwedge M = \bigwedge\{a \to m \mid m \in M\}$$

and $\{a \to m \mid m \in M\} \in \bigcup_{i\in J} S_i$ by (S2). Thus, $\{\bigwedge M \mid M \subseteq \bigcup_{i\in J} S_i\}$ is a sublocale, the smallest one among those containing all the S_i.

It remains to be proved that $\mathsf{S}(L)$ satisfies the coframe identity

$$(\bigcap_{i\in J} S_i) \vee T = \bigcap_{i\in J}(S_i \vee T).$$

The inclusion $\subseteq$ is obvious. Hence, consider an $x \in \bigcap_{i\in J}(S_i \vee T)$. Then for every i there are $s_i \in S_i$ and $t_i \in T$ such that $x = s_i \wedge t_i$. Set $t = \bigwedge_{i\in J} t_i$. We have

$$x = \bigwedge_{i\in J}(s_i \wedge t_i) = \bigwedge_{i\in J} s_i \wedge \bigwedge_{i\in J} t_i = (\bigwedge_{i\in J} s_i) \wedge t \leqslant s_i \wedge t \leqslant s_i \wedge t_i = x$$

so that $x = s_i \wedge t$ for all i. Then, by 4.3(H6), all the $t \to s_i$ coincide; denote by s the common value. Since $s = t \to s_i \in S_i$, $s \in \bigcap S_i$ and we conclude by 4.3(H5) that $x = t \wedge s_i = t \wedge (t \to s_i) = t \wedge s \in (\bigcap_{i\in J} S_i) \vee T$. □

5.4.2 Notes The fact that the system of all "generalized subspaces" ordered by inclusion is a coframe is in fact a pleasing one [218, 219]. Classically, the system of all subspaces of a space is a Boolean algebra and hence both a frame and a coframe. But the latter is how one works with it: namely, one typically works with intersections, unions and the difference $A \smallsetminus B$, that is with the co-Heyting operation, seldom with the Heyting arrow $A \to B = (X \smallsetminus A) \cup B$.

5.5 Congruences and nuclei From the frame perspective we can view subobjects of L as (equivalence classes of) onto frame homomorphisms, that is, as congruences on L. For a congruence E we have the quotient map

$$q_E\colon L \to L/E;$$

taking the adjoint localic map $(q_E)_*$ we can obtain the associated sublocale as $(q_E)_*[L/E]$. This may look complicated, but it is not, because constructing quotients of frames is surprisingly simple. Since we will need the procedure also for other purposes it is in order to present an outline.

5.5.1 Quotients of frames Consider any binary relation R on a frame L. An element $s \in L$ is *saturated* (more precisely, *R-saturated*) if

$$\forall a, b, c, \quad aRb \quad \Rightarrow \quad a \wedge c \leqslant s \text{ iff } b \wedge c \leqslant s$$

(using the Heyting arrow we have an equivalent, much more elegant, but not always expedient formula

$$\forall a, b, \quad aRb \quad \Rightarrow \quad a \to s = b \to s.\ )$$

It is immediate that meets of saturated elements are saturated, and not hard to see that for any $a \in L$ and s saturated, $a \to s$ is saturated. Hence the set

$$L/R = \{s \mid s \text{ is } R\text{-saturated}\}$$

is a sublocale.

Since meets of saturated elements are saturated, we have the least saturated one majorizing x,

$$\nu_R(x) = \bigwedge\{s \text{ saturated} \mid x \leqslant s\}$$

and $L/R = \{x \mid x = \nu_R(x)\}$. The characteristic properties of the map $\nu = \nu_R : L \to L$ are that

$$x \leqslant y \Rightarrow \nu(x) \leqslant \nu(y),\ x \leqslant \nu(x),\ \nu\nu(x) = \nu(x) \text{ and } \nu(x \wedge y) = \nu(x) \wedge \nu(y).$$

Such maps are called *nuclei*, and each nucleus $\nu\colon L \to L$ results from a saturation procedure (see, e.g., [220]).

Further, the map ν restricted to $\overline{\nu}\colon L \to L/R$ is a frame homomorphism and one has

5.5.2 Theorem *For aRb, $\overline{\nu}(a) = \overline{\nu}(b)$, and for every frame homomorphism $h\colon L \to M$ such that aRb implies $h(a) = h(b)$, there is a unique $\overline{h}\colon L/R \to M$ such that $\overline{h} \circ \overline{\nu} = h$. Moreover, $\overline{h}$ is the restriction of h on L/R.*

(See, e.g., [220].)

Hence, $\overline{\nu}\colon L \to L/R$ is the quotient by the congruence generated by R, and we see that it is represented as a sublocale of L.

To summarize: We have one-to-one correspondences between $\mathsf{S}(L)$, the set $\mathfrak{C}(L)$ of congruences on L, and the set of nuclei $\mathcal{N}(L)$. They are given by

$$S \mapsto \nu_S = (x \mapsto \bigwedge\{s \in S \mid x \leqslant s\}), \qquad \nu \mapsto S_\nu = \{a \mid \nu(a) = a\}$$

$$E \mapsto L/E, \qquad S \mapsto E_S = \{(x, y) \mid \nu_S(x) = \nu_S(y)\}$$

$$E \mapsto \nu_E = (x \mapsto \bigvee Ex), \qquad \nu \mapsto E_\nu = \{(x, y) \mid \nu(x) = \nu(y)\}$$

(the reader wishing for details can consult, e.g., [220]).

Note that the nucleus ν_R above is the ν_E associated with the congruence generated by the relation R (the smallest congruence containing R).

The translations of sublocales to congruences and nuclei are antitone in the natural orders. Hence we have

$$\mathfrak{C}(L) \cong \mathcal{N}(L) \cong \mathsf{S}(L)^{\mathrm{op}}$$

and $\mathfrak{C}(L)$ resp. $\mathcal{N}(L)$ are frames.

5.5.3 Typically we will need only the nucleus $\nu_S\colon L \to L$ associated with a sublocale S, that is,

$$\nu_S(x) = \bigwedge\{s \in S \mid x \leqslant s\} = \min(\uparrow x \cap S).$$

The following equation is very useful.

Proposition *If $s \in S$ then for every $a \in L$*

$$a \to s = \nu_S(a) \to s.$$

Proof Trivially, $\nu(a) \to s \leqslant a \to s$. On the other hand, by 4.3(H9), $a \leqslant (a \to s) \to s \in S$, hence $\nu(a) \leqslant (a \to s) \to s$ and by the exchange rule in 4.1, $a \to s \leqslant \nu(a) \to s$. □

5.5.4 The following immediate consequence should be recognized as a standard property of sublocales, but it is seldom (if ever) mentioned.

Corollary *Sublocales are precisely the complete-Heyting subalgebras. That is, they are precisely the subsets closed under all meets and the binary operation $a \to b$.*

(Indeed, obviously each sublocale is a $(\bigwedge, \to)$-subalgebra; on the other hand, let S be such a subalgebra of L and let $x \in L$ be general. Then $a = \nu_S(x)$ is in S and we have $x \to b = a \to b \in S$.)

5.6 Remarks on the dual situation When working with coframes one has the frame concepts naturally dualized.

Thus, *coframe homomorphisms* preserve all meets and all finite joins.

A coframe is naturally endowed with a *co-Heyting operation*, the *difference*[2] $a \smallsetminus b$ satisfying the formula

$$a \smallsetminus b \leqslant c \quad \text{iff} \quad a \leqslant b \vee c$$

and similarly like in A.4.1.2, *if b is complemented, and only in such case, the difference can be written as $a \smallsetminus b = a \wedge b^*$.*

We have *colocalic maps*[3] (left) adjoint to coframe homomorphisms. They are characterized, in among the join-preserving maps, by the formulas

(a) $f(x) = 0$ only if $x = 0$, and
(b) $f(b \smallsetminus h(a)) = f(b) \smallsetminus a$

(recall 4.4).

The natural subobjects, *sub-colocales*, are subsets S closed under joins and containing all the $s \smallsetminus x$ with $s \in S$ (recall 5.2).

---

[2]For instance, in the Boolean algebra of all subsets of a set, or of all subspaces of a space, it is the standard difference $A \smallsetminus B$.

[3]An example is the standard image function $f[-]\colon \mathfrak{P}(X) \to \mathfrak{P}(Y)$ between power sets, induced by a mapping $f\colon X \to Y$ and adjoint to the preimage function $f^{-1}[-]$. For a more general fact see 7.4.4 below.

6 Special Sublocales

6.1 Open and closed sublocales *Open* resp. *closed* sublocales are defined by the formulas

$$\mathfrak{o}(a) = \{x \mid a \to x = x\} = \{a \to x \mid x \in L\} \quad \text{resp.} \quad \mathfrak{c}(a) = \uparrow a$$

(the equality in the definition of $\mathfrak{o}(a)$ follows from 4.3(H7)). Checking that they are really sublocales is straightforward.

6.1.1 Closure By (S1) for every sublocale S, $\bigwedge S \in S$. Thus, we have a very simple formula for *closure*, the smallest closed sublocale containing S,

$$\overline{S} = \uparrow \bigwedge S.$$

For example, $\overline{\mathfrak{o}(a)} = \mathfrak{c}(a^*)$ (since the smallest element of $\mathfrak{o}(a)$ is $a \to 0 = a^*$).

6.1.2 Notes

1. We use the symbols $\mathfrak{c}(a)$, $\uparrow a$ interchangeably: the former typically when we wish to emphasize the relation with $\mathfrak{o}(a)$, the latter in computing.
2. The terminology is well founded. Open resp. closed sublocales correspond in spaces precisely to open and closed subspaces, and in the Isbell's pioneering article [152] to the *open and closed parts*.
3. *The frame* $\mathfrak{o}(a)$ *is isomorphic to* $\downarrow a$. Indeed, we have the onto frame homomorphism $\widehat{a} = (x \mapsto a \wedge x) : L \to \downarrow a$. The associated localic map f satisfies the equivalence $x \wedge a \leqslant y$ iff $x \leqslant f(y)$. Since $x \wedge a \leqslant y$ iff $x \leqslant a \to y$ we conclude that $f(y) = a \to y$ and the associated sublocale (isomorphic with $\downarrow a$, since f—adjoint to an onto mapping—is one-to-one) is $f[\downarrow a] = \{a \to y \mid y \leqslant a\} = \{a \to y \mid y \in L\} = \mathfrak{o}(a)$.)
4. When representing subobjects of frames by congruences, the open $\mathfrak{o}(a)$'s correspond to the *open congruences*

$$\Delta_a = \{(x, y) \mid x \wedge a = y \wedge a\}$$

and the closed $\mathfrak{c}(a)$'s correspond to the *closed congruences*

$$\nabla_a = \{(x, y) \mid x \vee a = y \vee a\}.$$

Regarding nuclei, the open $\mathfrak{o}(a)$'s resp. the closed $\mathfrak{c}(a)$'s correspond to the

$$\nu_{\mathfrak{o}(a)} = (x \mapsto (a \to x)) : L \to L \quad \text{resp.} \quad \nu_{\mathfrak{c}(a)} = (x \mapsto a \vee x) : L \to L.$$

(First, $\nu_{\mathfrak{o}(a)}(x) = \bigwedge\{a \to y \mid x \leqslant a \to y,\ y \in L\}$; if $x \leqslant a \to y$, then $a \to x \leqslant a \to (a \to y) = a \to y$, hence $\nu_{\mathfrak{o}(a)}(x) \geqslant a \to x$, and on the other hand, $a \to x \leqslant a \to x$ and by 4.3(H7) $a \to (a \to x) = a \to x$ proving the other inequality. The second is straightforward.)

6.2 Proposition *$\mathfrak{o}(a)$ and $\mathfrak{c}(a)$ are complements of each other.*

Proof If $x \in \mathfrak{o}(a) \cap \mathfrak{c}(a)$ we have $a \leqslant x = a \to x$, hence $a = a \wedge a \leqslant x$ and $x = a \to x = 1$ by 4.3(H2). On the other hand, each $x \in L$ is by 4.3(H8) equal to $(a \to x) \wedge (a \vee x) \in \mathfrak{o}(a) \vee \mathfrak{c}(a)$. □

6.3 Theorem *We have the following formulas*

$$\mathfrak{o}(0) = \mathsf{O}, \quad \mathfrak{o}(1) = L, \quad \mathfrak{o}(a \wedge b) = \mathfrak{o}(a) \cap \mathfrak{o}(b) \quad \textit{and} \quad \mathfrak{o}(\bigvee_{i \in J} a_i) = \bigvee_{i \in J} \mathfrak{o}(a_i),$$

$$\mathfrak{c}(0) = L, \quad \mathfrak{c}(1) = \mathsf{O}, \quad \mathfrak{c}(a \wedge b) = \mathfrak{c}(a) \vee \mathfrak{c}(b) \quad \textit{and} \quad \mathfrak{c}(\bigvee_{i \in J} a_i) = \bigcap_{i \in J} \mathfrak{c}(a_i).$$

Proof We will prove the formulas for $\mathfrak{c}$, those for $\mathfrak{o}$ will then follow by De Morgan formula. They are simple observations:

$\uparrow 0 = L$, $\uparrow 1 = \{1\} = \mathsf{O}$, $x \geqslant \bigvee a_i$ iff $x \geqslant a_i$ for all i, and finally, $x \geqslant a \wedge b$ iff $x = (x \vee a) \wedge (x \vee b)$, that is, if $x \in \uparrow a \vee \uparrow b$. □

6.3.1 It follows immediately from 6.2 and 6.3 that, for any $a, b \in L$, $\mathfrak{c}(b) \subseteq \mathfrak{o}(a)$ iff $a \vee b = 1$, and $\mathfrak{c}(b) \supseteq \mathfrak{o}(a)$ iff $a \wedge b = 0$.

6.4 Theorem *A general sublocale can be represented by open and closed sublocales as follows*

$$S = \bigcap\{\mathfrak{c}(\nu_S(x)) \vee \mathfrak{o}(x) \mid x \in L\} = \bigcap\{\mathfrak{c}(y) \vee \mathfrak{o}(x) \mid \nu_S(x) = \nu_S(y)\}.$$

Proof

I. If $a \in S$, then for arbitrary x, $x \to a \in S$. Hence by 4.3(H8)

$$a = (a \vee \nu(x)) \wedge (\nu(x) \to a) = (a \vee \nu(x)) \wedge (x \to a) \in \mathfrak{c}(\nu(x)) \vee \mathfrak{o}(x).$$

On the other hand, if $a \in \bigcap\{\mathfrak{c}(\nu(x)) \vee \mathfrak{o}(x) \mid x \in L\}$, then, in particular, $a \in \mathfrak{c}(\nu(a)) \vee \mathfrak{o}(a)$ and hence $a = x \wedge (a \to y)$ with $x \geqslant \nu(a)$. Since $a \leqslant a \to y$ we have $a \leqslant y$, hence $a \to y = 1$, so that $a = x \geqslant \nu(a)$ and $a = \nu(a)$, that is, $a \in S$.

II. In view of I, since $\nu(\nu(x)) = \nu(x)$, it suffices to show that if $\nu(x) = \nu(y)$, then $S \subseteq \mathfrak{c}(y) \vee \mathfrak{o}(x)$.

Hence let $a \in S$. We have $a = (a \vee y) \wedge (y \to a) = (a \vee y) \wedge (\nu(y) \to a) = (a \vee y) \wedge (\nu(x) \to a) = (a \vee y) \wedge (x \to a) \in \mathfrak{c}(y) \vee \mathfrak{o}(x)$. □

6.5 Open and closed sublocales in sublocales Let S be a sublocale of a frame L. As we have already observed, it is itself a frame. Denote by $\mathfrak{o}_S(a)$ and $\mathfrak{c}_S(a)$ the open and closed sublocales in the frame S. They behave precisely like open and closed subsets in subspaces. Namely, we have

Proposition *For any $a \in L$, $\mathfrak{o}(a) \cap S = \mathfrak{o}_S(\nu_S(a))$ and $\mathfrak{c}(a) \cap S = \mathfrak{c}_S(\nu_S(a))$; hence, if a is in S, $\mathfrak{o}_S(a) = \mathfrak{o}(a) \cap S$ and $\mathfrak{c}_S(a) = \mathfrak{c}(a) \cap S$.*

Proof $x \in \mathfrak{o}(a) \cap S$ iff $x \in S$ and $a \to x = x$, which by A.5.5.3 is the same as $\nu_S(a) \to x = x$ (the Heyting operator in S is the same as in L).

$x \in \mathfrak{c}(a) \cap S$ means that $a \leqslant x \in S$ which is equivalent with $\nu_S(a) \leqslant x \in S$, that is, with $x \in \mathfrak{c}_S(\nu_S(a))$. □

6.6 Two-element ("one-point") sublocales Let p be a prime of L. For any $x \in L$, either

$$x \to p = 1 \quad \text{or} \quad x \to p = p. \tag{$*$}$$

(Indeed, if $x \to p \neq 1$, that is, $x \nleqslant p$, we have $x \to p \leqslant p$ since $x \wedge (x \to p) \leqslant p$ by 4.3(H5), and hence $x \to p = p$ by 4.3(H3).)

Proposition *Let $a \in L$, $a \neq 1$. Then $\{a, 1\}$ is a sublocale of L iff a is prime.*

Proof The implication "$\Leftarrow$" is obvious by $(*)$. Conversely, let $\{a, 1\}$ be a sublocale. If $x \wedge y \leqslant a$, then $x \leqslant y \to a \in \{a, 1\}$, and if $y \nleqslant a$, that is, $y \to a \neq 1$, then $x \leqslant y \to a = a$. □

Note Since the *void* (generalized) subspace is represented by a *one-element* set $\{1\}$, and also in view of A.3.4.2, it is natural to refer to the $\{p, 1\}$ as the *one-point sublocales.*

6.7 Boolean sublocales For each $a \in L$,

$$\mathfrak{b}(a) = \{x \to a \mid x \in L\}$$

is the *smallest sublocale containing* a: a is in $\mathfrak{b}(a)$ as $a = 1 \to a$; $\mathfrak{b}(a)$ is closed under meets by (op-distr) and 4.3(H7) yields (S2). Moreover, by 4.3(H10)

$$\mathfrak{b}(a) = \{x \in L \mid x = (x \to a) \to a)\}$$

and therefore

each $\mathfrak{b}(a)$ is a Boolean algebra

(since a frame is a Boolean algebra iff $x^{**} = x$ for all $x \in L$ and the pseudocomplement in $\mathfrak{b}(a)$ is given by $x^* = x \to a$).

Proposition *A sublocale S of L is a Boolean algebra iff $S = \mathfrak{b}(a)$ for some $a \in L$.*

Proof Let S be a Boolean sublocale and consider $a = \bigwedge S = 0_S$. Of course, $\mathfrak{b}(a) \subseteq S$ since $\mathfrak{b}(a)$ is the smallest sublocale containing a. On the other hand, each $s \in S$ is complemented in S, and hence

$$s = s^{**} = (s \to 0_S) \to 0_S = (s \to a) \to a \in \mathfrak{b}(a). \qquad \square$$

6.7.1 Booleanization In particular we have the *Booleanization* of L,

$$\mathfrak{B}(L) = \mathfrak{b}(0) = \{x^* \mid x \in L\} = \{x \in L \mid x = x^{**}\}.$$

Obviously (see the Proposition above) it is the *largest Boolean sublocale* of L. On the other hand, it is the *smallest dense sublocale* of L (a dense sublocale of L contains the bottom 0 and hence all the $x^* = x \to 0$),[4] a phenomenon that is specifically point-free.

Although the construction is fairly canonical, it is not functorial (to achieve functoriality one has to reduce the system of morphisms—see, e.g., [50]).

The construction of Booleanization preceded point-free topology by decades. It plays a role in logic as a link between the Brouwerian and the classical logic [116].

7 Images and Preimages

7.1 If $f: L \to M$ is a localic map, then the set-image of a sublocale $S \subseteq L$, $f[S]$, is a sublocale of M. Indeed, since f preserves meets, $f[S]$ is closed under meets. Further, by 4.4, for an arbitrary $a \in M$ and $s \in S$, $a \to f(s) = f(h(a) \to s) \in f[S]$ because $h(a) \to s \in S$.

The situation with preimage is only slightly more complicated. First, observe that

the set-preimage $f^{-1}[T]$ of a sublocale T is closed under meets. $\qquad (*)$

Now we can easily prove

7.2 Theorem *Define the* (localic) preimage *of a sublocale $T \subseteq M$ as*

$$f_{-1}[T] = \bigvee\{S \in \mathsf{S}(L) \mid S \subseteq f^{-1}[T]\}.$$

Then we have monotone maps

$$\mathsf{S}(L) \underset{f_{-1}[-]}{\overset{f[-]}{\rightleftarrows}} \mathsf{S}(M)$$

[4] This fact is usually referred to as the Isbell's Density Theorem.

satisfying

$$f[S] \subseteq T \quad \textit{iff} \quad S \subseteq f_{-1}[T],$$

that is, $f[-]$ *is adjoint to the left and* $f_{-1}[-]$ *is adjoint to the right.*

Proof By the observation $(*)$ in 7.1 and the formula for join in the coframe of sublocales, $f_{-1}[T]$ is the largest sublocale contained in the set $f^{-1}[T]$. Thus, the adjunction follows from the standard

$$f[S] \subseteq T \quad \text{iff} \quad S \subseteq f^{-1}[T]$$

considering that S is a sublocale and hence $S \subseteq f_{-1}[T]$ iff $S \subseteq f^{-1}[T]$. □

7.3 Images and preimages of localic maps behave very much like images and preimages of continuous functions (see [104]). We have

Theorem *Let* $f: L \to M$ *be a localic map and let* h *be the associated (left adjoint) frame homomorphisms. Then*

1. *preimages of closed sublocales are closed, more precisely,*

$$f_{-1}[\mathfrak{c}(a)] = f^{-1}[\mathfrak{c}(a)] = \mathfrak{c}(h(a)),$$

2. *preimages of open sublocales are open, more precisely,*

$$f_{-1}[\mathfrak{o}(a)] = \mathfrak{o}(h(a)), \ \textit{and}$$

3. *for every sublocale* $S \subseteq L$, $f[\overline{S}] \subseteq \overline{f[S]}$.

Proof

1. Here the localic preimage even coincides with the set-preimage. The fact is trivial: by the adjunction $h(a) \leqslant x$ iff $a \leqslant f(x)$ we see that $x \in f^{-1}[\mathfrak{c}(a)]$, that is, $f(x) \in {\uparrow}a$, iff $x \in {\uparrow}h(a)$.
2. By 4.4(b), $f(h(a) \to x) = a \to f(x)$ and hence $\mathfrak{o}(h(a)) \subseteq f^{-1}[\mathfrak{o}(a)]$. Thus, we have to prove that every sublocale $S \subseteq f^{-1}[\mathfrak{o}(a)]$ is a subset of $\mathfrak{o}(h(a))$. Hence consider an $S \subseteq f^{-1}[\mathfrak{o}(a)]$ and $s \in S$. Then $(h(a) \to s) \to s$ is in S and $f((h(a) \to s) \to s) \in \mathfrak{o}(a)$ so that

$$\begin{aligned} f((h(a) \to s) \to s) = a \to f((h(a) \to s) \to s) &= \\ = f(h(a) \to ((h(a) \to s) \to s)) &= \\ = f((h(a) \wedge (h(a) \to s)) \to s) &= \\ = f((h(a) \wedge s) \to s) = f(1) &= 1 \end{aligned}$$

(the equalities, in this order, by definition of $\mathfrak{o}(a)$, 4.4(b), 4.3(H7), 4.3(H4), 4.3(H2), and preserving finite meets and hence 1). Thus, by 4.4(a), $(h(a) \to s) \to s = 1$, by 4.3(H2) $(h(a) \to s) \leqslant s$ and by 4.3(H3), $h(a) \to s = s$, that is, $s \in \mathfrak{o}(h(a))$.

3. If $x \geqslant \bigwedge S$, then $f(x) \geqslant f(\bigwedge S) = \bigwedge f[S]$. □

7.4 Preimage is a coframe homomorphism In the classical context, the preimage function $f^{-1}[-]$ between the power-sets (and hence between the Boolean algebras of subspaces) is a Boolean homomorphism. In this section we will prove that, similarly, the localic preimage function

$$f_{-1}[-]\colon \mathsf{S}(M) \to \mathsf{S}(L)$$

is a coframe homomorphism.

Fix a localic map $f\colon L \to M$ with $h\colon M \to L$ the adjoint frame homomorphism. To simplify the notation write φ for $f_{-1}[-]$. Thus, in this notation

$$\varphi(\mathfrak{c}(a)) = \mathfrak{c}(h(a)) \quad \text{and} \quad \varphi(\mathfrak{o}(a)) = \mathfrak{o}(h(a)). \tag{$*$}$$

Further denote by $\mathfrak{c}\mathsf{S}$ resp. $\mathfrak{o}\mathsf{S}$ the set of closed resp. open sublocales (in L or in M, it will be always obvious which of the cases it is), and, finally, set

$$\mathfrak{co}\mathsf{S} = \{\mathfrak{c}(a) \vee \mathfrak{o}(b) \mid a, b \text{ in the frame in question}\}.$$

Since φ is a right adjoint we have (using also 4.4(a))

7.4.1 Fact *φ preserves all meets, and $\varphi(\mathsf{O}) = \mathsf{O}$.*

7.4.2 Lemma *$\varphi(\mathfrak{o}(a)) \cap \varphi(\mathfrak{c}(b))$ is the complement of $\varphi(\mathfrak{c}(a) \vee \mathfrak{o}(b))$.*

Proof We have, by 7.4.1,

$$\begin{aligned}\varphi\big(\mathfrak{c}(a) \vee \mathfrak{o}(b)\big) \cap \varphi(\mathfrak{o}(a)) \cap \varphi(\mathfrak{c}(b)) &= \\ = \varphi\big((\mathfrak{c}(a) \vee \mathfrak{o}(b)) \cap \mathfrak{o}(a) \cap \mathfrak{c}(b)\big) &= \varphi(\mathsf{O}) = \mathsf{O}.\end{aligned}$$

Then, by distributivity and $(*)$,

$$\begin{aligned}&\varphi\big(\mathfrak{c}(a) \vee \mathfrak{o}(b)\big) \vee \big(\varphi(\mathfrak{o}(a)) \cap \varphi(\mathfrak{c}(b))\big) = \\ &\quad = \big(\varphi(\mathfrak{c}(a) \vee \mathfrak{o}(b)) \vee \varphi(\mathfrak{o}(a))\big) \cap \big(\varphi(\mathfrak{c}(a) \vee \mathfrak{o}(b)) \vee \varphi(\mathfrak{c}(b))\big) \supseteq \\ &\quad \supseteq \big(\varphi(\mathfrak{c}(a)) \vee \varphi(\mathfrak{o}(a))\big) \cap \big(\varphi(\mathfrak{o}(b)) \vee \varphi(\mathfrak{c}(b))\big) = \\ &\quad = \big(\mathfrak{c}(h(a)) \vee \mathfrak{o}(h(a))\big) \cap \big(\mathfrak{o}(h(b)) \vee \mathfrak{c}(h(b))\big) = L.\end{aligned}$$

□

7.4.3 Lemma *For $x_1, x_2 \in \mathfrak{co}\mathsf{S}$, $\varphi(x_1 \vee x_2) = \varphi(x_1) \vee \varphi(x_2)$.*

Proof If $x_1, x_2 \in \mathfrak{o}\mathsf{S}$ or $x_1, x_2 \in \mathfrak{c}\mathsf{S}$, it follows immediately from 7.3. Next, let $x_1 = \mathfrak{c}(a) \in \mathfrak{c}\mathsf{S}$ and $x_2 = \mathfrak{o}(b) \in \mathfrak{o}\mathsf{S}$. Then by 7.4.2 and Remark in A.2.1.1

$$\begin{aligned}\varphi(\mathfrak{c}(a) \vee \mathfrak{o}(b)) &= (\varphi(\mathfrak{o}(a)) \cap \varphi(\mathfrak{c}(b)))^{\#} = \\ &= \varphi(\mathfrak{o}(a))^{\#} \vee \varphi(\mathfrak{c}(b))^{\#} = \varphi(\mathfrak{c}(a)) \vee \varphi(\mathfrak{o}(b)).\end{aligned}$$

Finally, let $x_i = u_i \vee v_i$, $i = 1, 2$, with $u_i \in \mathfrak{c}\mathsf{S}$ and $v_i \in \mathfrak{o}\mathsf{S}$. Then we have

$$\begin{aligned}\varphi((u_1 \vee v_1) \vee (u_2 \vee v_2)) &= \varphi((u_1 \vee u_2) \vee (v_1 \vee v_2)) = \\ &= \varphi(u_1 \vee u_2) \vee \varphi(v_1 \vee v_2) = \varphi(u_1) \vee \varphi(u_2) \vee \varphi(v_1) \vee \varphi(v_2) = \\ &= \varphi(u_1) \vee \varphi(v_1) \vee \varphi(u_2) \vee \varphi(v_2) = \varphi(u_1 \vee v_1) \vee \varphi(u_2 \vee v_2).\end{aligned}$$ □

7.4.4 Theorem *Let $f : L \to M$ be a localic map. Then*

$$f_{-1}[-] : \mathsf{S}(M) \to \mathsf{S}(L)$$

is a coframe homomorphism (hence, in particular, it preserves complements). Consequently, the image function

$$f[-] : \mathsf{S}(L) \to \mathsf{S}(M)$$

is a colocalic map.

Proof We have already observed in 7.4.1 that $\varphi = f_{-1}[-]$ preserves zero and general meets. Thus let S, T be arbitrary sublocales of M. By 6.4 we have $S = \bigcap_i x_i$ and $T = \bigcap_j y_j$ with $x_i, y_j \in \mathfrak{co}\mathsf{S}$ and by 7.4.3 we obtain

$$\begin{aligned}\varphi(S \vee T) &= \varphi(\bigcap_i x_i \vee \bigcap_j y_j) = \varphi(\bigcap_{i,j}(x_i \vee y_j)) = \bigcap_{i,j}\varphi(x_i \vee y_j) = \\ &= \bigcap_{i,j}(\varphi(x_i) \vee \varphi(y_j)) = \bigcap_i \varphi(x_i) \vee \bigcap_j \varphi(y_j) = \varphi(\bigcap_i x_i) \vee \varphi(\bigcap_j y_j) = \\ &= \varphi(S) \vee \varphi(T).\end{aligned}$$ □

8 Binary Coproduct of Frames in a Broader Perspective

The reader can find constructions of coproducts of frames in the standard literature (e.g. [161, 220, 222, 236]). In this book we mostly need the binary case $L_1 \oplus L_2$ only. We will present its construction as a part of a more general procedure in a broader category. This will allow us to use the binary coproduct in cases that cannot be dealt with in the scope of frame homomorphisms only. Moreover, it will illustrate a certain general phenomenon connecting coproducts in categories enriched by extra operations distributing over the given structures (see 8.7 below).

8.1 We will be interested in

$$\bigvee\textbf{-Lat},$$

the *category of join-lattices* (in a sense extending the category of frames—which is naturally contained, but not as a full subcategory). It is a category of fundamental importance in various applications, see [173].

The objects of $\bigvee$**-Lat** are complete lattices and the morphisms $f: L \to M$ are the mappings preserving all joins (we will refer to them as *join-maps* or $\bigvee$*-maps*).

8.1.1 Taking quotients in $\bigvee$**-Lat** is even simpler than in frames (recall 5.5 above). We will discuss it in some detail.

Let L be a complete lattice and let $R \subseteq L \times L$ be a binary relation on L. An element $s \in L$ is *weakly R-saturated* if

$$\forall a, b \in L, \quad aRb \quad \Rightarrow \quad a \leqslant s \text{ iff } b \leqslant s. \tag{w-sat}$$

Then,

arbitrary meets of weakly saturated elements are weakly saturated.

Hence, if we define

$$\kappa(a) = \kappa_{wR}(a) = \bigwedge\{s \mid a \leqslant s, \ s \text{ weakly saturated}\}$$

we obtain a monotone mapping $L \to L$ such that $a \leqslant \kappa(a)$ and $\kappa\kappa(a) = \kappa(a)$. Set

$$L/_w R = \{x \in L \mid \kappa(x) = x\} = \{\kappa(x) \mid x \in L\}.$$

8.1.2 Recall the saturated elements from A.5.5.1, that is the s such that

$$\forall a, b, c \in L, \quad aRb \quad \Rightarrow \quad a \wedge c \leqslant s \text{ iff } b \wedge c \leqslant s. \tag{sat}$$

We see that

if the relation R is such that

$$aRb \quad \textit{implies} \quad (a \wedge c)R(b \wedge c) \textit{ for any } c,$$

then (*w-sat*) *is equivalent with* (*sat*).

In particular,

if R is a relation closed under meets on a frame L, then

$$L/_w R = L/R.$$

8.1.3 Proposition *$L/_w R$ is a complete lattice with joins $\bigsqcup_i a_i = \kappa(\bigvee_i a_i)$, and the mapping*

$$\kappa' = (a \mapsto \kappa(a)) \colon L \to L/_w R$$

is a join-map.

Proof $\kappa(0)$ is obviously the bottom of $L/_w R$. If $a_i \leqslant x \in L/_w R$, then $\bigvee a_i \leqslant x$ and consequently $\kappa(\bigvee a_i) \leqslant \kappa(x) = x$. Further, we have, for $a_i \in L$, $\kappa'(\bigvee a_i) \leqslant \kappa(\bigvee \kappa(a_i)) = \bigsqcup \kappa'(a_i) \leqslant \kappa'(\bigvee a_i)$. □

8.1.4 Theorem *We have the implication*

$$aRb \;\Rightarrow\; \kappa'(a) = \kappa'(b)$$

and if a join-map $f \colon L \to M$ satisfies

$$aRb \;\Rightarrow\; f(a) = f(b),$$

then there is exactly one join-map $\overline{f} \colon L/_w R \to M$ such that $\overline{f} \circ \kappa' = f$. Moreover, $\overline{f} = f_{|(L/_w R)}$.

Proof Since $a \leqslant \kappa(a)$ and $\kappa(a)$ is weakly saturated, $b \leqslant \kappa(a)$ and hence $\kappa(b) \leqslant \kappa(a)$; starting with $b \leqslant \kappa(b)$ we similarly obtain $\kappa(a) \leqslant \kappa(b)$.

Now let $f(a) = f(b)$ whenever aRb. Set

$$\sigma(x) = \bigvee\{y \mid f(y) \leqslant f(x)\}.$$

Since $f(x) = f(x)$ we have $x \leqslant \sigma(x)$, and

$$f(\sigma(x)) = \bigvee\{f(y) \mid f(y) \leqslant f(x)\} \leqslant f(x)$$

and hence $f(\sigma(x)) = f(x)$. Next, we see that $\sigma(x)$ is weakly saturated: indeed, if aRb and $a \leqslant \sigma(x)$, then $f(b) = f(a) \leqslant f(\sigma(x)) = f(x)$ and hence $b \leqslant \sigma(x)$.

Thus, $\kappa(x) \leqslant \sigma(x)$ and we conclude that $f(x) \leqslant f(\kappa(x)) \leqslant f(\sigma(x)) = f(x)$ and can set $\overline{f}(x) = f(x)$. □

8.2 For any poset X with zero consider the down-set lattice

$$\mathfrak{D}(X) = \{U \mid \emptyset \neq U = {\downarrow}U \subseteq X\}$$

(ordered by inclusion). Since the suprema in $\mathfrak{D}(X)$ are the unions (with the exception of $\sup\emptyset$ which is $\{0\}$) and the infima are the intersections, this lattice is complete, and indeed a frame.

We have mappings

$$\lambda = \lambda_X = (a \mapsto {\downarrow}a) \colon X \to \mathfrak{D}(X).$$

8.2.1 Lemma

(1) *For any complete lattice Y and monotone $f\colon X \to Y$ there is a unique join-map $\overline{f}\colon \mathfrak{D}(X) \to Y$ such that $\overline{f} \circ \lambda = f$, given by*

$$\overline{f}(U) = \bigvee\{f(u) \mid u \in U\}.$$

(2) *If X is a meet-semilattice, then λ is a meet-semilattice homomorphism.*
(3) *If X is a bounded meet-semilattice and Y a frame, and if f is a bounded meet-semilattice homomorphism, then $\overline{f}$ is a frame homomorphism.*

Proof (1) and (2) are straightforward.

(3): Since $\overline{f}$ is a join-map we have to show that $\overline{f}$ preserves meets. We have obviously $\overline{f}(X) = \overline{f}({\downarrow}1) = 1$. Further,

$$\begin{aligned}\overline{f}(U) \wedge \overline{f}(V) &= \bigvee\{f(u) \mid u \in U\} \wedge \bigvee\{f(v) \mid v \in V\} = \\ &= \bigvee\{f(u) \wedge f(v) \mid u \in U, v \in V\} = \bigvee\{f(u \wedge v) \mid u \in U, v \in V\} \leqslant \\ &\leqslant \bigvee\{f(w) \mid w \in U \cap V\} = \overline{f}(U \cap V) \leqslant \overline{f}(U) \wedge \overline{f}(V).\end{aligned}$$ □

8.3 The relation R On $\mathfrak{D}(X_1 \times X_2)$ consider the binary relation R consisting of all the pairs

$$\Big(\bigcup_{i\in J} {\downarrow}(a_i, b), {\downarrow}\big(\bigvee_{i\in J} a_i, b\big)\Big) \text{ with } a_i \in X_1 \text{ and } b \in X_2, \text{ and}$$

$$\Big(\bigcup_{i\in J} {\downarrow}(a, b_i), {\downarrow}\big(a, \bigvee_{i\in J} b_i\big)\Big) \text{ with } a \in X_1 \text{ and } b_i \in X_2$$

(also including the void J) and set

$$X_1 \otimes X_2 = \mathfrak{D}(X_1 \times X_2)/_w R = \mathfrak{D}(X_1 \times X_2)/R$$

(since R is closed under meets, recall A.8.1.2).

In the sequel we will work with this particular relation only. The resulting R-saturated down-sets $U \subseteq X_1 \times X_2$ are usually called *cp-ideals*.[5] It is easy to check that they are characterized by the following properties:

$$\begin{aligned}&\text{if } (a_i, b) \in U \text{ for all } i \in J \text{ then } (\textstyle\bigvee a_i, b) \in U \\ &\text{and if } (a, b_i) \in U \text{ for all } i \in J \text{ then } (a, \textstyle\bigvee b_i) \in U.\end{aligned} \qquad \text{(cp-ideal)}$$

[5]Short for "coproduct ideals"; the motivation for this expression will be apparent in 8.6 below.

Since it includes also the void J we have that

every R-saturated U contains $\mathsf{n} = \{(x, 0), (0, y) \mid x \in X_1, y \in X_2\}$.

8.4 Bi-$\bigvee$-maps; $X_1 \otimes X_2$ as a tensor product Let X_1, X_2 and Y be complete lattices. A mapping

$$f : X_1 \times X_2 \to Y$$

is a *bi-$\bigvee$-map* if each $f(x_1, -) : X_2 \to Y$ and each $f(-, x_2) : X_1 \to Y$ is a join-map.

8.4.1 Theorem *The mapping*

$$\mu = \kappa\lambda : X_1 \times X_2 \to X_1 \otimes X_2$$

is a bi-$\bigvee$-map, and for every bi-$\bigvee$-map $f : X_1 \times X_2 \to Y$ there is precisely one join-map $\tilde{f} : X_1 \otimes X_2 \to Y$ such that $\tilde{f} \circ \mu = f$.

Proof Consider the diagram

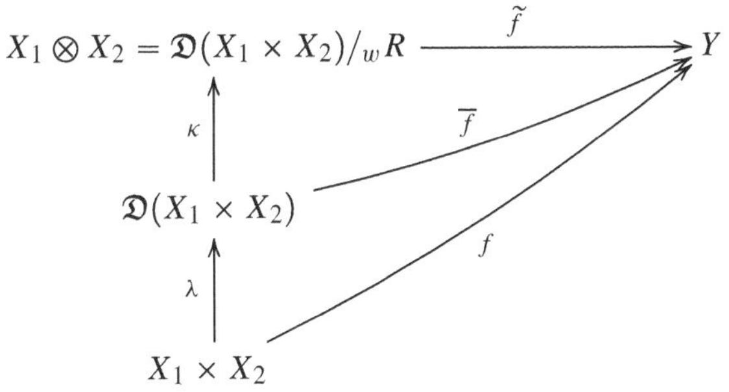

with $\overline{f}$ the join-map from 8.2.1. Since f is a bi-$\bigvee$-map we have

$$\overline{f}(\bigcup_{i \in J} {\downarrow}(a_i, b)) = \bigvee_{i \in J}(\overline{f}({\downarrow}(a_i, b))) = \bigvee_{i \in J} f(a_i, b) = f(\bigvee_{i \in J} a_i, b) = \overline{f}({\downarrow}(\bigvee_{i \in J} a_i, b))$$

and similarly $\overline{f}(\bigcup_{i \in J}({\downarrow}(a, b_i))) = \overline{f}({\downarrow}(a, \bigvee_{i \in J} b_i))$ and hence we obtain the $\tilde{f}$ by 8.1.4. □

8.5 Set $a \otimes b = \kappa({\downarrow}(a, b))$ and recall the

$$\mathsf{n} = \{(x, 0), (0, y) \mid x \in X_1, y \in X_2\}$$

from 8.3. Obviously

$X_1 \otimes X_2$ is join-generated by the $a \otimes b$.

We have

8.5.1 Proposition

(1) $a \otimes b = \downarrow(a, b) \cup \mathsf{n}$.
(2) $(\bigvee_{i \in J} a_i) \otimes b = \bigvee_{i \in J} (a_i \otimes b)$ *and* $a \otimes (\bigvee_{i \in J} b_i) = \bigvee_{i \in J} (a \otimes b_i)$.

Proof

(1) Any saturated U contains n and it is easy to check that $\downarrow(a, b) \cup \mathsf{n}$ is saturated.
(2) Since κ is a join-map we have

$$(\bigvee_{i \in J} a_i)\otimes b = \kappa(\downarrow(\bigvee_{i \in J} a_i, b)) = \kappa(\bigcup_{i \in J} \downarrow(a_i, b)) = \bigvee_{i \in J} \kappa(\downarrow(a_i, b)) = \bigvee_{i \in J} (a_i \otimes b).$$

□

Note The latter has nothing to do with any kind of distributivity in the lattices X_i's.

8.5.2 Proposition *Let* $f_i : X_i \to Y_i$, $i = 1, 2$, *be join-maps. Then the* $f_1 \otimes f_2$ *defined as* $\mu(\widetilde{f_1 \times f_2})$ *(the* $\widetilde{(-)}$ *from 8.4.1) is given by the formula*

$$(f_1 \otimes f_2)(x_1 \otimes x_2) = f_1(x_1) \otimes f_2(x_2).$$

Proof We have

$$(f_1 \otimes f_2)(x_1 \otimes x_2) = (f_1 \otimes f_2)(\kappa(\downarrow(x_1, x_2))) = (f_1 \otimes f_2)(\kappa(\lambda(x_1, x_2))) =$$
$$= \mu(f_1 \times f_2)(x_1, x_2) = \kappa(\downarrow(f(x_1), f(x_2))) = f_1(x_1) \otimes f_2(x_2).$$

□

8.6 Binary coproducts of frames Let $X_i = L_i$ be frames. To adjust the notation to the standard one, and (of course) indicating that we will obtain the coproduct ("sum"), we will write in this case

$$L_1 \oplus L_2 \quad \text{instead of} \quad L_1 \otimes L_2$$

and similarly $a \oplus b$ for the $a \otimes b = \downarrow(a, b) \cup \mathsf{n}$ from 8.5.1.

Consider the maps

$$\iota_1 = (a \mapsto a \oplus 1) \colon L_1 \to L_1 \oplus L_2, \quad \iota_2 = (a \mapsto 1 \oplus a) \colon L_2 \to L_1 \oplus L_2.$$

8.6.1 Theorem *The maps* $\iota_i : L_i \to L_1 \oplus L_2$ *are frame homomorphisms and constitute a coproduct in the category of frames.*

Proof $(a \oplus 1) \cap (b \oplus 1) = (a \wedge b) \oplus 1$ follows from the formula $x \otimes 1 = \downarrow(x, 1) \cup \mathsf{n}$, and the join is preserved by 8.5.1(2).

Now let $h_i : L_i \to M$ be frame homomorphisms. The mapping

$$g = ((a_1, a_2) \mapsto h_1(a_1) \wedge h_2(a_2)) : L_1 \times L_2 \to M$$

is obviously a bounded meet-homomorphism and hence by 3 in 8.2.1 we have a frame homomorphism $\overline{g} : \mathfrak{D}(L_1 \times L_2) \to M$ given by

$$\overline{g}(U) = \bigvee\{g(u) \mid u \in U\}.$$

We have

$$\overline{g}(\bigcup_{i\in J} \downarrow(a_i, b)) = \bigvee_{i\in J}(\overline{g}(\downarrow(a_i, b))) = \bigvee_{i\in J} g(a_i, b) = \bigvee_{i\in J}(h_1(a_i) \wedge h_2(b)) =$$
$$= (\bigvee_{i\in J} h_1(a_i)) \wedge h_2(b) = h_1(\bigvee_{i\in J} a_i) \wedge h_2(b) = g(\bigvee_{i\in J} a_i, b) = \overline{g}(\downarrow(\bigvee_{i\in J} a_i, b))$$

and hence $\overline{g}$ restricts by 5.5.2 to a frame homomorphism $h : L_1 \oplus L_2 \to M$ and we have $h(\iota_1(a)) = h(\kappa(\downarrow(a, 1))) = h_1(a_1) \wedge h_2(1) = h_1(a)$ and similarly $h\iota_2(a) = h_2(a)$. The unicity is obvious. □

Then, for any frame homomorphisms $f_i : L_i \to M_i, i = 1, 2$, we have a (unique) frame homomorphism $f_1 \oplus f_2 : L_1 \oplus L_2 \to M_1 \oplus M_2$ determined by $(f_1 \oplus f_2) \circ \iota_i = \iota_i \circ f_i$. In fact, we have more:

8.6.2 Proposition *Let L_i be frames and let $f_i : L_i \to M_i$, $i = 1, 2$, be join-maps. Then we have the $f_1 \otimes f_2 : L_1 \oplus L_2 \to M_1 \oplus M_2$ determined by*

$$(f_1 \otimes f_2) \circ \iota_i = \iota_i \circ f_i$$

(thus, $f_1 \otimes f_2$ is determined by the same expression as $f_1 \oplus f_2$ for f_i frame homomorphisms, even if the f_i do not preserve meets).

Proof Use the formula from 8.5.2. We have in particular

$$(f_1 \otimes f_2)(\iota_1(x)) = (f_1 \otimes f_2)(x \oplus 1) = f_1(x) \oplus f_2(1) = f_1(x) \oplus 1 = \iota_1(f_1(x))$$

and similarly $(f_1 \otimes f_2) \circ \iota_2 = \iota_2 \circ f_2$. □

8.7 Note: tensor products and coproducts Recall the standard construction of classical tensor products $A \oplus B$ of abelian groups A, B. What one does:

(1) first, one takes the cartesian product of the underlying sets, $A \times B$;
(2) then, the free abelian group $F(A \times B)$ generated by this set,
(3) and finally one takes the quotient $F(A \times B)/\sim$ by the congruence $\sim$ determined by

$$(a_1, b) + (a_2, b) \sim (a_1 + a_2, b) \quad \text{and} \quad (a, b_1) + (a, b_2) \sim (a, b_1 + b_2).$$

Now recall what one does when constructing the $X \otimes Y$:

(1) first, one takes the cartesian product of the underlying posets, $X \times Y$;
(2) then, the down-set join-lattice $\mathfrak{D}(X \times Y)$ generated by this set,
(3) and finally one takes the quotient $\mathfrak{D}(X \times Y)/\sim$ by the congruence $\sim$ determined by

$$\bigvee_{i \in J} {\downarrow}(x_i, y) \sim {\downarrow}(\bigvee_{i \in J} x_i, y) \quad \text{and} \quad \bigvee_{i \in J} {\downarrow}(x, y_i) \sim {\downarrow}(x, \bigvee_{i \in J} y_i).$$

Taking into account that $\mathfrak{D}$ is a construction of a free join-lattice over a poset (resp. of a free frame over a bounded semilattice) we observe a striking similarity (the steps (3) are adjustments of the additive structures), which becomes even more striking if we compare 8.6 with the fact that

tensor products of the underlying abelian groups underlie the binary coproducts in the category of commutative rings.

We have here a richer structure (commutative ring, frame) extending an "additive one" (commutative addition in abelian groups, join in join-lattices) by a new operation that distributes over the original structure in a certain way, and natural tensor products in the additive structure become coproducts in the enriched one. This is a general phenomenon. The reader can find the general categorical facts, e.g. in [46, 47]; for more precise and detailed analysis of the particular case of rings and frames see [222].

9 Free Frames and Defining Equations

9.1 Recall from 8.2 the down-set construction. For a bounded meet-semilattice S we consider the lattice

$$\mathfrak{D}(S) = \{A \mid \emptyset \neq A \subseteq S,\ {\downarrow}A = A\}$$

and the mapping (down-set lattice homomorphism)

$$\lambda_S = (a \mapsto {\downarrow}a) \colon S \to \mathfrak{D}(S).$$

This provides a *free functor from the category* **SLat$_{01}$** *to the category* **Frm** in the sense that for each frame L and each bounded frame homomorphism $f \colon S \to L$ there is precisely one frame homomorphism $h \colon \mathfrak{D}(S) \to L$ such that

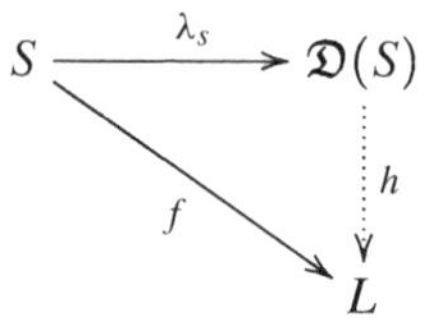

commutes.

9.2 If X is a set consider

$$\mathfrak{M}(X) = \{A \mid A \subseteq X,\ A \text{ finite}\} \cup \{0\}$$

ordered by *inverse* inclusion (thus, $A \wedge B = A \cup B$ and $1_{\mathfrak{M}(X)} = \emptyset$; 0 is an added least element) and the mapping

$$\mu_X = (x \mapsto \{x\})\colon X \to \mathfrak{M}(X).$$

Thus we have obtained a free functor $\mathfrak{M}\colon$ **Set** $\to$ **SLat$_{01}$**, again in the sense that for each bounded meet-semilattice S and each mapping $f\colon X \to S$ there is precisely one bounded semilattice homomorphism $h\colon \mathfrak{M}(X) \to S$ such that $h \circ \mu_X = f$ (defined by $h(\{x_1, \dots, x_n\}) = \bigwedge_k f(x_k)$.)

9.3 Combining the two constructions we obtain a free construction and a free functor $\mathbb{F}$ from the category **Set** of sets (and functions between sets) to **Frm**. To simplify the notation we will replace the

$$\lambda\mu(x) = {\downarrow}\{x\} = \{A \mid x \in A\}$$

by the symbol x and in the same sense we will use $x \wedge y$ or $\bigvee x_i$ for elements of $\mathbb{F}(X)$.

9.4 Working with defining relations When taking a quotient $\mathbb{F}(X)/R$ (recall A.5.5.1) we will describe the relation by a system of formal equalities, that is, if u, v are formulas to identify we write formally

$$u = v \quad \text{instead of} \quad uRv$$

(thus we represent the equivalence classes by their elements; this should not cause any more confusion than writing the fraction $\frac{1}{2}$ and meaning the class $\{\frac{k}{2k} \mid k \neq 0\}$).

9.4.1 How to define homomorphisms Note that in this convention the theorem on homomorphisms $L/R \to M$ represented as precisely those $h\colon L \to M$ for which aRb implies $h(a) = h(b)$ (from 5.5.2) results as the fact that we have the homomorphisms represented precisely by the mappings defined on the generators (and formally extended to the formulas in those generators) that respect all the defining equalities.

10 Point-Free Reals

10.1 We will use a frame of real numbers as constructed by Banaschewski [30] from a free construction on a certain set using defining relations (a similar construction can be found in [161]). Also we will introduce a modification of this construction more suitable for some of our purposes. Denote (as always) the set of rational numbers by $\mathbb{Q}$ and put

$$\mathcal{R} = \{p \cdot q \mid p, q \in \mathbb{Q}\}$$

($p \cdot q$ is just a way indicating an ordered pair p, q; we will have too many brackets and this will simplify the notation). The frame

$$\mathfrak{L}(\mathbb{R})$$

will be obtained from the free frame $\mathbb{F}(\mathcal{R})$ by the defining relations

(R1) $p \cdot q \wedge r \cdot s = (p \vee r) \cdot (q \wedge s)$,
(R2) $p \cdot q \vee r \cdot s = p \cdot s$ whenever $p \leqslant r < q \leqslant s$,
(R3) $p \cdot q = \bigvee\{r \cdot s \mid p < r < s < q\}$, and
(R4) $1 = \bigvee\{p \cdot q \mid p, q \in \mathbb{Q}\}$

(note that, by (R3), $p \cdot q = 0$ for $p \geqslant q$).

Note We keep in $\mathfrak{L}(\mathbb{R})$ the somewhat redundant $\mathbb{R}$ standardly used in the literature because of similar constructions concerning other spaces (for instance, the unit interval) where one wishes the results to be specified. Similarly in 10.3 below.

10.2 Lemma *For $r \cdot s$ in $\mathfrak{L}(\mathbb{R})$ set*

$$c(r \cdot s) = \bigvee_{p<r} p \cdot r \vee \bigvee_{q>s} s \cdot q$$

Then

(a) *for any $r' < r < s < s'$, $c(r \cdot s) \vee r' \cdot s' = 1$, and*
(b) $r \cdot s \wedge c(r \cdot s) = 0$.

Proof By (R2), for any $p < r$ and any $q > s$, $p \cdot q = p \cdot r \vee r' \cdot s' \vee s \cdot q \leqslant c(r \cdot s) \vee r' \cdot s'$. Since for any u, v there are $p < r$ and $q > s$ such that $u \cdot v \leqslant p \cdot q$ (by (R1)), then $c(r \cdot s) \vee r' \cdot s' = 1$ by (R4).

On the other hand, $r \cdot s \wedge c(r \cdot s) = 0$ by (R1), (R3) and distributivity. □

Note It is easy to see that $c(r \cdot s)$ is in fact the pseudocomplement of $r \cdot s$.

10.2.1 Proposition $\mathfrak{L}(\mathbb{R})$ *is completely regular.*

Proof If $p < r < s < q$ then, by 10.2, $r \cdot s \prec p \cdot q$. Since we can always interpolate $p < r' < r < s < s' < q$ we have in fact that $r \cdot s \prec\!\!\prec p \cdot q$. Since the elements $p \cdot q$ generate $\mathfrak{L}(\mathbb{R})$ by joins, the statement now follows from (R3). □

10.3 An alternative representation Instead of the generating set $\mathcal{R}$ take the set

$$\mathcal{R}_0 = \{p \cdot -, - \cdot p \mid p \in \mathbb{Q}\}$$

where $p\cdot -$, $-\cdot p$ are distinct symbols. The frame

$$\mathfrak{L}_0(\mathbb{R})$$

will be obtained from the free frame $\mathbb{F}(\mathcal{R}_0)$ by the defining relations

(r1) $p\cdot - \wedge -\cdot q = 0$ whenever $p \geqslant q$,
(r2) $p\cdot - \vee -\cdot q = 1$ whenever $p < q$,
(r3) for every $p \in \mathbb{Q}$, $p\cdot - = \bigvee_{r>p} r\cdot -$,
(r4) for every $p \in \mathbb{Q}$, $-\cdot p = \bigvee_{r<p} -\cdot r$,
(r5) $\bigvee_{p\in\mathbb{Q}} p\cdot - = 1$, and
(r6) $\bigvee_{p\in\mathbb{Q}} -\cdot p = 1$.

10.3.1 Theorem *The frames $\mathfrak{L}(\mathbb{R})$ and $\mathfrak{L}_0(\mathbb{R})$ are isomorphic. An isomorphism can be obtained translating $r\cdot s$ to $r\cdot - \wedge -\cdot s$, and on the other hand, $r\cdot -$ to $\bigvee_{u>r} r\cdot u$ and $-\cdot s$ to $\bigvee_{u<s} u\cdot s$.*

Proof Define $\alpha\colon \mathfrak{L}_0(\mathbb{R}) \to \mathfrak{L}(\mathbb{R})$ by setting

$$\alpha(r\cdot -) = \bigvee_{u>r} r\cdot u = \bigvee_{u\in\mathbb{Q}} r\cdot u \quad\text{and}\quad \alpha(-\cdot s) = \bigvee_{u<s} u\cdot s = \bigvee_{u\in\mathbb{Q}} u\cdot s$$

and $\beta\colon \mathfrak{L}(\mathbb{R}) \to \mathfrak{L}_0(\mathbb{R})$ by

$$\beta(r\cdot s) = r\cdot - \wedge -\cdot s.$$

Checking that thus defined maps respect the defining relations (recall A.9.4.1) is just a straightforward mechanical (and tedious) task. Hence we have here frame homomorphisms. We will prove they are inverse to each other.

We have

$$\alpha\beta(r\cdot s) = \alpha(r\cdot - \wedge -\cdot s) = \bigvee_{v\in\mathbb{Q}} r\cdot v \wedge \bigvee_{u\in\mathbb{Q}} u\cdot s =$$
$$= \bigvee\{r\cdot v \wedge u\cdot s \mid u, v \in \mathbb{Q}\} = \bigvee\{(r \vee u)\cdot(v \wedge s) \mid u, v \in \mathbb{Q}\}.$$

Now if $s \leqslant r$ (that is, $r\cdot s = 0$) we have all the $r\cdot v \wedge u\cdot s = 0$ since $v \wedge s \leqslant r \vee u$. For $r < s$ we have either $u \geqslant s$ resp. $v \leqslant r$ which makes $(r \vee u)\cdot(v \wedge s)$ zero, or $r \vee u \leqslant s$ and $r \leqslant v \wedge s$, hence in any case $(r \vee u)\cdot(v \wedge s) \leqslant r\cdot s$, and since $r\cdot s = (r \vee r)\cdot(s \wedge s)$ is one of the summands we conclude that $\alpha\beta(r\cdot s) = r\cdot s$.

Finally,

$$\beta\alpha(r\cdot -) = \beta(\bigvee_{v>r} r\cdot v) = \bigvee_{v>r} \beta(r\cdot v) =$$
$$= \bigvee_{v>r} (r\cdot - \wedge -\cdot v) = r\cdot - \wedge \bigvee_{v>r} -\cdot v = r\cdot - \wedge 1 = r\cdot -$$

and similarly $\beta\alpha(-\cdot r) = -\cdot r$. □

10.4 Theorem $\mathfrak{L}(\mathbb{R})$ *(and hence also* $\mathfrak{L}_0(\mathbb{R})$*) is isomorphic to* $\Omega(\mathbb{R})$.

Proof Define a homomorphism $h\colon \mathfrak{L}(\mathbb{R}) \to \Omega(\mathbb{R})$ by setting

$$h(p \cdot q) = (p, q).$$

The definition is correct since h obviously respects all the defining equalities (in fact, they are nothing else but imitating the behaviours of open intervals with rational ends). Since each (p, q) is an image and since the open intervals with rational ends are a basis of the topology of $\mathbb{R}$, h is onto.

By Lemma V.4.1.1 (and 10.2.1), to prove that it is also one-to-one it suffices to show it is codense. Hence let $h(a) = X$. Set $A = \{p \cdot q \mid p, q \in \mathbb{Q}, p \cdot q \leqslant a\}$ (so that $\bigvee A = a$). We have

$$X = h(\bigvee_{p \cdot q \in A} p \cdot q) = \bigcup_{p \cdot q \in A} (p, q).$$

Now let $r \cdot s$ be arbitrary. Since the closed interval $[r, s]$ is compact and $[r, s] \subseteq \bigcup_{p \cdot q \in A}(p, q)$, we have

$$[r, s] \subseteq (p_1, q_1) \cup \cdots \cup (p_n, q_n) \tag{$*$}$$

for some $p_k \cdot q_k \in A$. Consider any such finite cover with minimal n and order it so that $p_1 \leqslant p_2 \leqslant \cdots \leqslant p_n$. Then (p_k, q_k) and (p_{k+1}, q_{k+1}) cannot overlap because else $p_{k+1} < q_k$ and either $(p_{k+1}, q_{k+1}) \subseteq (p_k, q_k)$, or $q_k < q_{k+1}$ and $(p_k, q_k) \subseteq (p_k, q_{k+1})$ and the two $p_k \cdot q_k$ and $p_{k+1} \cdot q_{k+1}$ can be replaced either by $p_k \cdot q_k$ or $p_k \cdot q_{k+1}$, any of them in A. Thus we have the cover in $(*)$ disjoint, and since $[r, s]$ is connected, $n = 1$ and we have

$$(r, s) \subseteq [r, s] \subseteq (p_1, q_1)$$

and $r \cdot s \leqslant p_1 \cdot q_1 \in A$. Thus, A contains all the generators and hence $a = \bigvee A = 1$. □

10.4.1 Note Here we have an example of an advantage of the point-free techniques. Open sets in $\mathbb{R}$ are not quite as transparent as the system of all open intervals with rational ends (or, even, in the alternative case, on the $(r, +\infty)$ and $(-\infty, r)$ with rational r). It sometimes helps when we can define continuous maps as homomorphisms only on these transparent sets, with typically easy checking a few equations.

Bibliography

1. J. Adámek, H. Herrlich, G. E. Strecker. *Abstract and Concrete Categories: the Joy of Cats*. Pure and Applied Mathematics (New York). John Wiley & Sons Inc., New York, 1990.
2. A. D. Alexandroff. Additive set-functions in abstract spaces. *Mat. Sbornik N. S.* 8 (50), (1940) 307–348.
3. P. S. Alexandroff. Zur begründung der n-dimensionalen mengentheoretischen topologie. *Math. Ann.* 94 (1925) 296–308.
4. P. S. Alexandroff, H. Hopf. *Topologie*, vol. 1, Chelsea, New York, 1972.
5. C. E. Aull, W. J. Thron. Separation axioms between T_0 and T_1. *Indag. Math.* 24 (1963) 26–37.
6. F. Ávila, G. Bezhanishvili, P. J. Morandi, A. Zaldívar. When is the frame of nuclei spatial: A new approach. *J. Pure Appl. Algebra* 224 (2020) Art. 106302, 20 pp.
7. D. Baboolal, J. Picado, P. Pillay, A. Pultr. Hewitt's irresolvability and induced sublocales in spatial frames. *Quaestiones Math.* 43 (2020) 1601–1612.
8. R. Baire. *Leçons sur les fonctions discontinues, professées au collège de France*. Gauthier-Villars, Paris, 1905.
9. R. Balbes. Projective and injective distributive lattices. *Pacific J. Math.* 21 (1967) 405–420.
10. R. N. Ball. Distributive Cauchy lattices. *Algebra Universalis* 18 (1984) 134–174.
11. R. N. Ball, B. Banaschewski, T. Jakl, A. Pultr, J. Walters-Wayland. Tightness relative to some (co)reflections in topology. *Quaestiones Math.* 39 (2016) 421–436.
12. R. N. Ball, M. A. Moshier, A. Pultr. Exact filters and joins of closed sublocales. *Appl. Categ. Structures* 28 (2020) 655–667.
13. R. N. Ball, M. A. Moshier, J. L. Walters-Wayland, A. Pultr. Lindelöf tightness and the Dedekind-MacNeille completion of a regular σ-frame. *Quaestiones Math.* 40 (2017) 347–369.
14. R. N. Ball, J. Picado, A. Pultr. Notes on exact meets and joins. *Appl. Categ. Structures* 22 (2014) 699–714.
15. R. N. Ball, J. Picado, A. Pultr. On an aspect of scatteredness in the point-free setting. *Port. Math.* 73 (2016) 139–152.
16. R. N. Ball, J. Picado, A. Pultr. Some aspects of (non)functoriality of natural discrete covers of locales. *Quaestiones Math.* 42 (2019) 701–715.
17. R. N. Ball, A. Pultr. Maximal essential extensions in the context of frames. *Algebra Universalis* 79 (2018) Art. 32, 13 pp.
18. R. N. Ball, J. Walters-Wayland. C- and C^*-quotients in pointfree topology. *Dissertationes Mathematicae* (*Rozprawy Mat.*) 412 (2002) 1–62.

J. Picado, A. Pultr, *Separation in Point-Free Topology*,
https://doi.org/10.1007/978-3-030-53479-0

19. B. Banaschewski. *Untersuchungen über filterräume.* Doctoral dissertation, Universität Hamburg, 1953.
20. B. Banaschewski. Normal systems of sets. *Math. Nachr.* 24 (1962) 53–75.
21. B. Banaschewski. On Wallman's method of compactification. *Math. Nachr.* 21 (1963) 105–114.
22. B. Banaschewski. Frames and compactifications. In: *Proc. International Symp. on Extension Theory of Topological Structures and its Applications*, pp. 29–33. VEB Deutscher Verlag der Wissenschaften, 1969.
23. B. Banaschewski. *σ-frames.* Unpublished manuscript, 1980.
24. B. Banaschewski. Another look at the localic Tychonoff theorem. *Comment. Math. Univ. Carolinae* 29 (1988) 647–656.
25. B. Banaschewski. On pushing out frames. *Comment. Math. Univ. Carolinae* 31 (1990) 13–21.
26. B. Banaschewski. On proving the Tychonoff Product Theorem. *Kyungpook Math. J.* 30 (1990) 65–73.
27. B. Banaschewski. Compactifications of frames. *Math. Nachr.* 149 (1990) 105–115.
28. B. Banaschewski. Singly generated frame extensions. *J. Pure Appl. Algebra* 83 (1992) 1–21.
29. B. Banaschewski. *Completion in Pointfree Topology.* Lecture Notes in Mathematics and Applied Mathematics, vol. 2. University of Cape Town, 1996.
30. B. Banaschewski. *The Real Numbers in Pointfree Topology.* Textos de Matemática, vol. 12. University of Coimbra, 1997.
31. B. Banaschewski. The axiom of countable choice and pointfree topology. *Appl. Categ. Structures* 9 (2001) 245–258.
32. B. Banaschewski. A new aspect of the cozero lattice in pointfree topology. *Topology Appl.* 156 (2009) 2028–2038.
33. B. Banaschewski, G. Bruns. Injective hulls in the category of distributive lattices. *J. Reine Angew. Math.* 232 (1968) 102–109.
34. B. Banaschewski, T. Dube, C. R. A. Gilmour, J. Walters-Wayland. Oz in pointfree topology. *Quaestiones Math.* 32 (2009) 215–227.
35. B. Banaschewski, J. L. Frith, C. R. A. Gilmour. On the congruence lattice of a frame. *Pacific J. Math.* 130 (1987) 209–213.
36. B. Banaschewski, C. R. A. Gilmour. Stone-Čech compactification and dimension theory for regular σ-frames. *J. London Math. Soc. (2)* 39 (1989) 1–8.
37. B. Banaschewski, C. R. A. Gilmour. Pseudocompactness and the cozero part of a frame. *Comment. Math. Univ. Carolinae* 37 (1996) 577–587.
38. B. Banaschewski, C. R. A. Gilmour. Oz revisited. In: *Proceedings of the Conference Categorical Methods in Algebra and Topology* (ed. by H. Herrlich, H.-E. Porst), pp. 19–23. Math. Arbeitspapiere Nr. 54, Universität Bremen, 2000.
39. B. Banaschewski, C. R. A. Gilmour. Realcompactness and the cozero part of a frame. *Appl. Categ. Structures* 9 (2001) 395–417.
40. B. Banaschewski, C. R. A. Gilmour. Realcompact Alexandroff frames. *Topology Appl.* 161 (2014) 26–36.
41. B. Banaschewski, J. Gutiérrez García, J. Picado. Extended real functions in pointfree topology. *J. Pure Appl. Algebra* 216 (2012) 905–922.
42. B. Banaschewski, R. Harting. Lattice aspects of radical ideals and choice principles. *Proc. London Math. Soc. (3)* 50 (1985) 385–404.
43. B. Banaschewski, C. J. Mulvey. Stone-Čech compactification of locales I. *Houston J. Math.* 6 (1980) 301–312.
44. B. Banaschewski, C. J. Mulvey. Stone-Čech compactification of locales II. *J. Pure Appl. Algebra* 33 (1984) 107–122.
45. B. Banaschewski, C. J. Mulvey. The Stone-Čech compactification of locales III. *J. Pure Appl. Algebra* 185 (2003) 25–33.
46. B. Banaschewski, E. Nelson. Tensor products and bimorphisms. *Canad. Math. Bull.* 19 (1976) 385–402.

47. B. Banaschewski, A. Pultr. Distributive algebras in linear categories. *Algebra Universalis* 30 (1993) 101–118.
48. B. Banaschewski, A. Pultr. Paracompactness revisited. *Appl. Categ. Structures* 1 (1993) 181–190.
49. B. Banaschewski, A. Pultr. Variants of openness. *Appl. Categ. Structures* 2 (1994) 331–350.
50. B. Banaschewski, A. Pultr. Booleanization. *Cahiers Topologie Géom. Diff. Catég.* 37 (1996) 41–60.
51. B. Banaschewski, A. Pultr. Cauchy points of uniform and nearness frames. *Quaestiones Math.* 19 (1996) 101–127.
52. B. Banaschewski, A. Pultr. A new look at pointfree metrization theorems. *Comment. Math. Univ. Carolinae* 39 (1998) 167–175.
53. B. Banaschewski, A. Pultr. A constructive view of complete regularity. *Kyungpook Math. J.* 43 (2003) 257–262.
54. B. Banaschewski, A. Pultr. On weak lattice and frame homomorphisms. *Algebra Universalis* 51 (2004) 137–151.
55. B. Banaschewski, A. Pultr. A general view of approximation. *Appl. Categ. Structures* 14 (2006) 165–190.
56. B. Banaschewski, A. Pultr. Approximate maps, filter monad, and a representation of localic maps. *Arch. Math. (Brno)* 46 (2010) 285–298.
57. B. Banaschewski, A. Pultr. Pointfree aspects of the T_D axiom of classical topology. *Quaestiones Math.* 33 (2010) 369–385.
58. B. Banaschewski, A. Pultr. On covered prime elements and complete homomorphisms of frames. *Quaestiones Math.* 37 (2014) 451–454.
59. S. F. Barger. Hereditary sobriety and the cardinality of T_1 topologies on countable sets. *Quaestiones Math.* 20 (1997) 117–126.
60. J. Bénabou. *Treillis locaux et paratopologies*. Séminaire Ehresmann, 1re année, exposé 2. Paris, 1958.
61. G. Bezhanishvili, R. Mines, P. Morandi. Scattered, Hausdorff-reducible, and hereditarily irresolvable spaces. *Topology Appl.* 132 (2003) 291–306.
62. J. Blatter, G. L. Seever. Interposition of semi-continuous functions by continuous functions. In: *Analyse fonctionnelle et applications* (Comptes Rendus Colloq. d'Analyse, Inst. Mat., Univ. Federal Rio de Janeiro, Rio de Janeiro, 1972), pp. 27–51. Actualités Sci. Indust., No. 1367, Hermann, Paris, 1975.
63. F. Borceux. *Handbook of Categorical Algebra*. Encyclopedia of Mathematics and its Applications, vol. 52. Cambridge University Press, Cambridge, 1994.
64. N. Bourbaki. *Elements of Mathematics: General Topology*. Chapters 1–4, Springer, 1989.
65. G. Bruns. Darstellungen und erweiterungen geordneter mengen II. *J. Reine Angew. Math.* 210 (1962) 1–23.
66. G. Bruns, H. Lakser. Injective hulls of semilattices. *Canad. Math. Bull.* 13 (1970) 115–118.
67. J. R. Büchi. Representation of complete lattices by sets. *Portugal. Math.* 11 (1952) 151–167.
68. M. C. Charalambous. Dimension theory for σ-frames. *J. London Math. Soc.* 8 (1974) 149–160.
69. X. Chen. *Closed Frame Homomorphisms*. Doctoral dissertation, McMaster University, Hamilton, 1991.
70. X. Chen. On binary coproducts of frames. *Comment. Math. Univ. Carolinae* 33 (1992) 699–712.
71. X. Chen. On the paracompactness of frames. *Comment. Math. Univ. Carolinae* 33 (1992) 485–491.
72. M. M. Clementino. *Axiomas de Separação — Separação em Espaços Topológicos*. University of Coimbra, 1987.
73. M. M. Clementino. *Separation and Compactness in Categories*. Doctoral dissertation, University of Coimbra, 1991.
74. M. M. Clementino, E. Giuli, W. Tholen. A functional approach to general topology. In: *Categorical Foundations — Special Topics in Order, Algebra and Sheaf Theory* (ed. by M. C.

Pedicchio, W. Tholen), pp. 103–163. Encyclopedia of Mathematics and its Applications, vol. 97. Cambridge Univ. Press, Cambridge, 2004.
75. M. M. Clementino, J. Picado, A. Pultr. The other closure and complete sublocales. *Appl. Categ. Structures* 26 (2018) 891–906; errata ibid. 26 (2018) 907–908.
76. T. Coquand. Compact spaces and distributive lattices. *J. Pure Appl. Algebra* 184 (2003) 1–6.
77. A. S. Davis. Indexed systems of neighborhoods for general topological spaces. *Amer. Math. Monthly* 68 (1961) 886–893.
78. J. Dieudonné. Une généralisation des espaces compacts. *J. de Math. Pures et Appl.* 23 (1944) 65–76.
79. C. H. Dowker, D. Papert. Quotient frames and subspaces. *Proc. London Math. Soc. (3)* 16 (1966) 275–296.
80. C. H. Dowker, D. Papert. On Urysohn's lemma. In: *General Topology and its Relations to Modern Analysis and Algebra, II*, pp. 111–114. Proc. Second Prague Topological Sympos., 1966. Academia, Prague, 1967.
81. C. H. Dowker, D. Strauss. Separation axioms for frames. In: *Topics in Topology*, pp. 223–240. Proc. Colloq., Keszthely, 1972. Colloq. Math. Soc. Janos Bolyai, vol. 8, North-Holland, Amsterdam, 1974.
82. C. H. Dowker, D. Strauss. Paracompact frames and closed maps. In: *Symposia Mathematica, XVI*, pp. 93–116. Convegno sulla Topologia Insiemistica e Generale, INDAM, Rome, 1973. Academic Press, London, 1975.
83. C. H. Dowker, D. Strauss. Products and sums in the category of frames. In: *Categorical Topology* (ed. by E. Binz, H. Herrlich), pp. 208–219. Proc. Conf., Mannheim, 1975. Lecture Notes in Mathematics, vol. 540, Springer-Verlag, Berlin, 1976.
84. C. H. Dowker, D. Strauss. Sums in the category of frames. *Houston J. Math.* 3 (1977) 7–15.
85. C. H. Dowker, D. Strauss. T_1- and T_2-axioms for frames. In: *Aspects of Topology* (ed. by I. M. James, E. H. Kronheimer), pp. 325–335. London Math. Soc. Lecture Note Ser., vol. 93, Cambridge Univ. Press, Cambridge, 1985.
86. T. Dube. Submaximality in locales. *Topology Proc.* 29 (2005) 431–444.
87. T. Dube. An algebraic view of weaker forms of realcompactness. *Algebra Universalis* 55 (2006) 187–202.
88. T. Dube. A note on weakly pseudocompact locales. *Appl. Gen. Topol.* 18 (2017) 131–141.
89. T. Dube. Some connections between frames of radical ideals and frames of z-ideals. *Algebra Universalis* 79 (2018) Art. 7, 18 pp.
90. T. Dube, O. Ighedo. More on locales in which every open sublocale is z-embedded. *Topology Appl.* 201 (2016) 110–123.
91. T. Dube, S. Iliadis, J. van Mill, I. Naidoo. A pseudocompact completely regular frame which is not spatial. *Order* 31 (2014) 115–120.
92. C. Ehresmann. Gattungen von lokalen strukturen. *Jber. Deutsch. Math. Verein* 60 (1957) 59–77.
93. R. Engelking. *General topology.* Sigma Series in Pure Mathematics, vol. 6. Heldermann Verlag, Berlin, 1989.
94. M. Erné. The ABC of order and topology. In: *Category theory at work* (ed. by H. Herrlich, H.-E. Porst), pp. 57–83. Bremen, 1990. Res. Exp. Math., vol. 18, Heldermann, Berlin, 1991.
95. M. Erné. Adjunctions and Galois connections: origins, history and development. In: *Galois Connections and Applications* (ed. by K. Denecke, M. Erné, S. L. Wismath), pp. 1–138. Math. Appl., vol. 565, Kluwer Acad. Publ., Dordrecht, 2004.
96. M. Erné. Prime decomposition and pseudocomplementation. In: *Contributions to general algebra*, vol. 17, pp. 83–104. Heyn, Klagenfurt, 2006.
97. M. Erné. Distributors and Wallman locales. *Houston J. Math.* 34 (2008) 69–98.
98. M. Erné. Closure. In: *Beyond Topology* (ed. by F. Mynard, E. Pearl), pp. 163–238. Contemp. Math., vol. 486, Amer. Math. Soc., Providence, RI, 2009.
99. M. Erné. Algebraic models for T_1-spaces. *Topology Appl.* 158 (2011) 945–962.
100. M. Erné. The strength of prime separation, sobriety, and compactness theorems. *Topology Appl.* 241 (2018) 263–290.

101. M. Erné, M. Gehrke, A. Pultr. Complete congruences on topologies and down-set lattices. *Appl. Categ. Structures* 15 (2007) 163–184.
102. M. Erné, J. Koslowski, A. Melton, G. E. Strecker. A primer on Galois connections. In: *Papers on general topology and applications* (Madison, WI, 1991), pp. 103–125, Ann. New York Acad. Sci., vol. 704, New York, 1993.
103. M. Erné, J. Picado. Tensor products and relation quantales. *Algebra Universalis* 78 (2017) 461–487.
104. M. Erné, J. Picado, A. Pultr. Adjoint maps between implicative semilatices and continuity of localic maps. Preprint, 2020 (submitted).
105. L. Español, J. Gutiérrez García, T. Kubiak. Separating families of locale maps and localic embeddings. *Algebra Universalis* 67 (2012) 105–112.
106. K. Fan, N. Gottesman. On compactifications of Freudenthal and Wallman. *Indagat. Math.* 14 (1952) 504–510.
107. M. J. Ferreira. J. Gutiérrez García, J. Picado. Completely normal frames and real-valued functions. *Topology Appl.* 156 (2009) 2932–2941.
108. M. J. Ferreira. J. Gutiérrez García, J. Picado. Insertion of continuous real functions on spaces, bispaces, ordered spaces and point-free spaces — a common root. *Appl. Categ. Structures* 19 (2011) 469–487.
109. M. J. Ferreira, J. Picado, S. Pinto. Remainders in pointfree topology, *Topology Appl.* 245 (2018) 21–45.
110. M. P. Fourman. T_1-spaces over topological sites. *J. Pure Appl. Algebra* 27 (1983) 223–224.
111. J. L. Frith. *Structured frames.* Doctoral dissertation. University of Cape Town, 1987.
112. L. Gillman, M. Jerison. *Rings of continuous functions.* The University Series in Higher Mathematics. D. Van Nostrand Co., Inc., Princeton, 1960.
113. C. R. A. Gilmour. Realcompact spaces and regular σ-frames. *Math. Proc. Cambridge Philos. Soc.* 96 (1984) 73–79.
114. C. R. A. Gilmour. *Realcompact Alexandroff spaces and regular σ-frames.* Mathematical Monographs of the University of Cape Town, vol. 3. University of Cape Town, Department of Mathematics, 1985.
115. A. M. Gleason. Projective topological spaces. *Ill. J. Math.* 2 (1958) 482–489.
116. V. Glivenko. Sur quelques points de la logique de M. Brouwer. *Bull. Acad. R. Belg. Cl. Sci.* 15 (1929) 183–188.
117. J. Goubault-Larrecq. *Non-Hausdorff Topology and Domain Theory — Selected Topics in Point-Set Topology.* New Mathematical Monographs, vol. 22. Cambridge University Press, Cambridge, 2013.
118. A. Grothendieck, J. A. Dieudonné. *Eléments de Géometrie Algébrique, tome I: le Langage des Schémas.* Grundlehren der mathematische Wissenschaften, Number 166, Springer-Verlag, 1971. (Originally published by IHES in 1960.)
119. J. Gutiérrez García, T. Kubiak. A new look at some classical theorems on continuous functions on normal spaces. *Acta Math. Hung.* 119 (2008) 333–339.
120. J. Gutiérrez García, T. Kubiak. General insertion and extension theorems for localic real functions. *J. Pure Appl. Algebra* 215 (2011) 1198–1204.
121. J. Gutiérrez García, T. Kubiak, J. Picado. Lower and upper regularizations of frame semicontinuous real functions. *Algebra Universalis* 60 (2009) 169–184.
122. J. Gutiérrez García, T. Kubiak, J. Picado. Pointfree forms of Dowker's and Michael's insertion theorems. *J. Pure Appl. Algebra* 213 (2009) 98–108.
123. J. Gutiérrez García, T. Kubiak, J. Picado. Localic real-valued functions: a general setting. *J. Pure Appl. Algebra* 213 (2009) 1064–1074.
124. J. Gutiérrez García, T. Kubiak, J. Picado. An invitation to localic real functions. In: *Applied Topology: Recent progress for Computer Science, Fuzzy Mathematics and Economics*, pp. 119–129. Proc. WiAT'10, Gandia, 2010. Editorial Universitat Politècnica de València, 2010.
125. J. Gutiérrez García, T. Kubiak, J. Picado. Perfectness in locales. *Quaestiones Math.* 40 (2017) 507–518.

126. J. Gutiérrez García, T. Kubiak, J. Picado. On hereditary properties of extremally disconnected frames and normal frames. *Topology Appl.* 273 (2020) Art. 106978, 15 pp.
127. J. Gutiérrez García, T. Kubiak, J. Picado. Perfect locales and localic real functions. *Algebra Universalis* 81 (2020) Art. 32, 18 pp.
128. J. Gutiérrez García, I. Mozo Carollo, J. Picado, J. Walters-Wayland. Hedgehog frames and a cardinal extension of normality. *J. Pure Appl. Algebra* 223 (2019) 2345–2370.
129. J. Gutiérrez García, J. Picado. On the algebraic representation of semicontinuity. *J. Pure Appl. Algebra* 210 (2007) 299–306.
130. J. Gutiérrez García, J. Picado. Rings of real functions in pointfree topology. *Topology Appl.* 158 (2011) 2264–2278.
131. J. Gutiérrez García, J. Picado. On the parallel between normality and extremal disconnectedness. *J. Pure Appl. Algebra* 218 (2014) 784–803.
132. J. Gutiérrez García, J. Picado, M. A. de Prada Vicente. Monotone normality and stratifiability from a pointfree point of view. *Topology Appl.* 168 (2014) 46–65.
133. J. Gutiérrez García, J. Picado, A. Pultr. Notes on point-free real functions and sublocales. In: *Categorical Methods in Algebra and Topology*, pp. 167–200. Textos de Matemática, vol. 46. University of Coimbra, 2014.
134. H. Hahn. Über halbstetige und unstetige Funktionen. *Sitzungsberichte Akad. Wien Abt. IIa* 126 (1917) 91–110.
135. Y. H. Han, S. S. Hong. Separated locales. *Kyungpook Math. J.* 26 (1986) 113–117.
136. F. Hausdorff. *Grundzüge der mengenlehre.* Veit & Co., Leipzig, 1914.
137. F. Hausdorff. Gestufte Räume. *Fund. Math.* 25 (1935) 486–502.
138. W. He. Compact regular reflections of locales. *Acta Math. Sinica* (*Chin. Ser.*) 42 (1999) 441–444.
139. W. He. Remarks on completely regular Lindelöf reflection of locales. *Appl. Categ. Structures* 13 (2005) 71–77.
140. W. He, M. Luo. A note on proper maps of locales. *Appl. Categ. Structures* 19 (2011) 505–510.
141. W. He, M. Luo. Completely regular proper reflection of locales over a given locale. *Proc. Amer. Math. Soc.* 141 (2013) 403–408.
142. W. He, R. X. Sun. A remark on: "Wallman compactification of locales" [Houston J. Math. 10 (1984), no. 2, 201–206] by P. T. Johnstone. *J. Math. Res. Exposition* 28 (2008) 605–608.
143. R. W. Heath, E. A. Michael. A property of the Sorgenfrey line. *Compositio Math.* 23 (1971) 185–188.
144. H. Herrlich. A concept of nearness. *General Topology and Appl.* 4 (1974) 191–212.
145. H. Herrlich. Topological structures. In: *Topological structures*, pp. 59–122. Proc. Sympos. in honour of Johannes de Groot (1914–1972), Amsterdam, 1973. Math. Centre Tracts, vol. 52, Math. Centrum, Amsterdam, 1974.
146. H. Herrlich. *Topologie II: Uniforme Räume.* Heldermann Verlag, Berlin, 1988.
147. H. Herrlich, A. Pultr. Nearness, subfitness and sequential regularity. *Appl. Categ. Structures* 8 (2000) 67–80.
148. E. Hewitt. A problem of set-theoretic topology. *Duke Math. J.* 10 (1943) 309–333.
149. E. Hewitt. Rings of real-valued continuous functions, I. *Trans. Amer. Math. Soc.* 64 (1948) 45–99.
150. K. H. Hofmann, J. D. Lawson. The spectral theory of distributive continuous lattices. *Trans. Amer. Math. Soc.* 246 (1978) 285–310.
151. O. Ighedo, M. Mugochi. On some parallelism between complete regularity and zero-dimensionality. *Quaestiones Math.* 41 (2018) 423–435.
152. J. R. Isbell. Atomless parts of spaces. *Math. Scand.* 31 (1972) 5–32.
153. J. R. Isbell. Function spaces and adjoints. *Math. Scand.* 36 (1975) 317–339.
154. J. R. Isbell. Product spaces in locales. *Proc. Amer. Math. Soc.* 81 (1981) 116–118.
155. J. R. Isbell. Graduation and dimension in locales. In: *Aspects of Topology* (ed. by I. M. James, E. H. Kronheimer), pp. 195–210. London Math. Soc. Lecture Note Ser., vol. 93. Cambridge Univ. Press, Cambridge, 1985.

156. J. R. Isbell. First steps in descriptive theory of locales. *Trans. Amer. Math. Soc.* 327 (1991) 353–371; errata ibid. 341 (1994) 467–468.
157. J. R. Isbell. Some structure of Borel locales. *Proc. Amer. Math. Soc.* 126 (1998) 2477–2479.
158. J. R. Isbell, I. Kříž, A. Pultr, J. Rosický. Remarks on localic groups. In: *Categorical Algebra and its Applications* (ed. by F. Borceux), pp. 154–172. Lecture Notes in Mathematics, vol. 1348. Springer-Verlag, Berlin, 1988.
159. P. T. Johnstone. Conditions related to De Morgan's law. In: *Applications of Sheaves* (ed. by M. P. Fourman, C. J. Mulvey, D. S. Scott), pp. 479–491. Lecture Notes in Mathematics, vol. 753, Springer, Heidelberg, 1979.
160. P. T. Johnstone. Tychonoff's theorem without the axiom of choice. *Fund. Math.* 113 (1981) 21–35.
161. P. T. Johnstone. *Stone Spaces*. Cambridge Studies in Advanced Mathematics, vol. 3. Cambridge University Press, Cambridge 1982.
162. P. T. Johnstone. The point of pointless topology. *Bull. Amer. Math. Soc. (N.S.)* 8 (1983) 41–53.
163. P. T. Johnstone. Wallman compactification of locales. *Houston J. Math.* 10 (1984) 201–206.
164. P. T. Johnstone. Almost maximal ideals. *Fund. Math.* 123 (1984) 197–209.
165. P. T. Johnstone. Fibrewise separation axioms for locales. *Math. Proc. Cambridge Philos. Soc.* 108 (1990) 247–256.
166. P. T. Johnstone. The art of pointless thinking: a student's guide to the category of locales. In: *Category Theory at Work* (ed. by H. Herrlich, H.-E. Porst), pp. 85–107. Proc. Workshop Bremen 1990. Research and Exposition in Math., vol. 18, Heldermann Verlag, Berlin, 1991.
167. P. T. Johnstone. Elements of the history of locale theory. In: *Handbook of the History of General Topology* (ed. by C. E. Aull, R. Lowen), vol. 3, pp. 835–851. Kluwer Acad. Publ., Dordrecht, 2001.
168. P. T. Johnstone. *Sketches of an Elephant: a Topos Theory Compendium*. Oxford Logic Guides, vol. 44. The Clarendon Press, Oxford University Press, Oxford, 2002.
169. P. T. Johnstone. Complemented sublocales and open maps. *Ann. Pure Appl. Logic* 137 (2006) 240–255.
170. P. T. Johnstone, M. C. Pedicchio. Remarks on continuous Mal'cev algebras. *Rend. Istit. Mat. Univ. Trieste* 25 (1993) 277–297.
171. P. T. Johnstone, S.-H. Sun. Weak products and Hausdorff locales. In: *Categorical Algebra and its Applications*, pp. 173–193. Lecture Notes in Mathematics, vol. 1348. Springer-Verlag, Berlin, 1988.
172. A. Joyal. *Nouveaux fondaments de l'analyse*. Lecture notes, Montreal (manuscript), 1973 and 1974.
173. A. Joyal, M. Tierney. *An Extension of the Galois Theory of Grothendieck*. Mem. Amer. Math. Soc., vol. 309. Amer. Math. Soc., Providence, RI, 1984.
174. T. Kaiser. A sufficient condition of full normality. *Comment. Math. Univ. Carolinae* 37 (1996) 381–389.
175. P. Kalemba, S. Plewik. On regular but not completely regular spaces. *Topology Appl.* 252 (2019) 191–197.
176. M. Katětov. On H-closed extensions of topological spaces. *Časopis Pěst. Mat. Fys.* 72 (1947) 17–32.
177. M. Katětov. On real-valued functions in topological spaces. *Fund. Math.* 38 (1951) 85–91; errata ibid. 40 (1953) 203–205.
178. J. L. Kelley. The Tychonoff product theorem implies the axiom of choice. *Fund. Math.* 37 (1950) 75–76.
179. J. L. Kelley. *General Topology*. The University Series in Higher Mathematics, Van Nostrand, 1955.
180. I. Kříž. A constructive proof of the Tychonoff's theorem for locales. *Comment. Math. Univ. Carolinae* 26 (1985) 619–630.
181. I. Kříž. *Factorization of frames and its application in "pointless topology"*. Doctoral Dissertation, Charles University, 1988.

182. I. Kříž, A. Pultr. Systems of covers of frames and resulting subframes. *Rend. Circ. Mat. Palermo* (2) *Suppl.* 14 (1987) 353–364.
183. I. Kříž, A. Pultr. A spatiality criterion and an example of a quasitopology which is not a topology. *Houston J. Math.* 15 (1989) 215–234.
184. W. Kotzé, T. Kubiak. Insertion of a measurable function. *J. Austral. Math. Soc. Ser. A* 57 (1994) 295–304.
185. T. Kubiak. *On Fuzzy Topologies.* Doctoral Dissertation, Uniwersytet im. Adama Mickiewicza, Poznań, 1985.
186. T. Kubiak. On extremally disconnected subspaces. *Fasc. Math.* 19 (1990) 143–145.
187. T. Kubiak. Second open question. In: *Applications of category theory to fuzzy subsets* (ed. by S. E. Rodabaugh, E. P. Klement, U. Höhle), pp. 349. Kluwer, Dordrecht, 1992.
188. T. Kubiak. Separation axioms: extension of mappings and embedding of spaces. In: *Mathematics of Fuzzy Sets: Logic, Topology and Measure Theory* (ed. by U. Höhle, S. E. Rodabaugh), pp. 433–479. The Handbooks of Fuzzy Sets Series, vol.3. Kluwer, Dordrecht, 1999.
189. Y.-M. Li, Z.-H. Li. Constructive insertion theorems and extension theorems on extremally disconnected frames. *Algebra Universalis* 44 (2000) 271–281.
190. Y. Liu, M. Luo. T_D property and spatial sublocales. *Acta Math. Sinica (N.S.)* 11 (1995) 324–336.
191. S. Mac Lane. *Categories for the Working Mathematician.* Graduate Texts in Mathematics, vol. 5, Second edition. Springer-Verlag, New York, 1998.
192. D. S. Macnab. Modal operators on Heyting algebras. *Algebra Universalis* 12 (1981) 5–29.
193. H. M. MacNeille. *Extensions of Partially Ordered Sets.* Doctoral Dissertation, Harvard University, 1935.
194. H. M. MacNeille. Partially ordered sets. *Trans. Amer. Math. Soc.* 42 (1937) 416–460.
195. J. Madden. κ-frames. *J. Pure Appl. Algebra* 70 (1991) 107–127.
196. J. Madden, A. T. Molitor. Epimorphisms of frames. *J. Pure Appl. Algebra* 70 (1991) 129–132.
197. J. Madden, J. Vermeer. Lindelöf locales and realcompactness. *Math. Proc. Cambridge Philos. Soc.* 99 (1986) 473–480.
198. G. Manuell. Strictly zero-dimensional biframes and a characterisation of congruence frames. *Appl. Categ. Structures* 26 (2018) 645–655.
199. N. Marcus. The Wallman compactification of frames. *Quaestiones Math.* 21 (1998) 109–116.
200. J. Martinez, E. R. Zenk. Epicompletion in frames with skeletal maps, II. *Appl. Categ. Structures* 17 (2009) 467–486.
201. J. C. C. McKinsey, A. Tarski. The algebra of topology. *Ann. Math.* 45 (1944) 141–191.
202. K. Menger. Topology without points. *Rice Institute pamphlet* 27 (1940) 80–107.
203. K. Morita. On the simple extension of a space with respect to a uniformity I-IV. *Proc. Japan Acad.* 27 (1951) 65–72; 130–137; 166–171; 632–636.
204. M. A. Moshier, J. Picado, A. Pultr. Generating sublocales by subsets and relations: a tangle of adjunctions. *Algebra Universalis* 78 (2017) 105–118.
205. M. A. Moshier, A. Pultr, A. L. Suarez. Exact and strongly exact filters. *Appl. Categ. Structures* 28 (2020) 907–920.
206. I. Mozo Carollo. A lattice-theoretic approach to arbitrary real functions on frames. *Quaestiones Math.* 41 (2018) 319–347.
207. A. Mysior. A regular space which is not completely regular. *Proc. Amer. Math. Soc.* 81 (1981) 852–853.
208. L. Nachbin. *Topology and Order.* D. van Nostrand, Princeton, 1965.
209. S. B. Niefield, K. I. Rosenthal. Spatial sublocales and essential primes. *Topology Appl.* 26 (1987) 263–269.
210. D. Papert, S. Papert. *Sur les treillis des ouverts et paratopologies.* Séminaire Ehresmann (1re année, exposé 1). Paris, 1958.
211. J. Paseka. T_2-separation axioms on frames. *Acta Univ. Carolinae Mat. Phys.* 28 (1987) 95–98.
212. J. Paseka. Lindelöf locales and $\mathbb{N}$-compactness. *Math. Proc. Cambridge Philos. Soc.* 109 (1991) 187–191.

213. J. Paseka. Paracompact locales and metric spaces. *Math. Proc. Cambridge Philos. Soc.* 110 (1991) 251–256.
214. J. Paseka. Products in the category of locales: which properties are preserved? *Discrete Math.* 108 (1992) 63–73.
215. J. Paseka, P. Sekanina. A note on extremally disconnected frames. *Acta Univ. Carolin. Math. Phys.* 31 (1990) 75–84.
216. J. Paseka, B. Šmarda. T_2-frames and almost compact frames. *Czechoslovak Math. J.* 42 (1992) 297–313.
217. J. Picado. A new look at localic interpolation theorems. *Topology Appl.* 153 (2006) 3203–3218.
218. J. Picado, A. Pultr. Sublocale sets and sublocale lattices. *Arch. Math. (Brno)* 42 (2006) 409–418.
219. J. Picado, A. Pultr. *Locales mostly treated in a covariant way*. Textos de Matemática, vol. 41. University of Coimbra, 2008.
220. J. Picado, A. Pultr. *Frames and Locales: Topology without points*. Frontiers in Mathematics, vol. 28, Springer, Basel, 2012.
221. J. Picado and A. Pultr. (Sub)Fit biframes and non-symmetric nearness. *Topology Appl.* 168 (2014) 66–81.
222. J. Picado, A. Pultr. Notes on the product of locales. *Math. Slovaca* 65 (2015) 247–264.
223. J. Picado, A. Pultr. More on subfitness and fitness. *Appl. Categ. Structures* 23 (2015) 323–335.
224. J. Picado, A. Pultr. New aspects of subfitness in frames and spaces. *Appl. Categ. Structures* 24 (2016) 703–714.
225. J. Picado, A. Pultr. A Boolean extension of a frame and a representation of discontinuity. *Quaestiones Math.* 40 (2017) 1111–1125.
226. J. Picado, A. Pultr. Axiom T_D and the Simmons sublocale theorem. *Comment. Math. Univ. Carolinae* 60 (2019) 541–551.
227. J. Picado, A. Pultr, A. Tozzi. Locales. In: *Categorical Foundations — Special Topics in Order, Algebra and Sheaf Theory* (ed. by M. C. Pedicchio, W. Tholen), pp. 49–101. Encyclopedia of Mathematics and its Applications, vol. 97. Cambridge Univ. Press, Cambridge, 2004.
228. J. Picado, A. Pultr, A. Tozzi. Ideals in Heyting semilattices and open homomorphisms. *Quaestiones Math.* 30 (2007) 391–405.
229. J. Picado, A. Pultr, A. Tozzi. Joins of closed sublocales. *Houston J. Math.* 45 (2019) 21–38.
230. T. Plewe. Higher order dissolutions and Boolean coreflections of locales. *J. Pure Appl. Algebra* 154 (2000) 273–293.
231. T. Plewe. Quotient maps of locales. *Appl. Categ. Structures* 8 (2000) 17–44.
232. T. Plewe. Sublocale lattices. *J. Pure Appl. Algebra* 168 (2002) 309–326.
233. T. Plewe, A. Pultr, A. Tozzi. Regular monomorphisms of Hausdorff frames. *Appl. Categ. Structures* 9 (2001) 15–33.
234. A. Pultr. Pointless uniformities I. Complete regularity. *Comment. Math. Univ. Carolinae* 25 (1984) 91–104.
235. A. Pultr. Remarks on metrizable locales. *Rend. Circ. Mat. Palermo (2)* 6 (1984) 247–258.
236. A. Pultr. Frames. In: *Handbook of Algebra* (ed. by M. Hazewinkel), vol. 3, pp. 791–857. North-Holland, Amsterdam, 2003.
237. A. Pultr, A. Tozzi. Notes on Kuratowski-Mrówka theorems in point-free context. *Cahiers Topologie Géom. Différentielle Catég.* 33 (1992) 3–14.
238. A. Pultr, A. Tozzi. Equationally closed subframes and representation of quotient spaces. *Cahiers Topologie Géom. Différentielle Catég.* 34 (1993) 167–183.
239. A. Pultr, A. Tozzi. Separation axioms and frame representation of some topological facts. *Appl. Categ. Structures* 2 (1994) 107–118.
240. A. Pultr, A. Tozzi. A note on reconstruction of spaces and maps from lattice data. *Quaestiones Math.* 24 (2001) 55–63.
241. A. Pultr, J. Úlehla. Notes on characterization of paracompact frames. *Comment. Math. Univ. Carolinae* 30 (1989) 377–384.

242. G. N. Raney. A subdirect-union representation for completely distributive complete lattices. *Proc. Amer. Math. Soc.* 4 (1953) 518–522.
243. J. Rosický, B. Šmarda. T_1-locales. *Math. Proc. Cambridge Philos. Soc.* 98 (1985) 81–86.
244. E. V. Schepin. Real-valued functions, and spaces close to normal. *Sib. Math. J.* 13 (1972) 1182–1196.
245. G. Schlitt. ℕ-*Compact Frames and Applications.* Doctoral dissertation, McMaster University, Hamilton, 1990.
246. G. Schlitt. The Lindelöf-Tychonoff theorem and choice principles. *Math. Proc. Cambridge Phil. Soc.* 110 (991) 57–65.
247. V. Šedivă. On collectionwise normal and hypocompact spaces. *Czechoslovak Math. J.* 9 (84) (1959) 50–62 (in Russian).
248. R. A. Sexton, H. Simmons. Point-sensitive and point-free patch constructions. *J. Pure Appl. Algebra* 207 (2006) 433–468.
249. R. A. Sexton, H. Simmons. An ordinal indexed hierarchy of separation properties. *Topology Proc.* 30 (2006) 585–625.
250. N. A. Shanin. On separation in topological spaces. *C. R. (Dokl.) Acad. Sci. URSS (N.S.)* 38 (1943) 110–113.
251. N. A. Shanin. On the theory of bicompact extensions of topological spaces. *C. R. (Dokl.) Acad. Sci. URSS (N.S.)* 38 (1943) 154–156.
252. Z. Shmuely. The structure of Galois connections. *Pacific J. Math.* 54 (1974) 209–225.
253. R. Sikorski. A theorem on extensions of homomorphisms. *Ann. Soc. Pol. Math.* 21 (1948) 332–335.
254. H. Simmons. A framework for topology. In: *Logic Colloq. '77*, pp. 239–251. Stud. Logic Foundations Math., vol. 96. North-Holland, Amsterdam-New York, 1978.
255. H. Simmons. The lattice theoretic part of topological separation properties. *Proc. Edinburgh Math. Soc.* 21 (1978) 41–48.
256. H. Simmons. Spaces with Boolean assemblies. *Colloq. Math.* 43 (1980) 23–29.
257. H. Simmons. Regularity, fitness, and the block structure of frames. *Appl. Categ. Structures* 14 (2006) 1–34.
258. H. Simmons. A coverage construction of the reals and the irrationals. *Ann. Pure Appl. Logic* 145 (2007) 176–203.
259. H. Simmons. A curious nucleus. *J. Pure Appl. Algebra* 214 (2010) 2063–2073.
260. A. Simpson. Measure, randomness and sublocales. *Ann. Pure Appl. Logic* 163 (2012) 1642–1659.
261. M. Singal, S. Arya. On almost regular spaces. *Glasnik. Matematicki Ser. III* 4 (1969) 89–99.
262. M. Singal, A. Singal. Almost normal and almost completely regular spaces. *Glasnik Mat. Ser. III* 5 (1970) 141–152.
263. M. Singal, A. Singal. Mildly normal spaces. *Kyungpook Math. J.* 13 (1973) 27–31.
264. B. Šmarda. Completely normal locales. *Acta Univ. Carolin. Math. Phys.* 31 (1990) 101–104.
265. L. A. Steen, J. A. Seebach. *Counterexamples in Topology.* Springer-Verlag, 1970, 2nd ed. 1978.
266. M. H. Stone. Boolean algebras and their application in topology. *Proc. Nat. Acad. Sci. USA* 20 (1934) 197–202.
267. M. H. Stone. The theory of representations for Boolean algebras. *Trans. Amer. Mat. Soc.* 40 (1936) 37–111.
268. M. H. Stone. Boundedness properties in function-lattices. *Canad. J. Math.* 1 (1949) 176–186.
269. D. Strauss, F. Zhang. Separated frames. *Trans. Tianjin Univ.* 4 (1998) 74–77.
270. S. H. Sun. On paracompact locales and metric locales. *Comment. Math. Univ. Carolinae* 30 (1989) 101–107; errata ibid. 30 (1989) 817–818.
271. W. J. Thron. Lattice-equivalence of topological spaces. *Duke Math. J.* 29 (1962) 671–679.
272. H. Tietze. Über Funktionen, die auf einer abgeschlossenen menge stetig sind. *J. Reine Angew. Math.* 145 (1915) 9–14.
273. H. Tong. Some characterizations of normal and perfectly normal spaces. *Duke Math. J.* 19 (1952) 289–292.

274. A. Tychonoff. Über die topologische Erweiterung von Räumen. *Math. Ann.* 102 (1930) 544–561.
275. P. S. Urysohn. Über die Mächtigkeit der zusammen hängenden Mengen. *Math. Ann.* 94 (1925) 262–295.
276. R. Vaidyanathaswamy. On the lattice of open sets of a topological space. *Proc. Indian Acad. Sci. (Sect. A)* 16 (1942) 379–386.
277. J. J. C. Vermeulen. *Constructive Techniques in Functional Analysis*. Doctoral dissertation, University of Sussex, 1987.
278. J. J. C. Vermeulen. Some constructive results related to compactness and the (strong) Hausdorff property for locales. In: *Category Theory* (ed. by A. Carboni, M. C. Pedicchio, G. Rosolini), pp. 401–409. Lecture Notes in Mathematics, vol. 1488. Springer-Verlag, Berlin, 1991.
279. J. J. C. Vermeulen. Proper maps of locales. *J. Pure Appl. Algebra* 92 (1994) 79–107.
280. S. Vickers. *Topology via Logic*. Cambridge Tracts in Theoretical Computer Science, vol. 5. Cambridge University Press, Cambridge, 1989.
281. H. Wallman. Lattices and bicompact spaces. *Proc. Nat. Sci. U.S.A.* 23 (1937) 164–165.
282. H. Wallman. Lattices and topological spaces. *Ann. Math.* 39 (1938) 112–126.
283. G. T. Whyburn. Open and closed mappings. *Duke Math. J.* 17 (1950) 69–74.
284. J. T. Wilson. *The assembly tower and some categorical and algebraic aspects of frame theory*. Doctoral Dissertation, Carnegie Mellon University, 1994.

List of Symbols

J. Picado, A. Pultr, *Separation in Point-Free Topology*, https://doi.org/10.1007/978-3-030-53479-0

Index

J. Picado, A. Pultr, *Separation in Point-Free Topology*,
https://doi.org/10.1007/978-3-030-53479-0

Printed by Printforce, the Netherlands